Z1584
a995

AF863180

RESEARCH IN PHILOSOPHY & TECHNOLOGY

Volume 3 • 1980

Board members of the Society for Philosophy & Technology and editorial board members for *Research in Philosophy & Technology*

Joseph Agassi
Boston University
Tel Aviv University

Jean-Claude Beaune
University of Clermont, France

Edmund Byrne
Indiana/Purdue University Indianapolis

George Bugliarello
Polytechnic Institute of New York

Stanley Carpenter
Georgia Institute of Technology

Daniel Cérézuelle
University of Bordeaux

Melvin Kranzberg
Georgia Institute of Technology

Joseph Margolis
Temple University

Robert McGinn
Stanford University

Alex Michalos
University of Guelph

Friedrich Rapp
Technical University, Berlin

Egbert Schuurman
Free University, Amsterdam
Technical University, Delft

Henryk Skolimowski
University of Michigan

RESEARCH IN PHILOSOPHY & TECHNOLOGY

Editor: PAUL T. DURBIN
Philosophy Department and
Center for Science and Culture
University of Delaware

Review and Bibliography Editor: CARL MITCHAM
St. Catharine College
St. Catharine, Kentucky

OFFICIAL ANNUAL PUBLICATION OF THE SOCIETY FOR PHILOSOPHY & TECHNOLOGY

VOLUME 3 • 1980

JAI PRESS INC.
Greenwich, Connecticut

Copyright © 1980 JAI PRESS INC.
165 West Putnam Avenue
Greenwich, Connecticut 06830

All rights reserved. No part of this publication may be reproduced, stored on a retrieval system, or transmitted in any form or by any means, electronic, mechanical, filming, photocopying, recording or otherwise without prior permission in writing from the publisher.

ISBN NUMBER: 0-89232-102-4

Manufactured in the United States of America

CONTENTS

LIST OF CONTRIBUTORS

George J. Agich	School of Medicine, Southern Illinois University
Ronald E. Benson	Committee for the Humanities in Georgia
John Boli-Bennett	Stanford University
Albert Borgmann	University of Montana
J. van Brakel	State University of Utrecht, The Netherlands
Stanley R. Carpenter	Georgia Institute of Technology
Daniel Cérézuelle	University of Bordeaux
Thomas Conry	Environmental Action Foundation, Washington, D.C.
Paul T. Durbin	Center for Science and Culture University of Delaware
Jacques Ellul	University of Bordeaux
H. Tristram Engelhardt, Jr.	Kennedy Institute, Georgetown University
K.S. Shrader-Frechette	University of Louisville
Bernard Gendron	University of Wisconsin—Milwaukee
Michael Goldman	Miami University, Ohio
J. Bryan Hehir	Office of International Justice and Peace, U.S. Catholic Conference
David Lovekin	Sauk Valley College

Alex C. Michalos	University of Guelph
Henryk Skolimowski	University of Michigan
Katherine Temple	Catholic Worker, New York
David C. Thomasma	Center for the Health Sciences, University of Tennessee
Willis H. Truitt	University of South Florida
Langdon Winner	University of California—Santa Cruz

INTRODUCTION

Together, *Research in Philosophy & Technology,* volumes 1 and 2, constitute the best currently available introduction to the philosophy and technology movement as it exists world-wide. Volume 2 in particular corrected a lack of balance in the first volume, emphasizing European contributions of all sorts: Josef Banka and Socialist thinkers; Hans Lenk and Gunther Ropohl and the flourishing German tradition, especially that related to the VDI; Daniel Cérézuelle and other French writers; Hendryk Van Riessen and the Dutch neo-Calvinist theologian-philosophers of technology. (Many countries, including Japan, are still to be heard from, it must be admitted; and the Socialist countries—considering the vitality of their tradition—are still badly underrepresented.) In addition, volume 2, Part III, attempted to demonstrate some of the roots of the philosophy and technology movement and of the problems it addresses. Add to this the impressive collection of young thinkers (Albert Borgmann, Edmund Byrne, Stanley Carpenter, Phillip Fandozzi, Bernard Gendron, Robert

McGinn, Carl Mitcham, Frederick Rossini, Kristin Shrader-Frechette, Michael Zimmerman) and the token representation of "name" writers (Joseph Agassi, Joseph Margolis, Kai Nielsen representing Jürgen Habermas, and Willis Truitt speaking for the neo-Marxists)—from either volume 1 or volume 2—and the coverage for both the United States and the world is impressive. In short, the Society for Philosophy & Technology, with its annual publication in this series, can be taken to be well established.

Where do we go from here? With the foundations of the institution laid, it is time to get on with the building. What that means—among other things—is: (1) We will continue to promote conferences and symposia, both on our own and in conjunction with other societies, and to publish the best papers presented there. (2) We expect now to welcome more and more "name" figures into our pages; for instance, we are including in this volume an impressive symposium on Jacques Ellul, including his response to the (mostly friendly) critics. (3) We also anticipate more symposia devoted to topical issues—for instance, the *praxis* symposium printed here as Part IV. And of course (4) we will continue, indeed expand, our bibliographical efforts under the direction of Carl Mitcham. This year's bibliography has been completely restructured as well as expanded, which explains in part the decision to print it separately as Volume 4.

Our ship now launched, it remains for the scholarly community—in the future as well as the present—to adjudge its seaworthiness. We think the venture is worthwhile, and we welcome others to join us in the important task of philosophical assessment of technology and its impact on society.

ACKNOWLEDGMENTS

Thanks are again due to several institutions and individuals: to the Center for Science and Culture at the University of Delaware and its able secretary, Bootie Clark, who once again prepared the final manuscript; to Jim Grote who continues to aid Carl Mitcham in his valiant efforts; to the National Science Foundation which awarded a grant to support their efforts; and to JAI Press, especially Herb Johnson and Gayle Jerman, who have welcomed ours as one in their impressive collection of scholarly series.

PART I:
SELECTED SYMPOSIA FROM SECOND UNIVERSITY OF DELAWARE PHILOSOPHY AND TECHNOLOGY CONFERENCE, JUNE 1977

SYMPOSIUM ON THE CITIZEN AND TECHNOLOGICAL DECISION MAKING

PUBLIC INTEREST: RIGHTS FOR WHAT? RESPONSIBILITIES TO WHOM?

Thomas Conry

I want to point out some recent facts which I will subsequently try to weave together in the context of this speech:

(1) The clean air act was brutally gutted on the House floor by a utility-automaker-union coalition. Charles Warren, chairman of the Council on Environmental Quality, wondered if the environmentalists could mount a write-in campaign to save the bill in the Senate. As a group they could not (as it happened the Senate did restore much of the bill).
(2) One pro-environmentalist congressperson stated that the environmental movement couldn't carry a glass of milk in Congress.
(3) A number of federal critics, previously working outside the bureaucracy with citizen interest groups, are now within. Examples include Gus Speth at the

Research in Philosophy & Technology, Volume 3, 1980, pages 5–28.
Copyright © 1980 by JAI Press Inc.
All rights of reproduction in any form reserved.
ISBN: 0-89232-102-4

Council on Environmental Quality and Rick Cotten at the Department of Health Education and Welfare, both formerly with the Natural Resources Defense Council. A former Sierra Club lawyer, James Moorman is now at the Justice Department. Still others are being considered for government positions: Tom Cochran and Tony Roisman, presently at the Natural Resources Defense Council—both of whom orchestrated support for Carter's energy speech.

(4) Unions have lost a number of significant battles on the floors of Congress, and are now reassessing their policies and strengths.

(5) Some unions are now voting against pro-environment bills, essentially backing industry. Examples include the clean air act, nuclear energy, water projects and strip mining. Unions in other trades and other countries have not always taken the same stances.

(6) Charles Warren has admitted that the Council on Environmental Quality's labor links are the weakest of all—being almost nonexistent. This despite the fact that he has been able to build good environmental contacts.

These items are cut from the same piece of cloth. I want to pull the public or citizen interest thread and analyze the connections.

Public interest's past was mostly environmental and remains so today with around 3,000 environmentally related groups in the country. Most of the original battles were fought for entities without voice or legal standing: trees, streams, wilderness, wildlife. In this process common, characteristic principles of the environmental groups evolved: the value of life, ecosystems' multiplicity, and future generations.

Today's situation finds the issues that are of importance to the public and its interest groups more complicated—no longer being soot black or cloud white. Development of solar and alternative energy sources, environmental health, energy conservation, low-scale technology and waste resource conservation are topics at hand. The potentially massive amounts of federal funding, windfall profits and long-term impact on entire areas of science and its application fuel intense battles. Moreover, fuzzy or nonexistent lines of support or constituency on specific issues make successful conflict resolution tough.

Such confusion only heightens the potential for personal profit. Thus the consequences of mistakes or losses can be serious.

The currency of these struggles is power. Individuals or groups without it are the losers; those with it the winners; those with only small change ineffectual. At present much of the power in Washington is through money, sometimes in unfathomable amounts. Although a six-digit checking account balance immediately gives one recognized status and power, money is not the only access to power. Constituency, contacts, coordination with other groups or networks and credibility also open doors, achieve goals, sway votes or change behavior. In addition, working through grass roots constituency develops lines of support, and assures that issues will be addressed from less of a special-interest, private-gain view. Money alone does not.

Public interest has typically not operated on a sheer money basis, but rather through organizing around legitimate issues, many times displaying effective understanding of political battle tactics. Vocal and informed constituencies have been formed that have expressed their wrath through phone-ins and write-ins or more recently on the clean air act: a cough-in by the National Wildlife Federation's membership. But decisions and topics now before us are many times more subtle; the chance for personal gain, unfortunately, many times greater.

For instance, the implications of subtle wording differences or amendments may seem insignificant at first glance. Thus a proposed change allowing federal clean air limits to be exceeded for only 12 days a year would have seriously reduced air quality in parts of the country.

Small changes in exposure levels of workers or public can mean life or death to hundreds of citizens. Nuclear technology as only a last resort in the President's energy plan may actually delay or possibly kill solar or alternative energy research. Massive, high-technology waste-recycling units may ultimately work against recycling, energy conservation and employment.

The chances of raising the public to the levels of outrage as with DDT or the extinction of certain species is seldom possible, probably not desirable and can backfire as in the cancer-saccharin debacle. Education to the issues of cancer rather than use of fear might have avoided that problem. Thus a change in role and tactics, grounded in the basic ethics of public interest is urgently necessary as the issues become more technological, the groups proliferate, and governmental input on the part of citizens and their interest groups becomes institutionalized.

At the most basic level, I want to look at ethics because they speak to actions, highlight values, elucidate decisions and provide direction. One's ethics can be arrived at by viewing rights and concomitant responsibilities. Roles, actions and the use of power may be assessed, means judged and worthiness analyzed. It is precisely at this point, the infusion of ethics into the public interest issues, that public interest should be viewed, evaluated, judged and modified. It is here that some fail. Groups without ethical touchstones run the very real danger of being dizzied in Washington's rarefied air—losing perspective, becoming their own constituency and rationale for existence. It seems that much of Washington's federal bureaucracy would function unaffected if the 50 states no longer existed—so would such public interest individuals.

The basic rights of public interest groups are much the same as those of individuals. Many of these rights were set out at the recent conference on Toxic Substances and Trade Secrecy. The major rights determined are:

(1) To exist: Public interest groups share a recognized legitimacy greater than before, but far from universal.

(2) To gain access to information: Decisions on the dangers of exposure of certain chemicals, or the determination of conflicts of interest on the part of scientists, officials or agencies require adequate information.
(3) To communicate the issues through the media.
(4) To be free from harassment: One right still unrecognized by some agencies.
(5) To redress: Groups should be able to petition regulatory agencies and have legal standing in court.

But rights imply responsibilities and functions to perform. For my purposes these would be carried out in the arena of politics, technology, science and local communities. Defining some of the responsibilities may in fact be easier than defining public interest. Some of the specific accompanying responsibilities are:

(1) To stay informed of the issues at a grass-roots level, communicating with a constituency, and reflecting their concerns nationally when required.
(2) To act competently, authoritatively and accurately on issues.
(3) To follow issues through, avoiding the shallowness caused by continual pursuit of media exposure.
(4) To further the public interest by providing experience to those new to and attracted by public interest issues.
(5) To respect opponents, constituents, members and co-workers.
(6) To be accessible and open to the public. Sunlight remains the best cure for diseases that spawn in the darkness of secrecy.
(7) To educate the public to the importance and implications of the issues as well as current research and the means to apply it. Research application has been a real shortcoming of scientists. Examples abound of research lying dormant on lab or library shelves. Some solar research has never seen the light of day because of the lack of application efforts and information dissemination. Public interest should provide this linkage.
(8) To work in the public interest, not narrow special or self-interest. To this extent, revolving doors between public interest groups, government, and industry or political springboarding is suspicious and easily wrecks hard-earned credibility.
(9) To cooperate with other public interest groups whenever possible. Power divided is power lost, particularly now that lobbying and influence peddling has achieved new levels of sophistication. In sharing efforts, groups could become more widely known and their particular expertise clearer. Limited resources would be used more effectively.

Individuals are usually responsible to someone for something. In the case of public interest it is a constituency, members, or other groups, particularly at the local level. A vigorous set of grass roots supplies life and energy to a group, provides sources to draw from and give to. No matter how complex, how technical the issues involved, they must be communicated, reflected, reformed and focused.

Many times in espousing distance and separate identities from existing power groups (the feds, industry and unions) the lack of roots of the individuals and potentially their organizations stand quite starkly. Moreover, some actually seem to be pushing their own wares on government, later to become bureaucrats. Furthermore, in making what seem to be adolescent assertions, by fighting everything that exists, some groups are losing sight of similar groups facing similar problems, with similar responsibilities. In this case I am specifically referring to unions. Unions have spent much of their pasts gaining recognition, building grass roots, and garnering power. At present they are working at educating members to issues that affect them and have obvious concern with environmental health. Moreover, their members research future technologies, manufacture present technologies and suffer the consequences of past technologies. The similarities of unions and public interest groups give further insight into their functions, their pasts and futures. With that in mind, some questions arise over current conflicts.

Why are union–public interest relations so poor? There seems to be a flow of people from public interest into government and industry; why not unions? As mentioned in the beginning of this paper, communications with unions are wretched. Could this reflect on public interest, on their actions, showing them to be out of phase with the responsibilities just stated?

Could the fact that unions have been falling into issues of narrow self-interest strike a fearful chord of similarity in public interest groups?

Are the unions looked upon as lacking the necessary prestige for such important public interest individuals to consider working for them?

Is the constituency of unions too clear, exposing the rootlessness of some public interest groups?

I really believe that public interest has much to learn from union mistakes and successes. Cooperation and close relations between these groups could help public interest avoid the major pitfalls that unions have had, and help unions out of narrow, self-interest positions which in the long run also hurt the unions. Furthermore, as technological issues arise and decisions are required, a strong relation between the two will aid the decision process and check basic responsibilities.

At present, industry is all too aware of the poor relations between unions, public interest and government. The "jobs vs. environment" phrase has been swallowed by many, while the facts do not substantiate the dichotomy. Until communication improves, industry will be able to continue to exploit the falsehood, weakening potentially powerful foes.

A RECONSIDERATION OF THE IDEA OF A SCIENCE COURT

Alex C. Michalos

In August 1976 the U.S. Task Force of the Presidential Advisory Group on Anticipated Advances in Science and Technology published an interim report on something they called "The Science Court Experiment."[1] It struck me immediately as a potentially sound idea needlessly and perhaps hopelessly encumbered by some pretty poor philosophy. It seemed to be exactly the kind of idea that ought to receive but has not yet received serious attention in the philosophic literature on science and technology.

My plan is to present an outline of the proposed Science Court Experiment, and then to offer replies to several prima facie plausible objections to it. Although proponents of the Court may occasionally suspect that with friends like me they really don't need any enemies, I must admit (insist) that I'm engaged in a salvaging effort. I still think that we have witnessed the birth of a good idea. It just needs a little love and attention from the proper nursemaids.

The following outline has been written and circulated by the chief architect of the Science Court Experiment, Arthur Kantrowitz.

OUTLINE OF THE SCIENCE COURT EXPERIMENT

This is an abstract of the procedures with a fuller explanation to be found in the Interim Task Force Report (*Science*, August 20, 1976). One procedure, #15, has been added since the Interim Task Force Report.

Initiation

1. The experiment will begin with agreements between the Science Court Administration (SCA), the host institution and the funding institution concerning the plans for the experiment.
2. The first step will consist of contacts with regulatory agencies and others to search for a suitable issue. The first issue will be presumed to be bipolar and will be presumed to be a value-laden decision which must be made by the agency and in which the scientific facts are apparently in doubt.
3. An agreement with the regulatory agency to supply necessary legal powers for uncovering nonpublic information if necessary.

Organization

4. The SCA, equipped with an issue and funds, will seek Case Managers (CMs) for each side of the issue. The work of the CMs will be funded by the SCA.
5. The SCA will issue suggested procedures, suggested referees, and a panel of prospective judges.
6. The CMs will review these suggestions. The experiment will go forward only after both CMs agree on the referee, the judges and the procedures. A signed agreement to proceed will commit the SCA, the funding agency, the regulatory agency, the CMs, the judges, and the referee.

Suggested Procedures

7. The CMs formulate a series of factual statements which they regard as most important to their cases. Such statements must be results or anticipated results of experiments or observations of nature. The statements should be ranked in order of importance assigned by the CM.
8. The judges examine each statement to determine that it is a relevant scientific fact.
9. The CMs then exchange statements. Each side is invited to accept or challenge each of the opponents' statements.
10. The list of statements accepted by both sides will constitute the first output from the S.C.
11. Challenged statements are first dealt with by a mediation procedure in which attempts are made to narrow the area of disagreement or to negotiate a revised statement of fact that both CMs can accept.
12. The mediated statements are added to the S.C.'s output. Those statements which remain challenged are then subjected to an adversary procedure.
13. CMs prepare substantiation papers on statements remaining challenged and transmit these to the judges and the opposing CMs starting with the first (most important) challenged statement.
14. The substantiation is cross-examined by opposing CMs and judges and contrary evidence is presented and cross-examined.
15. A second attempt to negotiate a mediated statement is made and if successful this statement is added to the Science Court's output.
16. If this is not successful the judges write their opinions on the contested statement of fact.
17. This procedure is repeated for each of the challenged statements.
18. The accepted statements plus the judges' statements constitute the final output of the procedure. (See Figure 1.)

Figure 1. Output Decision-Points of the Science Court.

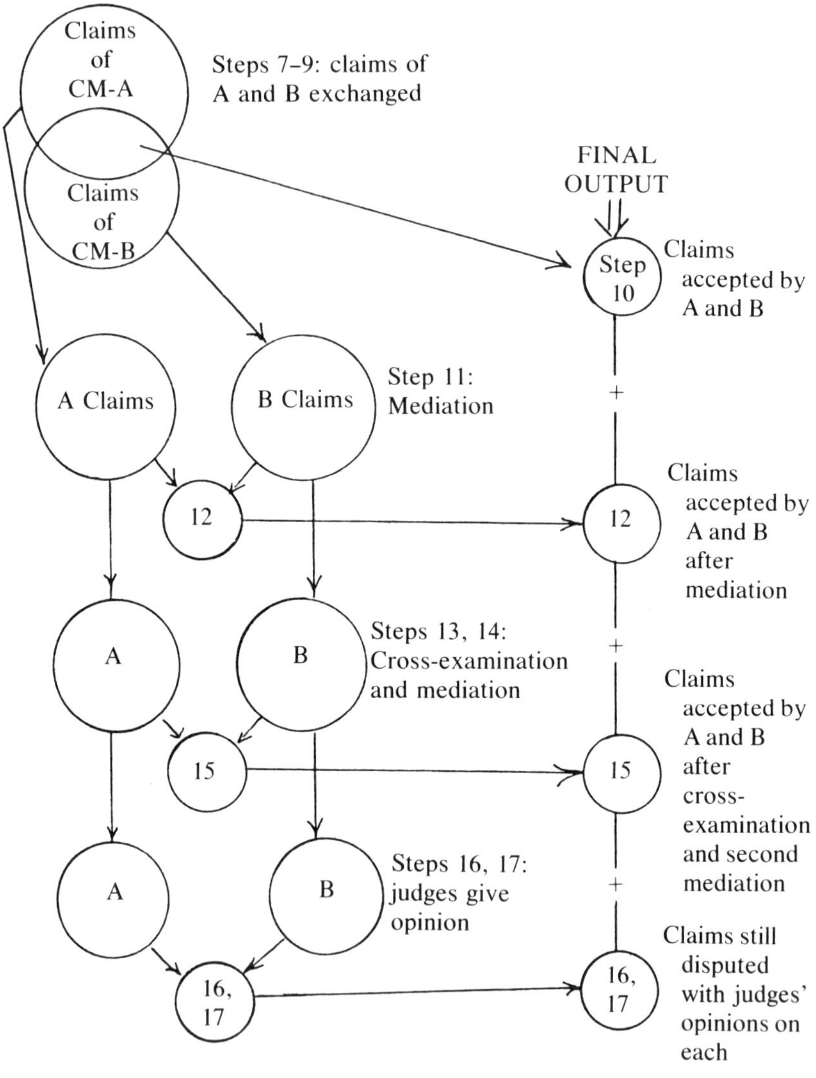

In the remaining paragraphs I will largely ignore the sections of the outline headed "initiation" and "organization," and focus attention on the section headed "suggested procedures." For ease of reference I have numbered all the objections, but the order is not significant.

Obj. 1. The term "Science Court" is a misnomer because

a) most of the examples used to illustrate the need for and operation of the court are drawn from technology and medicine (e.g., banning saccharin, Red Dye No. 2, legalizing marijuana).

b) use of the term "court" implies a verdict or a value-laden decision.[2]

Reply: There's no need for argument here on either score. We can think of the term "science" as merely an abbreviation for "science, technology and medicine." This may trouble linguistic purists but it need not trouble us. The procedure described by the Court's proponents may be applicable to problems that straddle disciplines or fields of study. For example, the question of whether or not saccharin is carcinogenic may be regarded as belonging to science, medicine, medical science or even technology. As long as we have an abundance of prima facie suitable cases for such an institution, we need not worry much about borderline cases.

The question of the term "court" implying "a verdict or a value-laden decision" is only a symptom of a deeper problem that proponents of the Science Court have with values. So instead of attacking the symptom, I think we should go right after the disease.

Obj. 2. The Science Court cannot work because facts cannot be separated from values.[3]

Kantrowitz's writings are speckled with assertions that in one way or another require a separation between facts and values. For examples, he claims, "The Science Court will restrict itself to stating facts,"[4] "Scientists take great pains to separate the factual statements they make from any of their personal values,"[5] "We can do a better job of separating facts from values if we address ourselves to that goal,"[6] "How will we actually go about eliminating values and irrelevancies from factual statements...?"[7] "I personally do not believe it is possible for scientists to have deeply held moral and political views about a question and simultaneously maintain complete objectivity concerning its scientific components."[8] "It is the old issue of the separation of facts from values, and I submit that this separation of facts from values can always be made."[9] "It bears repeating that the Science Court will stop at a statement of the facts and will not make value-laden recommendations."[10]

Reply: These are certainly ideas whose time has come, somewhere around 1935, I should imagine (for those of us with short memories). This is the stuff of which Vienna Circles are made. But a circle by any other name finally leads nowhere.

It's difficult to know where to begin arguing with someone who seems to be committed to the view that evaluations are worthless or, what is almost as bad, that allegedly value-free claims are somehow better than value-laden claims. On the face of it people with such views seem to have no appreciation for logical absurdity in the strict sense of self-contradictoriness. After all, it is patently self-contradictory to affirm, for example, that evaluations are worthless, but this evaluation of the worthlessness of all evaluations is true. If this value judgment is true then the principle of the worthlessness of evaluations is false. Similarly, if it is true that allegedly value-free claims are somehow better than value-laden claims then one who says so would be better advised not to say so. He ought to know better on his own principles. In short, it is plainly logically impossible to say anything good or bad about anything without saying something that is value-laden. It is just literally absurd to pooh-pooh values in the interests of good science, good methodology, good sense, good manners or good God almighty.

Obviously, I have come not to praise this feature of the Science Court, but to bury it. However, I intend to breathe some new life into the Court out of the ashes of this fact-value problem. The reply I want to offer to the fact-value dichotomy objection is this: It's a red herring. The distinction between facts and values is not essential to the main purposes of the Court and it's a pity it has been given the center of the stage. The main purpose of the Court, as I see it, is to arrive at a *timely and authoritative agreement* about issues broadly classifiable as scientific. The idea is to design a procedure that will give elected representatives access to the most reliable and valid information relevant to such issues as soon as possible. Although it has been called a court, it is plainly in the family of special task forces, commissions and boards of inquiry. Just as we often describe such institutions as ad hoc fact-finding bodies, it is natural to think of the mandate of a Science Court in the same way. Nevertheless, just as task forces, commissions and boards of inquiry tend to focus on what is important, whether it's overtly value-free or value-laden, we ought to think of the Science Court in the same way.

Apart from the fact that the objection is misleadingly beside the point of the main purpose of the Court and the fact that it's logically impossible to say anything good about anything without an evaluation, it is also proper to maintain a certain amount of humility in the presence of notoriously thorny epistemological problems. From an epistemological point of view, there is no way of distinguishing what is *generally agreed to be* the case from what *is* the case. Logically or conceptually, of course, there is a difference. It's roughly the difference between epistemology and ontology, i.e., the difference between warranted assertion and the real furniture of the world. It is virtually a truism that the most soundly warranted of empirical claims may turn out to be false. But there is nothing analogous to this epistemological truism at the level of ontology. What is indeed the case cannot turn out to be not the case in the same way, at the same time and so on. This is not to deny the possibility of change, just the possibility of things existing with contradictory attributes. Whenever we arrive at what we are inclined to say *is* the case, I think it finally has to be admitted that we have merely arrived at an epistemological resting place. What we know is always subject to error, correction, criticism and so on. If what we think we know is never shown to be in error then it may be fairly said that *for all we know,* we know what is the case—not merely what is generally agreed to be the case. Still, there's no great triumph here since we began with "for all we know," a nice epistemological caveat.

If you agree that in the last analysis we can't distinguish facts from what are generally agreed to be the facts, then I think you have a good reason to regard the fact-value dichotomy as largely inconsequential. From an epistemological point of view one is in exactly the same boat in the case of values that one is in in the case of facts—assuming one could distinguish them. The most one can hope to do is get general agreement about this or that evaluation. There is no way to cross the logical gap between this epistemological position and the ontological ground one hankers for. Just as there is no way to demonstrate that what we all agree is the case is the case, there is no way to demonstrate that what we all agree is good is good in and of itself. Maybe there are naked facts and naked "objective" values in the world to be discovered, but there does not seem to be any way to reach them. Logically, I just don't see any way to bridge the gap from epistemology to ontology, and I don't see that we are at all better or worse off with values than we are with facts—supposing they could be distinguished. Thus, it would seem to be a waste of time and effort to make a big deal out of determining facts only and scrupulously avoiding values. It would be wiser just to try to figure out what we all think is true, well supported or worthy of belief, whatever its ontological makeup happens to be. Instead of raising misleading criticisms like "That's a value judgment!" as many scientifically trained people are inclined to do, one ought to say "That's unwarranted!" Then the air is cleared immediately for rational debate over the reasons underlying one's claims, instead of being swiftly engulfed in the metaphysical fog of yet another unproductive fact-value debate.

Obj. 3. A Science Court would overemphasize facts and underemphasize important values, thus becoming a misleading and possibly harmful institution.[11]

Reply: In view of my remarks in response to objection 2, little more has to be said here. The fact-value dichotomy is inessential to the Court, and what is finally emphasized or ignored will be almost entirely a function of problem-specification, search procedures and personnel. These elements will require much more attention than the question of what is factual and what is not.

Obj. 4. The proposed Court procedure is unworkable because good people will not be willing to "risk interruptions in their careers and possible damage to their reputations to join in a procedure that is as yet untried and undeveloped."[12]

Reply: Kantrowitz's own reply to this objection is that "the motivation to participate must stem from an enthusiasm for bringing before the public and policy-making officials the best possible formulation of the state of scientific fact when such formulations are needed for urgent policy questions."[13] Presumably the "enthusiasm" is based on patriotism and pride.

I have never seen an investigation of the quality of commission and task force researchers compared to run-of-the mill researchers. The fact that there have been some splendid commission and task force reports shows that it is possible to get good people to shift their research interests in the public interest. Whenever a new bandwagon is formed, it's likely that a lot of people are inclined to join because they have nothing to lose, either because they are so close to the bottom or to the top in some self-selected hierarchy. But this is completely compatible with the possibility of getting first-rate people to venture into new areas.

Obj. 5. The procedure will be unworkable because it involves a winning and a losing side, and no one of any stature is going to risk public exposure as a loser.

Reply: Insofar as the scientific enterprise is regarded as an ongoing and self-regulating activity, no researcher has to be characterized as a loser. Instead, they may enjoy the luxury of thinking of themselves and even being thought of by others as participants in a more or less creative battle against ignorance. That's not only a wonderful image to face in the mirror every morning, it's also a terrific ego defense mechanism. It stands in sharp contrast to the image of an advocate throwing all his professional weight into a fight and being personally publicly pulverized. Nevertheless, two important points should be kept in mind.

First, it should be remembered that three out of the four decision or output points in the procedure do not identify a winner or loser. In step 10 a "list of statements accepted by both sides will constitute the first output." In step 12 "mediated statements are added to the . . . output." In step 15 "a second attempt to negotiate a mediated statement is made and if successful . . . is added to the . . . output." Finally, "the judges write their opinions on the contested statement of fact." By the time this final output emerges everyone should have had enough clear or quasi-successes to be able to stand a little reversal.

In the second place, it should be possible to orchestrate the proceedings of the Court so that it is clear that all participants are engaged in research in the public interest. Logically I'm sure that's possible, but of course the force of the objection was not logical but psychological. Unfortunately, the empirical evidence for the objection is not available yet.

Obj. 6. The Court procedure is unworkable because all the knowledgeable personnel will be committed to one side or the other of any live issue.[14]

Reply: This objection fails to distinguish the types of knowledgeable personnel required by the Court, namely, case managers (advocates), referees and judges. Case managers should be knowledgeable advocates for a particular position. They should combine a Popperian zeal for criticism with a Carnapian penchant for justification. In the best of all possible worlds they would be as critical of their own views as they are of those of the other side. In such a world much of the force of the fifth objection would be dissipated. The trouble is, of course, that the real world may be a long way from this ideal fantasy land. However, we shouldn't try to slay more than one dragon at a time. The main problem with respect to biased advocates was considered in the previous objection, and that's where the matter has to rest.

So far as referees and judges are concerned, bias is intolerable. The job of the referee is to see that the decision procedures agreed upon by both sides are scrupulously followed, and the Task Force has even suggested that referees should be "advised by legal counsel."[15] The job of the panel of judges is, again, to "write their opinions on the contested statement of fact."[16]

Judges and referees should be scientists in the broad sense of the term used here, but not specialists in the particular issues to be debated by the Court. Since contemporary science and practically every other field today tends to be dominated by specialists, it shouldn't be difficult to find people who are very good at something that is structurally, logically, methodologically and/or conceptually similar to issues going before the Court. For example, it should be easy to find chemists, epidemiologists or pharmacologists willing and able to serve as judges in cases involving the carcinogenic effects of saccharin, even though their own research has not dealt with that particular chemical.

Obj. 7. The Court is unnecessary because we already have too many commissions, task forces and special boards.

Reply: I really like the spirit of this objection, and I don't think it represents a serious problem. Although many advisory bodies come into being and seem to persist long after there is anything useful for them to do, some don't. In Canada, for example, royal commissions on the drug industry, taxation and farm machinery have come and gone. The U.S. Commission on Obscenity and Pornography had a short life. So we might think of the Science Court as analogous to such short-run commissions of inquiry rather than to standing committees or boards. Indeed, if we think of the Court primarily as a *procedure* rather than an *institution,* much less a structure literally set in concrete, we probably come closer to giving it an adequate characterization. It's entirely possible that the Science Court procedure could be used as the operating procedure of a commission, task force or board of inquiry. Since it's roughly a mixture of Delphi, adversary and mediation tactics, several of its features would be suitable for other boards of inquiry and possibly even familiar to board members.

For the record, I should mention that Kantrowitz has claimed that "it would be expected that periodically repeated procedures would be necessary with the advance of scientific knowledge, in contrast with 'single jeopardy' in some legal court proceedings."[17] Although this suggests a standing body or institution of some sort, I think the force of most of his writing is on the side of a decision-making procedure. He has also claimed that at the Leesburg Colloquium on the Court "the consensus was that we should really be planning a procedure rather than an institution."[18]

Obj. 8. The Court is unnecessary at best and harmful at worst because its output will only be used if it's compatible with policies preferred by those in power.

Reply: This is an objection that applies as well to virtually all research institutions. Even if one accepts its premise as generally true, it's difficult to accept its conclusion. Insofar as the objection applies to all research institutions, the result would be the dissolution of research in the interests of preventing its misuse. That may be wise in specific instances, but it can't be plausibly recommended as a general policy. As a general policy it's roughly equivalent to insisting that wisdom is always on the side of flying by the seat of your pants, a pretty silly proposition, to be sure.

Obj. 9. The Court will be self-defeating because it is based on an adversary procedure, and such procedures have as their aim not the discovery of truth but the resolution of disputes without violence.[19]

Reply: This is a false dichotomy. The pursuit of truth and the resolution of disputes are neither exclusive nor exhaustive aims of courts of law or of the proposed Science Court. If we have learned anything since the Popper-Carnap controversy in the fifties it is that truth or proximity to the truth are not necessary or sufficient for the acceptability of all scientific hypotheses in all circumstances. Our scientific journals are stuffed with trivially true hypotheses tortuously defended with exotic experiments and magnificently sophisticated statistical machinery. It is clear, however, that very often a wide variety of considerations like relevance, explanatory power, internal and external coherence, and so on must be weighed in order to determine the most acceptable hypothesis at a given point in time. So the model of a court procedure for decision-making in the realm of science broadly conceived strikes me as eminently reasonable. Indeed, as I have argued at length elsewhere, I believe rational decision-making generally is more appropriately characterized by the model of a judge weighing considerations in a trial than of a druggist weighing chemicals in a laboratory. The richness of most decision-making situations just tends to be obliterated in the druggist model.

It is also important to remember that the idea of a Science Court has emerged as a response to contemporary conflicts in science broadly conceived. It is precisely because we have had and will continue to have so many serious disagreements over scientific-technological issues that the idea of the Court has had initial plausibility. Nuclear energy, solar energy, SST, space shuttles, Red Dye No. 2, monosodium glutamate, saccharin—the list is virtually endless. Clearly we cannot subject our people to a continuous barrage of quasi-popular, media-polished claims and counterclaims about all the benefits and costs of this or that scientific theory or technological device. The result would be confusion at best and decision-making controlled by the loudest voices, at worst. The best hustlers might lead us to the worst calamities. The idea of having an adversary procedure with rules of evidence and fairness instead of a steady diet of media events seems to be just what's required for the kinds of conflict resolution envisioned by the Court's proponents.

Obj. 10. The adversary procedure proposed for the Court will be self-defeating because it will tend to polarize issues and exclude intermediate positions.[20]

Reply: Such a procedure could have the effects mentioned in this objection, but there is no good reason to think that it must have those effects. It's true that in the design of the original procedure it is assumed that the issue investigated will be bipolar. I assumed that this was a matter of simplicity and convenience, rather than a limitation in principle. In principle and in fact ordinary courts are not limited to the adjudication of bipolar issues. So the Science Court doesn't have to be limited that way either. Furthermore, since each case manager will be producing and reviewing several relevant statements, I doubt that the exclusion of intermediate positions is likely. If I had to bet, I would bet that the procedure will be threatened more by an overabundance of more or less relevant views than by polarization around a couple. One of the most difficult problems of group decision-making is getting members to collectively pursue a managable set of aims, rather than individually taking off in different directions. So I think the constraints built into the Science Court procedure will be remedial for the sort of behavior most likely to occur.

Obj. 11. The Court will turn out to be self-destructive because lawyers will eventually have more to say than scientists about its proceedings and then respectable scientists will abandon it.[21]

Reply: Since the vast majority of participants in the Court are going to be scientists, I doubt that lawyers will eventually have more to say. However, supposing that respectable scientists do abandon it, a couple different things can happen. On the one hand, the machinery could continue to crank out garbage, which in the worst of all possible worlds would be joyfully consumed by everybody with disastrous results. On the other hand, the machinery could crank out garbage which is immediately identified by its smell and thoughtfully committed to the nearest trash can. Hopefully, we would be nearer to the best than to the worst of possible worlds, but there's no guarantee of that. Still, these alternative scenarios are logically possible consequences of most decision procedures. So they don't represent special problems for the Science Court.

Obj. 12. The Court will operate as a loyal agent of whoever sponsors it; so its output is bound to be biased.[22]

Reply: I have no illusions about the ability of people to avoid bias in their everyday affairs or their scientific research. But a Science Court would not be any more exposed to this sort of aberration than any other research institution. On the contrary, the adversary procedure involved in the Court may contribute to the reduction of bias, as its proponents originally hoped.

Obj. 13. Judgments of the Court would be misleading guides to public policy because they would overemphasize technical and underemphasize all other dimensions of problems.[23]

Reply: This objection is a variant of the third, where the issue was put in terms of facts versus values. But this time it seems as if what's being resisted is the very idea of a division of labor in the review of scientific issues. Presumably such a division is precisely one of the central aims of Science Court proponents. In defense of the latter, it ought to be emphasized that Science Courts are not in any way exclusive of other decision-making institutions. There is no reason in principle why such Courts could not work in tandem or simultaneously with commissions, task forces, boards of inquiry, civil servants, elected representatives or anyone else. For this or that specific problem, the Court may provide an excellent auxiliary approach toward a solution.

The question is: Would such a suggestion fail by giving up more than proponents of the Court want and less than the opponents want? For all I know, the intellectual arteries of both sides may have hardened to the point where proponents and opponents see no need or possibility of accommodating the other side. Opponents may feel that the kind of technical decision-making apparently envisioned by Court proponents is really illusory. In economic jargon, they may believe that there are no real externalities; there are only costs and benefits which for one reason or another one refuses to face. Proponents may deny just that, or they may accept it and insist that there are good reasons for scientists or people with other special interests and expertise to treat some effects as external. In other words, they may just think, rightly I suspect, that not everyone can or should try to attend to every aspect of every problem. Some division of labor is necessary. In particular cases it may be difficult to decide how to cut the pie, but I don't think this operational problem is sufficient to reject the very idea of having a division. After all, lots of things are perfectly acceptable in principle although they are often tough to operationalize; a just society, for example.

Obj. 14. The Court would be self-defeating because its decisions would "seem to carry more authority than a conventional publication," and it would be extraordinarily difficult to alter them.[24]

Reply: This is another objection that seems to be based on a disagreement of principle rather than of any particular feature of the Court itself. One of the benefits of the Court, according to its proponents, is that it would produce authoritative decisions about scientific issues. The point is to create an institution that would bring diffuse sore spots to a head. The objection seems to be tantamount to the view that the proposed cure would be worse than the disease. If I can squeeze my metaphor a bit further, the objection seems to assume that scientific sore spots are self-limiting anomalies which ought to be allowed to run their course.

I don't think the objection comes to grips with the problem that Science Court proponents are trying to solve. The problem is that in the world of ordinary affairs decisions often have to be made sooner rather than later. Some of these decisions require scientific information that may be widely dispersed in the literature, very complicated and loaded with contradictions. If any old staff assistant is sent for the latest word in some area, he or she is likely to return with only half the story, and the result could be disastrous for public policy. The Science Court is intended to help reduce the risk of some such catastrophes. It represents a major screening device for scientific decision-making in the interests of preventing major and catastrophic consequences. The expected benefits of providing this type of public service seem to outweigh the expected costs of introducing some rigidity into the scientific enterprise.

Obj. 15. The Court would be unworkable because we have no criterion for deciding when a case has been conclusively established.[25]

Reply: Presumably this is a by-product of the view I have already granted that empirical claims are always in principle corrigible. However, the fact that it's always logically possible to falsify such claims is not a reason to doubt them or to say that they have not been established. For example, it would be silly for anyone to doubt that everyone dies, although it's logically possible that someone won't. Just as we have excellent reasons for believing that everyone dies, although it might be difficult to articulate the general criteria that have to be satisfied in order to reach that warranted certainty, we may have excellent reasons for believing more esoteric things. We manage to muddle through epistemological problems with the same facility that we muddle through political problems. Our paths are seldom logically tidy, but the human race would probably have gone the way of the dinosaur by now if our paths weren't more or less correct, more often than not.

Obj. 16. The establishment of a Court would undermine efforts to persuade working scientists to be more concerned with their broader social responsibilities.

There is a significant tendency for all human beings to try to put restrictions on the numbers and kinds of things for which they may be held responsible. In other words, almost everyone likes to pass the buck. So, whenever anyone suggests the need for a new institution, structure or organization designed to put an end to buck-passing, one may be justifiably skeptical. In the case of the Science Court, the fear is that while some of us will be occupied with the business of creating a purification system for scientific information, the vast majority of working scientists will still be busy creating an increasingly unmanageable cesspool. The problem is not structural or systemic, but personal. If individual scientists and their sponsors had the pursuit of truth and the avoidance of falsehood as their primary aim, there would be no need for a Science Court. Furthermore, the more we talk about the latter, the more the focus of attention is shifted from real to cosmetic solutions.

Reply: As you might imagine, the author of the previous paragraph is fairly sympathetic to this objection. But he doesn't, i.e., I don't, think it's decisive against the idea of a Science Court. It cuts too deep and it poses a false dichotomy. If it were accepted as a decisive consideration in rejecting the idea of a Science Court, it would be difficult to consistently avoid appealing to it whenever any new structural changes are proposed. In some measure I think it ought to be used fairly generally, but as a rule of thumb rather than a knock-down argument. One ought to be skeptical about organizational or institutional solutions to problems, all right, but healthy skepticism shouldn't prevent the consideration of every case on its own merits. Apart from that, however, one really doesn't have to choose a priori between designing systems or structures that facilitate one's aims and recommending responsible individual behavior. There can certainly be a conflict of interests between such designs and recommendations, but even *within* science and *within* the realm of morality there can be such conflicts. Sometimes it's just impossible to satisfy two scientific aims or two moral aims at the same time, but the existence of such conflicts doesn't demonstrate that, a priori, one must choose between any two scientific aims or any two moral aims. Particular conflicts of interest have to be resolved on the basis of the particular benefits and costs involved case by case, the hard way.

Obj. 17. The Court will be undermined by those who will use it to stall for time and those who will use it to hurry decisions that require more time.[26]

It is likely that representatives of industries that have sunk a lot of research and development resources into some new product will be anxious to have market clearance for it. So they are likely to press for a hearing by a Science Court. Once a hearing is granted, they are bound to have much more information about their product than anyone else. They will have a definite edge over competitors at the start of the hearing. Hence, the sooner they can force a decision, the more probable it is that their initial advantage will be maintained and that they will have their way. On the other hand, competitors who find themselves on the short end of such a battle may subvert the proceedings by delaying them until relevant facts emerge that tip the scales in their own favor.

Reply: Although one can solve the problem of delays with time limits, one can't solve the problem of rushed decisions that way. It is also not plausible to expect that every side of every issue will be defended with exactly the same amount of resources. So I think one has to assume that some issues will come before the Court with a handicap and others will come with an edge. The question is then: Can one compensate for such anomalies? Clearly some useful steps can and are taken in the interests of giving people an equal opportunity to defend their positions in ordinary courts. However, virtually every public opinion poll, specialist and official commission that has addressed the problem of equality in North American courts has concluded that rich people are better served than poor people. So, while one may design compensating devices for the resource-poor participants in a Science Court, it's not likely that they will be able to make full compensation.

Nevertheless, I believe the safeguards built into the Science Court procedure give us more protection than available alternatives. This is basically the result of the hybrid nature of the Court. Bits of Delphi, mediation and adversary procedures have been blended together for maximum strength. Besides, it should be noticed that the third item of the proposal calls for the "necessary legal powers for uncovering nonpublic information if necessary" and the fourth item stipulates that case managers are to be funded by the Science Court Administration. If reputable judges can be found who have no special ax to grind and no particular bias in favor of anything but scientific probity, I don't see how we can go wrong. If one side has an enormous advantage over another when the hearings begin, that would certainly be taken into account in the final report. There is no reason why we cannot require the reporting of such essential facts in every case. Indeed, there is no reason why judges could not dismiss a case on the grounds that the sides are so poorly balanced that a fair, unbiased hearing could not take place. This would make much more sense than going through the whole exercise and then announcing that there was, after all, no way to make it fair.

Obj. 18. The proceedings of the Court will be biased by the resources of the advocates, their style, cleverness, determination and information.[27]

Even if all advocates are committed to nothing but scientific probity, they may be more or less persuasive than they should be. If it is assumed that there are some more or less neutral facts to be discovered, then in principle at least it's possible for advocates to prove too little or too much. Apart from this unwanted result, a stronger advocate may overwhelm a weaker one. The stronger advocate's report may receive more attention and dominate the view of the final report.

Reply: Broadly speaking, I suspect that the human resources referred to in this objection are normally distributed. That is, I think the chances of picking individuals with more similar than dissimilar relevant characteristics are far greater than the chances of picking individuals with the opposite characteristics. Even assuming that the draws are not random but very selective, I would expect to see the same sorts of advocates chosen. Certainly in the long run, if there is a long run, there will tend to be some convergence on people with characteristics that seem to yield generally satisfactory results in the Science Court environment. While one should expect more disparity in the short than in the long run, I can't see any good reason to expect much disparity in either case.

I think it should be emphasized again that judges have the option of dismissing cases in which there is little or no hope of a fair hearing. That may be putting a heavier burden on the judges than they ought to have to bear, but I don't see any escape from it. In ordinary courts I doubt that cases are ever dismissed on the grounds that one of the parties involved has resources that are vastly superior to those of the other. I imagine that the party with fewer resources just loses in one way or another. By refusing to take on such cases, the Science Court would be a bit better off than ordinary courts. Moreover, the public interest would be better served by having a court that occasionally dismissed cases than by not having a court at all. In the former situation we are informed that there is apparently no fair way to judge the issue, while in the latter situation we don't have such information. Thus, in the latter situation, which is what we have now, decisions may be made anyhow on the basis of a hopelessly biased hearing. Whether or not they are made depends on how this or that particular commission or board operates. On the contrary, the Science Court is being designed with the option of dismissal-due-to-inherent-unfairness being built into it.

Obj. 19. The Court procedure tends to reinforce the misleading impression of science as a field in which complete certainty is possible.[28]

Reply: This is roughly the opposite of the problem mentioned in Objection 15. There the worry was that we could not establish claims. Here the worry is that we will create a demand or expectation for certainty that we cannot deliver. Since the very point of using the Science Court procedure is to try to reach some sort of a consensus of views about issues that patently defy complete certainty, that are too complicated, fuzzy, open-ended and so on to be nailed down once and for all, I doubt that the Court will create a misleading impression. I believe that such an impression has been created by schoolteachers and texts, news media and some practicing scientists, and that these same people and institutions will have to play a major role in setting the record straight. The only way the Court might inhibit this activity is by creating the impression that usually there is no need for a Science Court because scientific problems tend to be clear, simple and solvable with complete certainty. That is, by insisting that the Court is designed to handle extraordinary problems, we have left the door open for the inference that scientific problems typically have characteristics just opposite of those brought before the Court. So the existence of the Court might lead to the problem cited in Objection 19 but in a roundabout fashion. I suppose it might; but I doubt that it would create more than a ripple in a sea of misleading impressions. What's more, there will be strong forces operating in the opposite direction. Presumably participants in the Court procedure will be fully aware of the uncertainties with which all scientific investigation is entangled.

Obj. 20. A Science Court would be dangerously elitist, operated by and for recognized authorities in various fields.[29]

All the nice talk about reputable scientists pursuing the truth in an unbiased way is really sop. A lot of money is going to be involved in the operation of the Court, perhaps one to ten million dollars.[30] No one is going to give that kind of money to Joe Blow. Indeed, proponents of the Court have never suggested that the Court should be run by anyone but extraordinary scientists. In the words of one participant in the Colloquium on the Science Court, "What we have had here for two days is a gathering of important leaders agreeing that important decisions should be made by important leaders. . . . These leaders have told us why the judges have to be important leaders. Dr. Kantrowitz . . . said that, in the great public debates, too many scientists are going around the nation spreading wrong information. The science court is supposed to stifle those voices with its authority. . . . What is sought here is stipulation as the facts."[31]

Reply: This is roughly the opposite of Objection 4, that "the best and the brightest" will not come forward to serve the public interest. Here the fear is that "the best and the brightest" will come forward, all right, but not necessarily to serve the public interest. In fact, there is a strong suspicion that they will merely serve their own interests. Again, I don't think there will be a serious problem of finding good people. However, the interesting thing about this objection is its fairly overt suggestion that one should not try to find exceptional people. Put in its boldest form, the suggestion would appear to be that scientific issues ought to be settled in a strictly democratic fashion, i.e., in a public debate among all interested parties. Although I find populism generally very attractive, I'm pretty sure it would be self-defeating if applied to many scientific issues. For example, how much light could an ordinary citizen be expected to shed on the question of whether or not saccharin is carcinogenic?

Perhaps the objector would want to allow only specialists of one kind or another to be advocates or judges, but have all participants drawn at random. If we didn't have some minimum requirements for specialists to satisfy in order to qualify for the pool, we could get some real duds. So we would have to have minimum requirements. For all I know a set of advocates and judges picked at random from such a pool of well-qualified people might produce excellent results. For purposes of experimentation, it would be worth a try. The risks of picking people with disparate resources and abilities might be increased with this procedure, but I don't know how much difference that will make. My guess is that if the advocates are worth their salt, they are going to consult recognized authorities and draw upon all available resources anyhow. So backing off from the attempt to select "the best and the brightest" immediately in favor of this more randomized procedure may not have a significant effect on results.

Obj. 21. Participation on other committees would be neglected because it would not have the prestige or other payoffs promised by participation on the Science Court.[32]

To take a single example, the U.S. "National Academy of Sciences has about six hundred permanent committees, each with probably twenty members."[33] If people become enamored with the Court, it may be impossible to find people to serve on all these committees. On top of that, there are thousands of other committees that require scientific consultants.

Reply: I doubt that any well-qualified scientist is going to sit around waiting for the phone to ring. Besides, there are many more thousands of scientists than committees looking for scientists.

Conclusion

Having presented the basic objections to the idea of a Science Court and plausible replies to them, I think the ball is now in someone else's court. The idea seems to have survived some serious criticisms and to merit further attention. I hope, therefore, that the U.S. Task Force will be able to get the "experiment" off the drawing boards in the near future. I would also like to see some efforts in this direction in Canada.[34]

NOTES

1. Task Force of the Presidential Advisory Group on Anticipated Advances in Science and Technology, "The Science Court Experiment: An Interim Report," *Science* (August 1976), pp. 653–656.
2. A. Kantrowitz, "The Science Court Experiment: Criticisms and Responses," *Bulletin of the Atomic Scientists* (April 1977), p. 44.
3. E. Callen, "The Science Court," *Science* (September 1976), pp. 948–951; L. Lipson, "Technical Issues and the Adversary Process," *Science* (November 1976), p. 890.
4. Kantrowitz, *Bulletin* (see note 2), p. 44.
5. *Ibid.*, p. 46.
6. *Ibid.*
7. *Ibid.*, p. 47.
8. A. Kantrowitz, "Controlling Technology Democratically," *American Scientist* (September–October 1975), pp. 506–507.
9. *Ibid.*, p. 507.
10. Task Force, *op. cit.*, p. 653.
11. Kantrowitz, "The Science Court Experiment: Criticisms and Responses," p. 46.
12. *Ibid.*, p. 47.
13. *Ibid.*
14. Kantrowitz, "Controlling Technology Democratically," p. 506.

15. Task Force, *op. cit.*, p. 653.
16. Kantrowitz, "The Science Court Experiment: Criticisms and Responses," p. 45.
17. *Ibid.*, p. 44.
18. *Ibid.*
19. N. E. Abrams and R. Stephen Berry, "Mediation: A Better Alternative to Science Courts," *Bulletin of the Atomic Scientists* (April 1977), p. 51.
20. *Ibid.*; Kantrowitz, "Controlling Technology Democratically," p. 508; P. M. Boffey, "Experiment Planned to Test Feasibility of a 'Science Court,' " *Science* (July 1976), p. 129.
21. Abrams and Berry, *op. cit.*, p. 51.
22. *Ibid.*
23. *Ibid.*
24. *Ibid.*, p. 52, Boffey, *op. cit.*, p. 129.
25. Abrams and Berry, *op. cit.*, p. 52.
26. J. Primack in the U.S. Department of Commerce, *Proceedings of the Colloquium on the Science Court* (Washington, D. C., 1977), p. 233.
27. *Ibid.*
28. J. Holdren, "The Nuclear Power Controversy," *Proceedings of the Colloquium on the Science Court*, p. 177.
29. E. Callen in the *Proceedings of the Colloquium on the Science Court*, p. 217.
30. J. Primack, *op. cit.*, p. 237.
31. E. Callen, see note 3, above.
32. J. R. Killian, Jr., in the *Proceedings of the Colloquium on the Science Court*, p. 287.
33. R. O. Simpson in the *Proceedings of the Colloquium on the Science Court*, p. 37.
34. I would like to express my thanks to Dr. Kantrowitz for kindly providing me with or directing me to most of the material referred to in this paper.

AGAINST FEYERABEND: THE MEANING OF PROGRESS IN SCIENCE

Michael Goldman

Throughout history, even in this century, it has been possible to find intellectual attacks on science following in the wake of social dissatisfaction with the scientific and technological enterprise. In his article "Weimar Culture, Causality, and Quantum Theory, 1918–1927: Adaptation by German Physicists and Mathematicians to a Hostile Intellectual Environment," Paul Forman considers the capitulation of even the most secure science, physics, to the irrationalism which was prevalent in Weimar Germany.[1] He argues that the development of quantum physics, with its attribution of a fundamental "irrationality" to nature, was a result of this capitulation.

Forman sees in contemporary culture much of the same anti-scientific attitude, and his perceptions seem to me to be correct. Not only a significant portion of the general population, but quite a few philosophers as well appear to have abandoned the respect for science so characteristic of our culture in the post-World War II period. This change in attitude is reflected in the neo-Luddite and Zero Growth movements, the "back-to-nature" and "natural life" tendencies which are so very strong in America today. There is some evidence that these kinds of anti-science, anti-technology attitudes arise coincidentally with a more universal social pessimism, and that this in turn reflects the financial and economic conditions of the culture rather than its actual scientific or industrial potential. When funds and social resources which ought to be devoted to economic development are instead earmarked primarily for debt service payments, and when it is financially necessary to keep capital investment low and unemployment high, there is, not surprisingly, a plethora of theories and social rationalizations for denying the funds necessary for scientific and technological development. Such was the case in post-World War I Germany, and such is the case in the Western world today. These rationalizations take many forms, as I have suggested, but my concern in this paper is with the philosophical (specifically metaphysical and epistemological) presuppositions offered in defense of these attitudes. According to Forman, it was the *Lebensphilosophie* of Oswald Spengler, together with an emerging existentialism which served as the most forceful theoretical justifications for the anti-objectivism implicit and explicit in the attack on science in Weimar Germany. Today, it is the so-called "new" philosophy of science, initiated in part by Thomas Kuhn and his "historical" approach to science, and by Paul Feyerabend with his admittedly "anarchistic" epistemology, which serves as the intellectual cutting edge of the generally pessimistic, anti-rational, anti-scientific movements.

I do not mean for a moment to suggest that these intellectual exercises, now or then, are decisive or even important for understanding the attacks on science as these attacks actually happened. Rather, these exercises must be understood as providing "cover," in the intellectual community, for the attacks, and it is only for the purpose of "blowing" this cover that these subsequent remarks may be valuable.

The special strength of Feyerabend's position is that it proceeds from a systematic and devastating attack on the standard positivistic defense of the objectivity of science. Feyerabend shows that no methodology, no theory of truth, no epistemology so far proposed can adequately account for all of the manifestations of science in history, much less for the historical development of science. He is convinced on theoretical grounds that there is no such single epistemology or methodology to be found. Instead we must be content with a succession of methodologies, each one coherent with its subject matter, but each appropriate to a very limited domain. In one of his most recent works, a paper prepared for and presented to the Conference on the Epistemological Relations between the Sciences and the Humanities at Miami University, Feyerabend took his most radical position to date.[2] In that paper he urged the "democratization" of science, and proposed that a community or society have the power to allocate resources for research along whatever lines, using whatever methodologies, the majority wished. Thus astrology, magic, witchcraft, etc., not only might be, but, Feyerabend thinks, would be allocated large sums of money at the expense of the more traditional approaches to science. Since there are no grounds, according to him, for preferring traditional approaches to these more exotic ones, such a turn of events is to be applauded for its coherence with the democratic ideal.

Feyerabend can reach this kind of conclusion only because his conception of science is of an activity wholly alienated from the material, physical world in which it must exist. Science for him is an intellectual-spiritual activity, one in which the achievements of the scientist are exhaustively expressible in systems of explanation. In this, he is typical of most contemporary philosophers, whether they agree with him or not about specifics. To be sure he rejects most of the restrictions placed on such systems by his philosophical predecessors: formal restrictions concerning axiomatic form, deductive-nomological explanation, etc., and epistemological restrictions concerning verification, confirmation, meaning invariance, etc.; but the results of scientific inquiry remain in and for the understanding. I am not saying that Feyerabend denies the significance of experimental work, even experimental field work. But such work serves the purpose of helping to create an intellectually satisfying account of phenomena, a characterization which is less significant than it might seem when it is remembered that for Feyerabend phenomena are themselves determinate (not determining) elements of the theory spun to account for them!

If this were all that science had to offer—if the creation of intellectually satisfying systems to which allegiance could be pledged on aesthetic, religious, or any other egoistically determined grounds were the essence of scientific activity—then we would be justified in asking, why bother to allocate resources to science at all? Such a question is not, after all, unprecedented. The arts and the humanities are often called upon (especially in financially troubled times, but not only then) to justify their continued support by the "productive" community, and when they are, the usual justification given takes the form either of an admission to the charge of parasitism together with an appeal to the "ornamental" value of such humanistic and artistic activities, or of a vague defense in terms of usually undefined "spiritual" values, by which the community will be "uplifted."[3] Such "defenses" of the arts and the humanities are hardly inspiring, and in times of material scarcity would fall, with justification, upon deaf ears. They tend to affirm the accusation implicit in the question, and offer a sermon instead of an answer. If the arts and humanities are to be defended at all, it must be in terms of the necessary contribution they make to the development of a creative and productive civilization, a defense which is quite possible but beyond the scope of this paper. If Feyerabend is correct, however, he has legitimized a similarly uninspired response by the scientist when his or her request for funds is challenged.

But Feyerabend is not correct about science (just as the "defenders" of the arts and the humanities alluded to above are not correct about these enterprises either). Theoretical science is an absolutely essential part of the human strategy for survival, and it is in terms of its success at that task that specific methods in the sciences can be evaluated. It is of course true that the scientist often contents himself or herself with exploring phenomena in an effort to create a coherent explanatory system for them. But it is not true that all the significant work is finished when such a system has been articulated. On the contrary, those efforts at system building serve to guide mankind in its persistent and necessary effort to master its physical and social environment through technologies (and by "technologies" here I mean not just physical but social techniques as well). It is true that a given technology does not always require a decent theoretical explanation before it can be practically applied (and those people who demand such a theoretical understanding are often being deliberately obstructionist) but sometimes just such a theoretical understanding suggests specific new technologies. Furthermore a theoretical understanding of an existing technology makes possible the improvement or perfection of it, and thereby creates the possibility for qualitatively better modes of human existence. For instance, current theoretical work in non-linear processes, designed to understand and make explicit the properties of plasmas, is essential to the development of controlled nuclear fusion reactions, which promise to provide a virtually unlimited source of energy, hopefully in the near future. More generally, it is appropriate to conceptualize the totality of human labor, in its relationship to fundamental human survival needs, as falling into two categories. The first, and the one most familiar to the overwhelming majority of human beings, consists in the transformation of material (physical) nature: farmers, miners, auto workers, clothing manufacturers, etc., all work to alter nature in such a way as to make it amenable, directly or indirectly, to human consumption requirements. This sort of labor is sometimes called "cooperative" labor by Marxists, and I will occasionally use that term. (It is also true that some cooperative labor is absorbed in the production of things either not amenable to human requirements or even antithetical to them; but while this sort of labor, even where it exists, is dispensible, the kind I am considering here is not). A critically important feature of such labor is that while over the short run it can be repetitive and hence boring and unrewarding to the particular laborer (a feature which has gained particular attention in recent years), over the long run such labor can absolutely not be repetitive. No society can lock itself for long into non-evolving forms of labor, labor dependent upon fixed, unchanging technologies. For if it does so, it quickly runs into serious, often fatal problems. Resources appropriate to the fixed technology begin to run out, or at least to require more sophisticated extraction methods (for instance, the enormous technological requirements of the Alaska pipeline, the new deep-water off-shore drilling techniques, etc., are required merely to stay abreast of existing demands, not even to meet legitimate but unmet current needs); populations grow (or even diminish) so that more needs to be produced (or produced by fewer people; the best known example is food); diseases, ecological reactions, changes in weather patterns, etc., all mitigate against the possibility of fixed modes of nature-alteration, at least for any significant length of time. This suggests the necessity for the second broad category of human labor: the creative or "intellectual" laborer, and especially in this context, the scientist. We will again follow frequent Marxist usage and refer to this as "universal" labor. Such labor is essential if we are to discover the necessary new means of providing for our continued existence. I do not mean to suggest that creative scientific work must be done by a separate "class" of people, although in almost all societies this is presently the case; but I do submit that, as a percentage of the total labor time of the population, creative scientific work must exist. Again, it is possible for some of this time to be absorbed in the consideration of socially useless projects: war technologies for example or fruitless avenues of research; but again, not all of it can be absorbed in this way, lest the society find itself locked into technologies which have lost their power to provide for human needs.

It is sometimes argued, by those who would prefer to deindustrialize us (rather than bomb us) back to the Stone Age, that this is an unwarranted generalization, and that in fact the Stone Age offers a counterexample to the need for technological innovation. Such a counterexample is, however, problematical on several counts: (1) the fact is that Stone Age man *did* develop new technologies, although these developments were exceedingly slow in coming. It is not at all clear that these developments were unnecessary impositions on a secure and stable social and economic order. The fact that they did occur suggests that they may well have been necessary, although it does not, of course, prove it. (2) Even if the Stone Age example is a genuine counterexample to the universal generalization, the feature which makes it possible for it to be a counterexample is not relevant to any existing or foreseeable world: *viz.*, an exceedingly small population in an exceedingly large environment.

Creative science is necessary then, at the very least to maximize the possibility of rational response to crises, and ideally to give us the power to self-consciously control social destiny. Of course, it would be the worst sort of idealism to suppose that creative science *alone* makes this possible. Obviously scientific creation must be "materialized" by cooperative labor before it can meet human needs. Nonetheless scientific creation is at least a necessary condition for human survival. To this extent methodology in science has not been, and cannot ever be, wholly arbitrary. If astrology, alchemy, magic, witchcraft, etc. have for the most part disappeared, it is because they have proved of little value in the struggle for genuinely human existence. Their results simply could not be materialized through the efforts of productive labor. If these methodologies ever do emerge (as, for instance, alchemy might with the implementation of the fusion torch), it will be because they offer genuine hope of contributing to the effort of human beings to become free from material necessity through the mastery of life-enhancing technologies. It is simply not the case that there are *no* criteria for scientific achievement or *no* criteria for preferring science to something else. There may be no formal or merely "intellectual" criterion, but there is an unambiguous material criterion: the enhancement of human existence. To the extent that human life becomes less tenuous, less insecure through the power of productive labor to produce *with ever decreasing effort*[4] the material necessities of its reproduction, the scientific methodologies employed to maintain, extend, and improve the relevant technologies are confirmed.

If what I have said so far is generally correct, we can see that the criterion of scientific progress (as opposed to mere scientific change) is a fundamentally *moral* one, for the principle I am appealing to, the principle which one must accept in order to defend the case I am making is that the continued existence of the human species as a human species itself is the ultimate value. To the extent that it is possible to "prove" that it is an ultimate value, such an effort will be made below. First, though, it might be worthwhile to make a few observations designed to persuade the reader that human species existence is in fact a moral value, and indeed the value which serves as the condition for all other values. It is not likely that many people would doubt this, except for the fact that it is not the sort of value that individuals are often conscious of seeking to maximize, and hence not the sort of value philosophers or others are likely to focus on when considering individual moral choice. Instead, issues such as euthanasia, abortion, pornography, sexual relationships, etc. dominate contemporary attention. Nevertheless, it ought to be clear that the very possibility of individual moral choice depends first and foremost upon the proper social choice having been made: specifically, that social choice which organizes cooperative and universal labor in such a way as first, to maintain human life, and second to reduce the socially necessary labor time required for mere survival. Such a choice has as a consequence freeing individuals from material necessity to the extent that what we think of as individual freedom, and hence the possibility of individual moral action, emerges. In cultures which have not achieved levels of technological development which allow for the reduction of socially necessary labor time (and which are not blessed by "natural abundance"), one simply does not find "moral problems" in the sense of the term we are familiar with. All time is socially necessary labor time, and questions of individual morality never even arise, since there is no time unaccounted for in which to exercise moral freedom. Indeed the moral problems of concern to the contemporary Western philosopher are often luxury items produced by technologically created abundance. One obvious example of this is the set of issues loosely categorized under the heading of "Medical and Biological Ethics." Without the possibilities made available by sophisticated medical technologies these problems would not arise. Similarly the "problem" of pornography becomes a real one only with the availability of mass media. We ought not lose sight of this fact, for it points to the primacy of social morality in so far as the successful structuring of cooperative and universal labor for the enhancement and preservation of human species existence is the major problem of social morality. Only when this fundamental problem of social morality is solved can any other problem even emerge as a problem for the individual.[5] Since we have seen how science is essential to the solution of this problem of social morality and hence to the process of social survival, we have also seen how the correct choice of a science is fundamentally a moral choice.

Lest these remarks be misunderstood, let me say that I do *not* mean to suggest that cultures without sophisticated technologies are somehow "immoral." On the contrary, they are engaged constantly in the most moral of all human activities: cooperative efforts to ensure social survival. However, their realm of freedom is circumscribed by this social necessity, and the individual moral problems which can arise only outside that realm are subsequently absent. Also, it might well be argued, and I am not sure that I would disagree, that the most interesting moral problems are those that concern social cooperation for social survival. However, the opportunity for large numbers of people to engage in the self-conscious solution of these problems is contingent upon the presence of time made free by technological innovation (or natural abundance).

We are now in a position to state more formally the nature of the social contribution a methodology must make before we ought to afford it the title "science" and the honor and social support such a title usually brings. If we designate the amount of materially productive (i.e., cooperative) labor time required to produce the goods and services necessary to maintain the materially productive population at a certain standard of living as M, and view this as a portion of the total amount of materially productive labor time available at that standard of living, T, then the difference, T-M, represents the amount of materially productive labor time, S, available for the support of those activities which we usually designate as "progressive," including science, art, education, etc. Since, as we have seen, human existence requires the development of new technologies and creatively different modes of structuring social and physical environments, the value of S must always be maximized, both absolutely, but more importantly, as a percentage of T.[6] This is done, not surprisingly, through (1) scientific and technological innovation itself, which makes available technologies which reduce the amount of labor time necessary for the production of given goods and services and (2) the materialization of these innovations through the materially productive, cooperative labor of a growingly more sophisticated skilled labor force. To contend for the title "science," then, a methodology must make the claim, with plausible reasons, to increase the value of S/T; and to earn the title it must in fact increase S/T. One thing should be apparent: no single methodology must be the most appropriate for all times and situations. Feyerabend and Kuhn are quite correct when they point to "paradigm shifts" and the depth of change such shifts involve. But they are quite wrong in supposing that there is no "higher order" logic to those shifts. That "logic" is embedded in the struggle, often not self-conscious but becoming more and more so, to make life more secure by freeing us from material necessity. When Feyerabend says "Stone Age Man was already the fully developed *homo sapiens,* he was faced by tremendous problems, he solved them with great ingenuity. Science is always praised because of its achievements. So let us not forget that the inventors of myth invented fire, and the means of keeping it,"[7] he is not, as he seems to think, giving evidence for the superiority of myth over science. He is merely showing that all cultures which have survived or will survive have used whatever intellect they had in whatever ways were appropriate to the solution of life-threatening problems. If the best Stone Age Man could do with his myth-making was keep fire it certainly does not follow that we will transcend our need for new sources of energy by returning to myth-making. A "technology" cannot be divorced from the social situation which supports it. If our population levels were the same as Stone Age Man's, if our physical, emotional, and intellectual requirements were the same, then perhaps myth-making would suffice. But they are not.[8] Would Feyerabend take the side of Zeus against Prometheus?

There is a curious irony about a "proliferation" theory such as Feyerabend's. Scientific activity is not self-supporting: for the scientist to engage in serious and creative work, he or she must depend upon the largesse of others. In industrial societies this takes the form of allocating "social surplus"—that is, some of the material production not required for maintenance—to scientific research. (In capitalist industrial societies the social surplus is alienated from the materially productive class, thereby turning scientists into employees of capitalists, either directly through private research laboratories or indirectly through foundation grants or government grants. This does not change the essential dependency of scientists upon cooperative labor, but it has provoked many unfair and misguided "leftist" attacks on the enterprise of science itself.) If we want genuinely to "proliferate" theories, we have to maximize social surplus. But our ability to do that depends upon maximizing support for that one or those few methodological approaches most likely to increase our socially productive capacities. (And note here that "likelihoods" must rapidly become actualities before proliferation becomes possible). Proliferation of theory is a luxury purchasable only through support for real science. This is not to say that it is too expensive a luxury. Sometimes it is, perhaps up to now it always has been, but hopefully it won't always be. But we must never lose sight of the parasitical nature of most of these luxuries. Admittedly some forms of exotic research may bear fruit, and in times of plenty the small likelihood of such results may justify allocations to any and all projects. Again, in times of plenty exotic projects may be supported for aesthetic reasons: a more interesting if not more beautiful world certainly deserves to be encouraged, after human needs are satisfied and the medium range future fairly well assured.

It is also ironic that success in terms of the production of social surplus—or in other words success in solving the fundamental moral problem of all human civilization: how best to maximize the possibility of further existence—is the necessary condition not merely for the solution of all other ethical problems, but for their very emergence as real problems. What "freedoms" are to be tolerated within the bounds of social decency? What life styles ought bring forth censure? What sexual behavior is permissible? These and dozens of similar questions express problems which arise only as necessary labor time diminishes; only, in short, as we become more and more free from material necessity.

It is worth saying that the diminution of *necessary* labor time does not have to lead to the diminution of *total* labor time. It is sometimes claimed that it is through labor that human nature or the human essence most thoroughly manifests itself. It is beyond the scope of this paper to consider this question, but if we assume for the moment that this claim is true, it is clear that a world in which most labor was eliminated would not be a morally desirable world. However, those who argue that human essence is best expressed in human labor usually also argue that this is so only when the labor is freely chosen. Therefore, the elimination of *necessary* labor is to be applauded when by so eliminating it we create the possibility for freely chosen labor.[9]

Feyerabend's position, and the positions of those who urge for other reasons the abandonment of science, is, ultimately, materially self-contradictory. If in fact there were no criteria for choosing among methodologies, and if as a result we were to allocate the bulk of our social surplus to exotic projects which did not contribute to further social surplus (as Feyerabend is willing to countenance), there soon would be nothing left for further allocation. These exotic projects would disappear, as did their more prosaic counterparts, although unlike the latter which would have died as a result of an alienated epistemology, the former would have died as a result of social devolution. This is the ultimate *reductio ad absurdum* of Feyerabend's position, and it suggests the best proof we can offer for the claim that human species existence is the ultimate value: if we were to act on the denial of this principle, there would soon be nothing left to do the valuing.

It remains only to consider to what extent "democratic" procedures are appropriately applied to the allocation of resources to scientific projects. In this connection it is well to recall the commonly understood distinction between democracy as the expression of informed choice and democracy as the collection of whims and fancy. There is no objection at all—indeed there is, ultimately, the necessity—for decisions to be democratically made regarding the allocation of social surplus. But such decisions must be informed decisions made in light of an understanding of what hangs in the balance. To the extent that a population understands that, they will without doubt inform themselves about the alternatives, and the collective choice which results, while not necessarily the best, is likely to be close enough often enough to be tolerable. On the other hand, the kind of "democracy" which Feyerabend occasionally suggests[10] has little to do with informed choices, but is meant only to express the majority sentiment, feelings, whims, or prejudices. Those who "groove on astrology," are "into mysticism," "get off on the I Ching," or for that matter have a selfish interest in cancer research, simply vote their preferences with no moral obligation to themselves, much less to others, to consider the significance of their vote. Such "democracy" is indeed the kind we are most familiar with, but it absolutely is not the kind appropriate to a free society, or more importantly in this context, to a society which values self preservation.

NOTES

1. "Weimar Culture, Causality, and Quantum Theory, 1918–1927: Adaptation by German Physicists and Mathematicians to a Hostile Intellectual Environment" by Paul Forman in Russell McCormmach, ed., *Historical Studies in the Physical Sciences,* Volume 3 (University of Pennsylvania Press, 1971).

2. "Towards a Realistic Appraisal of Science," by Paul Feyerabend (mimeographed). To be completely fair to Feyerabend the following should be noted: (1) the paper to which I

refer has not, to the best of my knowledge, been published but exists in mimeographed form as it was read at the conference. It is possible that Feyerabend wishes to rethink some of his positions before commiting the paper to print; (2) my remarks speak to one thrust of his paper only, perhaps not the main one. Nevertheless, the aspects of his paper which I criticize here are important because they can be found in the culture as a whole, as well as in semi-popular books such as Theodore Roszak's *The Making of a Counter Culture* (Doubleday & Co., 1969). To find them also in Feyerabend demands that philosophical attention be paid to them.

3. This issue is discussed in Peter Schuller's essay "Philosophers and Parasites: The Fundamental Value of Theoretical Research" read to the American Philosophical Association Western Division meeting, April 29, 1976, New Orleans.

4. Social reproduction, the power of the species to perpetuate itself through time, always requires that at least some social effort (that is, some social labor time) be devoted to the production of items of mere maintenance. We have already seen that mere maintenance will not do: there must be social effort devoted to the development of new techniques as well as social surplus to materially implement those techniques. To the extent that a large percentage of labor time must be devoted to mere maintenance, less is available for progress, and hence human existence itself becomes more endangered, more tenuous. (This entire issue is discussed in detail in Lyn Marcus, *Dialectical Economics*, D. C. Heath, 1974).

5. Some of the recent writings on alienation hint that at least one of the manifestations of alienation in modern capitalist society is the inability of persons merely to understand, much less to seek to solve or have the power to implement a solution to the fundamental problem of social morality.

6. The reason an absolute increase in S is not decisive is that if the total population is quite large, thereby demanding a large M (and hence a rapid consumption of resources), a value of S larger than before might still not be enough, relative to the social needs, to eliminate crises.

7. Feyerabend, p. 21.

8. And if I had to choose between the rain dance technology Feyerabend praises (p. 24) and the desalination techniques proposed by Professor Lloyd Motz of Columbia as a solution to California drought, I'd be a fool to choose the rain dance. Perhaps rain dances work for Indian tribes, but few Californians live in one, nor could very many live in one even if they wanted to. I'm afraid that science is the way of our people.

9. In an earlier draft of this paper I asserted that the elimination of socially necessary labor time would lead to a (desirable) increase of "leisure" time. This was obviously wrong, since the very concept of "leisure" is tied to a social system in which work is an alienated activity. Nor should it be inferred from my remarks that socially necessary labor cannot also be freely chosen.

10. Again, to be fair to Feyerabend, he occasionally speaks about informed choice as well.

SYMPOSIUM ON ETHICS AND BIOMEDICAL RESEARCH

BIOMEDICAL RESEARCH, POLICY, AND PHILOSOPHY

George J. Agich

Of the issues involving medicine today, biomedical research policy appears as one area of central concern for philosophy. The work of the National Commission for the Protection of Human Subjects has focused philosophical attention on the issues of the rights of experimental subjects, the rights to informed consent and privacy as well as the ethics of the use of children, the retarded, prisoners, pregnant women and fetuses as experimental subjects and in psychosurgery. Recently, the recombinant DNA research issue has generated discussion not only of the National Institutes of Health "Guidelines for Research Involving Recombinant DNA Molecules,"[1] but also of the broad questions of the right or even the duty to prohibit possibly dangerous research, the obligations of scientists to the public, the use of means/ends thinking in scientific research, as well as the accountability of basic scientists. In addition,

Research in Philosophy & Technology, Volume 3, 1980, pages 39–75.
Copyright © 1980 by JAI Press Inc.
All rights of reproduction in any form reserved.
ISBN: 0-89232-102-4

philosophical topics relevant to biomedical research in general, e.g., the question of the nature of rights (the right to health care), persons, obligations, equality and equity in health care, paternalism and autonomy, as well as the nature and use of technology, are now regularly addressed.

In light of this, it is fair to say that biomedical research is well recognized as the locus of serious philosophical problems. While these problems are well recognized, other aspects of biomedical research policy are not. Not surprising, then, that certain aspects of biomedical research policy continue to need clarification. For instance, the definition of biomedical research is not itself entirely clear, especially when viewed in its historical context. Second, the relation of biomedical research to progress in medicine is more complex than is often appreciated. And third, the meaning of policy in the phrase "biomedical research policy" often involves a confusion of the notion of a priority structure in terms of which decisions and choices are made and the rules implementing those choices. This latter confusion is compounded by the uncritical reference of biomedical research to concepts of health and disease.

I. Cultural and Historical Background

Contemporary medicine is marked by a broad commitment to scientific method, a commitment especially prominent in United States public policy.[2] As early as the Federal Quarantine Act of 1878 the Federal government had been involved in supporting the health sciences. Under the Chamberlain-Kahn Act of 1918 the Public Health Service made grants to 25 institutions to study veneral disease. These grants set the precedent for what has become a multibillion-dollar National Institutes of Health Extramural Research Program. In 1922 a Special Cancer Investigation Laboratory was established at the Harvard Medical School by the Public Health Service. And the National Institute of Health was established through the National Cancer Act of 1937. Congress authorized the pluralization of "Institute" in the NIH title in 1948 when programs to explore heart and dental disease were added to the existing program of cancer research. From the end of World War II until the mid 1960s the Federal government liberally and directly stimulated the growth of biomedical research. During that period, almost as much money was appropriated as was needed to fund all the proposals judged to be scientifically sound by the NIH's Peer Review Groups. These Federal funds stimulated the establishment of a huge biomedical research endeavor both within academic research institutions and the National Institutes of Health. The impact of science on medicine as stimulated by Federal policy can be seen in the intimate relationship which has developed between our ideas of medical progress as the control and elimination of disease and the status now accorded to basic medical and clinical research. It has also influenced the structure of medical education and has helped to shape public expectations regarding medical and health care.

The belief that medical progress must be founded upon biomedical (or what was then simply basic life science) research was clearly articulated by Abraham Flexner in his momentous Report to the Carnegie Foundation published in 1910.[3] Flexner's recommendation that medical schools affiliate with universities and hospitals and that medical education be founded on the basic sciences turned out to be more important in the long run for medical education than his recommendation for a drastic reduction in the number of medical schools. These proposals not only established basic science education as an intrinsic and even fundamental part of medical education but contributed to the formation of an ideal role model for the teacher of medicine. Subsequent events which stimulated the realization of this role model include massive Federal support for basic medical research and the creation of the Medicare and Medicaid system which decreases the need for private patient referral from outside the university hospital. These influences have occasioned William Regelson to argue that the result has been the death of the Oslerian style of teaching humanistic medicine.[4]

Today the practicing physician is, outside the medical school, not the vibrant model within it. In his place is a new kind of academic physician who is predominately occupied with research rather than teaching or patient care. If this analysis is correct, then the impact of biomedical research on medical education is broader and perhaps more insidious than many have been inclined to believe.

Usually the influence of biomedical research is cast in terms of specific developments in medical technology, e.g., the continuous flow-blood analyzer, the computerized axial tomography (CAT) scanner, polio and rubella vaccines, renal dialysis and organ transplantation. The positive role of these capabilities in patient care continues to be stressed, even though they are increasingly subject to a line of criticism which refuses to take for granted that medical technology *eo ipso* improves the quality and style of medical care.[5] Yet, the belief that medical progress is tantamount to specific scientific and technical developments and that these signal better health care is widespread and fits well with our cultural perspective which sees science and technology as basic elements in social and historical advancement. To be sure, we are also aware that this cultural belief is not without its critics. Of particular note in the present context is Hans Jonas, who sees the power implicit in modern technology as requiring a new sense of responsibility which involves an "ethics of long-range responsibility."[6] It consists in our taking control of technological and scientific development by accepting accountability for future states rather than abrogating responsibility by assuming a limited concept of the causal efficacy of human decisions and actions. But this view has not been fully elaborated for the context of biomedical research, especially in light of the problems involving the definition of health and disease.

II. Biomedical Research and Progess

Hence, it is often uncritically assumed that biomedical research underpins and in some sense generates progress in medicine. The growth of medical technology is often taken as the conclusive index of this relationship. Yet, several points need to be clarified before this relationship can be evaluated.[7]

First, medical technologies are quite divergent in nature and are used for a variety of purposes. According to their nature, medical technologies include drugs, equipment, techniques (*actions* taken by health-care providers that do not require specialized equipment), and procedures (e.g., the implantation of a cardiac pacemaker). In addition, medical technologies can be classified according to their purposes: diagnostic (CAT scanner, continuous flow-blood analyzer), preventive (vaccine), therapeutic or rehabilitative, organizational (used in management and administration to insure the efficiency and effectiveness of medical services), and supportive (e.g., food and laundry services in hospitals).

Second, medical technologies are developed in a wide variety of ways. Some occur from the confluence of several lines of basic, applied, and clinical research (e.g., the cardiac pacemaker) where the logic of discovery is discernible only in retrospect. Others occur as *inventions* (e.g., the continuous flow-blood analyzer of Leonard Skeggs).

Third, modern technologies pose a variety of interrelated technical and social problems. Technical issues include those of safety and efficacy, and social problems include those resulting from features or effects of the technology itself (e.g., the challenge to society's traditional view of death and dying) or the economic burden that its use places on society.

In light of these features of medical technology, two issues emerge regarding the view that biomedical research underpins and in some sense generates progress in medicine. First, to think that biomedical research can be understood in terms of a causal relationship to progress in medicine is relatively naïve. Of course, developments do not occur without an appropriate context of knowledge.[8] Nevertheless, it is not in the least clear that the context of knowledge can unequivocally be credited to biomedical research. The invention of the continuous flow-blood analyzer and the CAT scanner are cases in point.[9] Second and more importantly, medical technology involves technical and social problems, the presence of which seriously challenges the accuracy of the term "progress." The relationship between biomedical research and medical technology is thus open to question precisely in terms of the idea of progress insofar as technical and social problems are engendered by new technologies.

"Progress" is clearly an evaluative term. The role of biomedical research in medical progress can be worked out only if we can satisfactorily define progress, but it is not clear that the definition can be provided philosophically. In fact, the meaning of progress is determined historically by reference to standards and expectations which are revised in light of new experience. So, by choosing some lines of research rather than others or by giving priority to research over implementing available technologies, both the state of medicine and the *criteria* for the evaluation of medical progress are determined. In this determination empirical and normative factors are intertwined, since "progress" includes both judgments about what we can do and what we ought to be able to do. For this reason the question of policy in biomedical research is centrally involved even for the philosophical analysis of medical technology and practice. Only at the policy level can these factors comprising the notion of progress be analytically addressed.

III. Biomedical Research

Donald Fredrickson, Director of the National Institutes of Health, has noted: "The reduction of life in all its complicated forms to certain fundamentals that can then be resynthesized for a better understanding of man and his ills is the basic concern of biomedical research."[10] This (in some sense predominant) view presupposes a specific orientation to knowledge. Knowledge or understanding is seen as analytical in nature; explanation and evaluation are conceived as analytical and perhaps reductive processes. Given this orientation, one might reasonably ask what in particular comprises the subject matter of biomedical research; that is to say, which analytical orientations are to figure most significantly in the determination of medical progress?

The family of related but different (and perhaps even divergent) activities which are covered by the name "biomedical research" is indicated in an appendix to the *Report of the President's Biomedical Research Panel* which addresses "the state of the science" and includes reports from 11 "interdisciplinary clusters."[11] A listing of these clusters gives one a sense of the enormous breadth of activities understood as biomedical research:

1. Biochemistry, molecular genetics, and cell biology
2. Immunology and microbiology
3. Neurosciences
4. Behavioral sciences

5. Social and behavioral development
6. Tissue and organ biology
7. Developmental biology
8. Nutrition
9. Communicative sciences
10. Epidemiology, biostatistics, and bioengineering
11. Pharmacology, substance abuse, and environmental toxicology

Besides the number of these research domains, their complexity in terms of subtopics is most impressive. Confronted with this great diversity, one might well ask whether it owes to the particular research activities themselves or to the overall goals or purposes of the research domains. That is to say, is the heterogeneity accidental, apparent only at the level of particular examples of basic research and explainable in terms of intermediary goals, or is it essential, determined or structured by different final purposes which guide individual projects or kinds of research?

Conceiving biomedical research as research into the mechanisms and processes by which human organisms become diseased and through which health can be restored or preserved does not eradicate important differences among the activities comprising biomedical research. The list of "interdisciplinary clusters" looks very unstructured. It is difficult to tell whether the items so listed are intended for a single purpose or whether they are individually intended as holders for a list of various enterprises, each serving a different purpose.

One way to solve this dilemma is to notice that the list of "interdisciplinary clusters" contains relatively clear examples of two major research approaches: the biological (including biochemistry, molecular genetics, and cell biology, immunology, etc.) and the behavioral (including social and behavioral development and the communicative sciences). Yet, even this broad distinction fails to account for nutrition, epidemiology, biostatistics, and bioengineering, or pharmacology, substance abuse, and environmental toxicology, because in these cases, the behavioral and biological approaches are collaborative not exclusionary. In this regard, one wonders whether we are to assume that these latter clusters are interdisciplinary in a heterogeneous sense including both the biological and the behavioral approaches, whereas the other clusters are interdisciplinary only in the sense of including various behavioral *or* biological disciplines but not a combination of the two approaches?

This question marks an impasse in our analysis of biomedical research.[12] From the perspective of current practice and policy it cannot be easily resolved since ambiguities in the implied definition of the goal of biomedical research as the elimination of disease are also involved. This prompts us, therefore, to consider more fully the suggestion that biomedical research be defined as *research into the preservation of health and the elimination of disease*. From what we have already seen, biomedical research is tacitly understood in a broad rather than a narrow way. Accordingly, if our definition is to be inclusive of what is ordinarily understood as biomedical research, it will have to be relatively broad in order to encompass recognized and even contested empirical instances. By the same token, the definition must not be too broad if it is to be useful.

IV. Concepts of Health and Disease

If we even tentatively define biomedical research as research into the preservation of health and the elimination of disease, we are immediately confronted with another set of problems. First, the notion of health or human well-being is itself ambiguous. It can be defined as broadly as the definition provided by the World Health Organization: "Health is a state of complete physical, mental, and social well-being and is not merely the absence of disease or infirmity."[13] As many have pointed out, this definition is far too broad since it includes both moral and political well-being and as such is inherently an ideal rather than an achievable goal.[14] It speaks less to how health is or could be conceived, but more to how it *should* be conceived in general rather than specific terms.

As we shall see, the meaning of policy importantly includes an empirical component as well as a normative one. In fact, as Eugene J. Meehan has argued, the empirical component of policy is prior insofar as there can be no normative judgment until options have been identified.[15] In this regard, one feature which distinguishes the activities of the health professions from those of education and moral discourse is their preoccupation with diagnosis, prognosis, and therapy; they deal primarily with empirical regularities rather than ideal states of affairs which in some sense should exist. This also reinforces a point made earlier, namely, that the concept of biomedical research must be able to include the most commonly accepted examples or else it will be seen as irrelevant. Thus, the major problem with the World Health Organization's definition is not that it is too broad, but that it is inherently an ideal, and a rather amorphous one at that, than an achievable goal. Hence, this definition does not help to illuminate the meaning of biomedical research for policy purposes.

Second, the notion of health or human well-being involves particular value judgments about what will count as proper human goals. These value judgments cannot be capricious or idiosyncratic preferences, though we must recognize the difficulty in justifying particular valuings even when generally held. In clarifying particular value judgments involved in the notion of human well-being, the notion of a normal human style of life is especially important. Yet, it should also be clear that "normal human life style" cannot be precisely defined since human life style is a complex notion depending not only on particular human preferences but on social and environmental factors as well. Since we cannot control *all* relevant factors—or even determine which social and environmental factors are relevant without already engaging in a selection and hence a valuational process—we are forced to examine the notion of normal human life style (and with it human well-being and particular value judgments) in a piecemeal fashion in order to spell out the meaning of health. Understandably, then, there will be a circularity to our definition and an openness to our concept of health. These are shaped by what we think is proper to human beings. Biomedical research will thus be defined in terms of these rather open-textured notions. This point can be made most easily by considering the concept of disease.

Engelhardt believes that the concept of health can best be laid out in a "backwards" fashion by analyzing the concept of disease (cf. n. 14, p. 125). Disease is usually defined in terms of two broad kinds of considerations: what is conducive or nonconducive to the long-range survival of the species and what augments or restricts the basis of the rational free agency. Since we cannot know what actually contributes or will contribute under altered future conditions to the long-range survival of the human species, the naturalistic definition of human disease in terms of species survival must remain speculative. Also, to hold this view commits one to saying that persons with sickle cell disease are not diseased since in the presence of falciparum malaria, the sickle cell trait would be adaptive for the species. Their suffering and death would be the price for group survival since the sickle cell trait confers resistance to malaria. On the other hand, the definition of disease in terms of rational free agency allows for a more open and immediate specification of what is proper to human beings: namely, rational free agency.

The important difference between these two ways of conceiving disease—in terms of species survival or adaptation and rational free agency—is that rational free agency allows for the possibility of human rational control and specification of particular disease states. Here, a disease would be any condition which would inhibit rational free agency or thwart its expression. Thus, schizophrenia or organic brain disease would illustrate the former, while any number of disorders causing pain, disfigurement, or loss of bodily capacity, as for instance renal disease, could possibly thwart the expression of rational free agency. To be sure, there are problematic cases such as homosexuality, color blindness, and perhaps aging, but these kinds of cases serve more to display the plastic and open character of disease concepts than they do to remove the conception of disease in terms of rational free agency to a speculative domain beyond rational discussion.

As indicated above, it is desirable that the concept of biomedical research be somewhat open and even circular insofar as it then can follow our common usage of the concepts of health as what is proper to human beings and the concept of disease. This circularity, however, is not vicious insofar as we are dealing with empirically determined regularities relative to normative standards rather than normative ideals alone. However, there are forms of the analysis of biomedical research which beg the question of the relationship between empirical and normative aspects in the concept of biomedical research.

For example, Lewis Thomas argues that biomedical research is simply research in which the science of biology is paramount.[16] Biology is paramount because of the belief, first, that medicine is not quite scientific enough to stand on its own, and second, to reflect the belief that the answers to medical problems will come from biology rather than from some other science. Accordingly, biomedical research is research conducted along the lines of the biological sciences. However, there are problems with both of these reasons.

First, although we might agree that medicine is not "scientific enough," where that phrase (still vaguely) means "not a science in the sense of physics and other natural sciences," it is not clear that this enables the inference that therefore medicine cannot "stand on its own." The second point, that answers to medical problems will ultimately come from biology rather than from some other science, can be criticized as being in effect a *stipulative* answer to the question of biomedical research. Such an answer, to be sure, does have adherents. Placed in historical context this view is understandable as the result of past policy decisions which supported and enforced such a view of things; and so, the tendency to define biomedical research exclusively as biological research points, once again, to the topic of policy.

V. Meaning of Policy

In clarifying the meaning of biomedical research policy, we must now face the difficulties present in such a discussion which surround the meaning of the term "policy."[17] In many discussions of the subject the meaning of "policy" is simply left undefined or assumed; in other cases policies are identified variously and not always consistently with the actions of persons or corporations, with laws or administrative regulations, with actions, ideals, and intentions, or with aspirations or official statements. Hence, there is a plethora of examples of policies, but no definition of "policy."

One way around this morass is to stipulate a meaning for "policy." Minimal requirements for such a definition include reference to significant dimensions of human activity which lie within the realm of human capacity (since we are basically concerned with what is realizable, not ideals) and that the term be general enough to be inclusive rather than exclusive of troublesome examples and current usage. Hence, policy can be defined as "a guide to action or choice."

Establishing a guide to action or choice involves the performance of two basic tasks. First, one must connect the present capacity of some specified actor (who may of course be a collectivity) to a set of alternative future outcomes. Second, one must order those alternative outcomes in terms of a set of hierarchically ordered preferences (already available or created for the occasion). Carrying out these tasks involves considering both empirical and normative factors. Of these, the empirical component is prior since normative judgment (i.e., choice) cannot be exercised until the options have been established.[18] So understood, a "policy" is simply the rule of action or instrument that applies a priority structure to a particular choice. It involves an empirical component, namely, a set of options regarding alternative future states as well as a normative component, the priority structure in terms of which particular choices are made. The distinction between policy as a rule of action and a priority structure is important. Unfortunately, they are often confused.

For instance, while the priority structure may remain constant—as with a child who wishes to increase his savings rather than buying a new toy; the rules of choice may vary: choose the largest coin, choose the brightest coin, choose the most valuable coin. Knowledge gained through experience may lead to the replacement of one rule by another, and this development is built into the procedure whereby policies or rules of action are determined and acted upon.

In complex issues such as biomedical research, then, a very large number of rules are involved and so there is more than ample room for disagreement and inconsistency. Hence, to speak of biomedical research policy is misleading, for "biomedical research policy" is complex, including, as we have seen, a variety of broad research areas or interdisciplinary clusters, various approaches, as well as particular lines of investigation, any of which might comprise a policy or family of rules of action. Hence, the evaluation of such policies must be made with the priority structure in mind, for without it talk of policy is incomplete. If we bear in mind that the priority structure for biomedical research involves concepts of health and disease which are relatively open, the complexity involved should become apparent. This brings us finally to the relationship of policies or rules of action to the priority structure for biomedical research in general.

VI. Conclusion

As we saw, biomedical research is an open-textured concept. In part, this is reflected in the concrete instances of biomedical research which we confronted in our reflection, but it was unclear whether the plethora of examples depends upon different goals or purposes guiding the research or not. This dilemma is not categorically resolved by defining biomedical research in terms of health and disease, because these concepts also are circular in nature.

Identification of certain states as disease states already implies the concept of health or what is proper for human beings. This implication is most often vague, at least in the sense of being inarticulate. Hence, one main task for research is not simply to uncover the mechanisms of an alleged disease process, but (perhaps in doing so) to justify its being labeled as disease. Uncovering a vector or positive agent (as in infectious disease), displaying a physiological process as pathological, discovering a way to cure or alleviate the symptoms, or ultimately explaining the phenomenon are common ways in which the use of disease labels is justified. In this regard, it is especially important to recognize that justification involves not simply the implementation of rules of action for research (e.g., investigating schizophrenia as biological rather than psychological or psychosocial), but the implementation *and* evaluation of the priority structure itself, since any priority structure is also contextual in nature.

For instance, to consider aging as a disease rather than a normal phenomenon impacts upon how we are likely to treat aging and the aged. The identification of aging as a subject for research already implies a priority structure regarding the evaluation of the phenomenon or set of phenomena we call aging. The task of research in this connection is to follow out particular rules of action (lines of investigation) *given* this priority structure. In doing so, a broad and multifaceted approach is preferred.

If biomedical research were conceptually restricted to investigating only physiological and pathophysiological processes, the behavioral, social, or psychological approaches to aging would be excluded. However, it should be clear upon reflection that such an exclusion would be capricious for two reasons. The common meaning of aging refers to behavioral, social, and psychological characteristics distributed throughout each age group. So it would be arbitrary to exclude these factors in defining research into aging solely in physiological and pathophysiological terms. But second, and more importantly, the identification of physiological and pathophysiological processes as significant items for research can occur only on the basis of judgments about what is desirable or not desirable, i.e., in terms of some notion about what comprises a normal human style of life. This judgment clearly involves behavioral, social, and psychological factors, and perhaps others as well. Hence, one can exclude these factors from biomedical research into aging only at the risk of being circular in a vicious sense.

Aging is not a simple phenomenon, but complex. Hence, it is an ideal to be able to correlate phenomena multifactorially, to include relevant biological, sociological, psychological, and behavioral characteristics. Empirically, research into aging will appear rather piecemeal since many as yet unconnected themes have to be pursued in order to gain a knowledge base from which to conceive what a composite picture would or should be. Hence, at the rule or action level of policy, research opportunities are followed as they present themselves and as they appear fruitful. Gaining experience and knowledge at this level provides the basis for the reassessment of the priority structure insofar as what we should mean by normal human life style is redefined as research progresses. A dialectical process is clearly at work in biomedical research policy.

Since the concept of biomedical research, like that of the concepts of health and disease, is open-textured, a potential field for human action is indicated. This field is not an abstractly structured field of options open to human choice, but is determined by concrete preferences. The shift from one set of preferences to another in the policy-making process reflects the experiential dialectics of gaining increased knowledge which in turn opens new options for us. Hence, not only is the priority structure relatively open, it is also fluid and necessarily so. Thus, in speaking of biomedical research policy and distinguishing between levels of rules of action and the priority structure within policy, we have suggested a way to begin the philosophical investigation of these issues. This investigation will have to distinguish more rigorously the logical stages in the dialectics of biomedical research policy. Such a task is uniquely appropriate to philosophy.

NOTES

1. *Federal Register,* 41 (July 7, 1976), 27902–27943 and "Recombinant DNA Research: Proposed Revised Guidelines," *Federal Register* 42 (September 27, 1977), 49596–49606.

2. The following historical overview is drawn from The Rand Corporation, "Policy Analysis for Federal Biomedical Research," in *Report of the President's Biomedical Research Panel,* Appendix B (Washington, D.C.: U.S. Government Printing Office, 1976), p. 73, and Donald S. Fredrickson, "Health and the Search For Knowledge," *Daedalus* (Winter, 1977), 160–61.

3. *Medical Education in the United States and Canada: A Report to the Carnegie Foundation for the Advancement of Teaching* (New York, 1910).

4. "The Weakening of the Oslerian Tradition," *JAMA* 239 (January 23, 1978), 317–19.

5. See for instance in this connection, Ivan Illich, *Medical Nemesis: The Expropriation of Health* (New York: Pantheon Books, 1976); Rick J. Carlson, *The End of Medicine* (New York: John Wiley and Sons, 1975); and Victor R. Fuchs, *Who Shall Live? Health, Economics and Social Choice* (New York: Basic Books, 1974).

6. "Technology and Responsibility: Reflections on the New Tasks of Ethics," *Social Research* 40 (Spring, 1973), 31–54 and "Responsibility Today: The Ethics of an Endangered Future," *Social Research* 43 (Spring, 1976): 77–97.

7. The source for the following discussion is United States Congress Office of Technology Assessment, *Development of Medical Technology: Opportunities For Assessment* (Washington, D.C.: U.S. Government Printing Office, 1976).

8. The phrase "appropriate context of knowledge" appears frequently in discussions of Federal biomedical research policy. It seems to signal what usually is an uncritical belief in science's ability to stimulate progress through research.

9. A brief discussion of these two technologies is to be found in *Development of Medical Technology: Opportunities For Assessment,* pp. 10–13.

10. "Health and the Search for Knowledge," p. 162.

11. *Appendix A: The Place of Biomedical Science in Medicine and the State of Science* (Washington, D.C.: U.S. Government Printing Office, 1976).

12. An impasse is marked insofar as the question of policy is involved. The impasse involves the confusion of the rules of action aspect of policy with the priority structure. For this discussion see Section V below.

13. Constitution of the World Health Organization (Preamble). *The First Ten Years of the World Health Organization* (Geneva: World Health Organization, 1958), p. 459.

14. "Human Well-Being and Medicine," in H. Tristram Engelhardt, Jr. and Daniel Callahan, eds., *Science, Ethics and Medicine* (New York: Institute of Society, Ethics and the Life Sciences, 1976), p. 122.

15. "Philosophy and Policy Studies," in Stuart Nagel, ed., *Policy Studies and the Social Sciences* (Lexington, Mass.: Lexington Books, 1975), p. 188.

16. "Future Directions in Biomedical Research," in *Beyond Tomorrow: Trends and Prospects in Medical Science* (A 75th Anniversary Conference, March 8, 1976; New York: The Rockefeller University, 1977), p. 111.

17. This discussion of the meaning of policy is drawn rather freely from the paper by Eugene J. Meehan cited in note 15 above.

18. We are stressing the empirical component with Meehan in order to underscore the point that the normative judgments involved must be realizable rather than ideal. A more rigorous analysis would have to show the interrelationship between facts and values in particular examples of biomedical research policy, but that is not our present task.

TECHNOLOGICAL INNOVATION AND HEALTH PLANNING POLICY

Ronald E. Benson

The improvement of health in the United States during the twentieth century has been dependent to a large extent on the achievements of biomedical research. Future developments in medical technologies hold further promise for raising the quality of health care. An assessment of the value of new medical technologies requires extensive research analysis, not only in the process of development and testing, but also with reference to diffusion and adoption. The specific context addressed here is capital expenditures for innovative equipment which are the object of review by health planning agencies. The issues are illustrated with the controversy occasioned by the introduction of computerized tomography for medical diagnosis.

Policy recommendations regarding the contributions of biomedical research to health planning are proposed following a discussion of some social implications implicit in the recent proliferation of computerized tomography. In American culture both the medical profession and the public generally tend to expect to achieve the highest possible quality of health care based on an uncritical acceptance of the technological imperative. These expectations, however, are being challenged by contemporary social goals to assure equal access for every citizen to health care and to effect cost containment in the face of soaring charges for medical care. Political and economic necessities of our era dictate that the allocation of scarce material and human resources may result in changes in some familiar aspects of an affluent American way of life.

I

Health is a relative state of complete mental and physical well-being for a person functioning as a whole. Good health is relative to such external circumstances as one's chronological age and the cultural setting which includes the technological capacity and the economy of the society. Health care is the provision of goods and services to maintain, improve and restore health. Health care is directed toward minimizing suffering, avoiding disease and disability, and preventing and treating illness.

Health is recognized as one of the highest goods of life. Some bare minimum level of health is required for life to continue. While happiness can be experienced without optimum health, the lack of adequate health can mercilessly threaten liberty and happiness as well as life itself.

According to American ideals, every person is of equal worth as a human being, even though obvious inequalities exist among persons. Many strengths and weaknesses are due to the accident of birth. No one chooses his genetic code. Natural characteristics are given, as it were, through a natural lottery. Loss of health remains a universal hazard that might strike anyone at any time from birth to death.

Good health is a goal toward which people aspire, but is a goal which for some will be impossible to achieve and maintain. Health is one of those conditions that rests as much on the chance of birth as it does on personal care and effort. It is unjust to cause (or even to permit) a person to suffer through no fault of the sufferer whenever it is possible to prevent this suffering, insofar as this does not increase the suffering of others. Certainly no person nor any government can guarantee a person good health. Government can, however, use its influence to stimulate increasingly effective ways to enhance the welfare of its citizens. Each person is worthy of being protected from suffering health tragedies that can be prevented or minimized by a reasonable level of health care. Since unlimited resources are not available to anyone individually nor everyone jointly, efficiency in the distribution of health care is a vital responsibility.

The view that there is a *public* responsibility for the health care of individual citizens is not a new idea which is alien to traditional American principles. In both the colonial period and the Revolutionary era, Americans were acutely aware of the value of health. Early settlers were well acquainted with the persistent hazards of disease and famine. The deadly pestilence of yellow fever resulted in the enactment of quarantine laws as early as the seventeenth century. In 1796 Congressman William Lyman of Massachusetts declared to the House of Representatives that "the right to the preservation of health is inalienable."[1] Two years later Congress imposed a compulsory tax on the employers of seamen to establish a system of hospitals and medical care along the Atlantic coast. Within a few years the Federal government assumed responsibility for the purity of vaccines and offered to provide free cowpox vaccine to any U.S. citizen to help combat the scourge of epidemics of contagious disease. The concept that medical care ought to be considered as a human right like food, clothing, and shelter was officially recognized in the *Federal Rules and Regulations* No. 7 of 1933 with reference to the program of the Federal Emergency Relief Administration.

Recognition of health care as a positive right has received increasing acceptance. On July 17, 1969, the House of Delegates of the American Medical Association approved a resolution which asserted that "it is the basic right of every citizen to have available to him adequate health care." Two years later the House of Delegates of the American Hospital Association approved a detailed Policy Statement on the Provision of Health Services whose opening paragraph declared that:

> As a nation we must provide better quality, more convenient health care for all the people, at a reasonable cost, and in a manner in keeping with human dignity. This must be done because we accept one basic, irreducible principle: Health care is an inherent right of each individual and of all people of the United States.[2]

Like the right to receive an education, an entitlement to health care is an ideal that is not specified in the Constitution. The social controversies in both areas center on the appropriateness and effectiveness of the means to accomplish the objective.

II

The National Health Planning and Resources Development Act of 1974 (P.L. 93-641) was signed into law on January 4, 1975. This act provides a structure for coordination, resource allocation, and regulation of the American health care delivery system.

The objectives of several prior Federal health care measures were incorporated into the new legislation. Through the Hill-Burton Act (P.L. 79-725, initiated in 1946 and subsequently amended), the Federal government provided support for planning, construction and modernization of hospitals and other health facilities. In 1965 the Heart Disease, Cancer and Stroke Amendments (P.L. 89-239) established Regional Medical Programs whose goal was to improve medical training and treatment of specific diseases. Federal support for health care planning in state and substate areas through councils, a majority of whose members were consumers of health care services, was included in the Comprehensive Health Planning and Public Health Services Amendments of 1966 (P.L. 89-749). Section 1122 of the 1972 Social Security Amendments (P.L. 92-603) requires review and approval of certain capital expenditures that result in reimbursement through Medicare and Medicaid. Reviews of capital expenditures mandated under section 1122 are commonly conducted by the state health planning agency in consultation with the local agencies for the appropriate geographical area.

One principal aim of the National Health Planning and Resources Development Act of 1974 is "the achievement of equal access to quality health care at a reasonable cost."[3] This legislation specifies ten national health priorities including multi-institutional coordination of services, an integrated health system in each geographical area, and the availability of primary care services to underserved populations. National guidelines for health planning policy in the form of precise standards and goals are being developed.

In order to implement health planning at the local level, the United States is divided into 212 health service areas. Ordinarily the population of an area ranges from 500,000 to three million. A health systems agency (HSA) is designated to perform planning functions in each health service area. Fifty-one to 60 percent of the health systems agency board members are consumers, including a broad representation from various community groups, in addition to a variety of providers of health care. The general purposes of a health systems agency are to improve the health of the residents, improve access to quality health care, help restrain increases in the cost of health services, and prevent unnecessary duplication of health resources. The most important means to accomplish these purposes is the development of a health systems plan (HSP) for the area. The goals stated in the health systems plan are translated into local objectives and priorities in an annual implementation plan.

An agency of state government is designated as the State Health Planning and Development Agency (SHPDA). The state agency is advised by a Statewide Health Coordinating Council. The Federal administrative function is directed by the Secretary of Health, Education and Welfare through ten regional offices. A major responsibility for health planning is delegated to the state and substate organizations which have citizen boards with a consumer majority.

The process of project review for changes in facilities, services and major equipment has evolved across the years from informal voluntary cooperation to legislative mandate with sanctions. The force of review and regulation of changes in health care delivery undoubtedly will be strengthened in the future. The State Health Planning and Development Agency (SHPDA) ordinarily assumes responsibility for conducting reviews under section 1122 of the Social Security Act and also administering a state certificate of need program for all new institutional health services proposed (P.L. 93-641, Title XV, Sec. 1523). At the present time 36 states have established certificate of need programs, but none is in full compliance with recently issued Federal regulations. States that do not adopt appropriate certificate of need programs by 1980 will risk loss of substantial Federal funds for health programs.

The certificate of need statute adopted by a state must meet minimum standards, although any state may establish a more strict requirement for reviewability than the Federal criteria.[4] The state must furthermore provide sanctions for reviewable expenditures which fail to receive a certificate of need that are effective enough to prevent the introduction of unauthorized services and facilities.

Health systems agencies serve a crucial role in the review of plans for capital expenditures by hospitals for new facilities, substantive changes in facilities, and purchase of expensive equipment. The recommendation of a health systems agency is reviewed by the state agency for final action. Both the state and local planning agencies evaluate a proposal with reference to the health systems plan prepared by the health systems agency for its health service area. While the plan must be consistent with national guidelines, it is important to note that the plan is developed by local people to serve the needs of their own area. The adequacy of the plan is dependent to a considerable degree on the involvement of the consumers and providers as plan goals are adopted at the local level. Short-term priorities and strategies are identified through the annual implementation.

The goal of improved health care is a population that is healthy and thereby free to enjoy the benefits of useful and happy lives. Health status is the consequence of all factors affecting personal health that can be approximated by quantifiable indicators, including natality, mortality, and morbidity rates. Success in working together to implement sound health planning will be manifested in the degree to which a higher health status for Americans is achieved.

The purpose of health planning is to encourage providers and consumers in each area around the country to form a partnership to create rational plans capable of coordinating services in an effective system of health care delivery. Improving the distribution of services can be one means of promoting justice and overcoming inequities.

III

Biomedical research has provided resources for innovations in medicine that have dramatically improved the quality of health care. Confidence in the results of research is reflected in the rapid increase of health research expenditures. National health research support in 1970 was $2,856,000,000, which is more than triple the level of support for 1960.[5] Presently two-thirds of the health research in the United States is financed by the government. This is in stark contrast to government support which accounted for only 7 percent of the total in 1940.[6]

New medical technologies are among the most tangible products of basic, applied and clinical biomedical research. Technology is the application of scientific knowledge to serve practical needs. Medical technology refers to "techniques, drugs, equipment, and procedures used by health care professionals in delivering medical care to individuals and the systems within which such care is delivered."[7] Technological innovation has transformed medical practice. Some familiar examples are cardiac pacemakers, renal dialysis, polio vaccines, ultrasonic imaging, artificial respirators, and antibiotics. The adoption of medical technologies is commonly associated with increased specialization and medical practice located in capital-intensive institutional settings. Technologies of medical practice may be categorized in terms of their physical nature as techniques, procedures, drugs, or equipment. The functions performed by medical technologies are preventive, therapeutic, organizational, supportive, or diagnostic.[8]

Diagnostic technologies represent a key element in contemporary medical practice. Numerous widely utilized tests are the fruits of technology. While these are important factors in the high quality of care, the increasing number of diagnostic tests is contributing to the escalating costs of medical care. Approximately half of the increase in expenditures for hospital care is related to changes in technologies.[9]

Computerized tomography is an example of a diagnostic technology which utilizes equipment requiring a major capital investment. The computerized axial tomographic (CAT) scanner is a technological innovation which has been hailed as the most significant advance in radiology since the discovery of X-rays. Basic research in mathematics, radiology, and computer technology laid the foundation for the development of computed tomography,[10] which was patented in 1969 by Godfrey Hounsfield of England. Earlier in the United States, similar research by Oldendorf and Cormack failed to attract the interest of others, or to secure either public or private funding for development of their work. In October 1971 at a London hospital, Ambrose became the first physician to use a prototype scanner in medical practice. The first units operational in the United States were installed by EMI, Ltd., the first at the Mayo Clinic in

June 1973 and the second a month later at Massachusetts General Hospital. Subsequently other manufacturers have successfully entered the market as well, although EMI continues to be the major producer.

Computed tomography reproduces cross-sectional images of internal body structures and is capable of distinguishing even the differentiation of soft tissues. An X-ray machine mounted on a gantry directs X-ray beams through the body. The X-ray particles that emerge after passing through the body are absorbed in a detector to indicate differences in tissue density. The gantry rotates around the body in a series of rapid steps. All the information from many X-rays is sequenced to reconstruct the images of the cross-sectional plane. The tomogram or image of the body plane has proven to be accurate and useful in medical diagnosis, especially for the brain. Sometimes the computed tomographic scanner is referred to as a "brain scanner" because of its early application and remarkable success in the diagnosis of neurological abnormalities.

The prevalent fears about indirect potential hazards, such as those expressed with regard to recombinant DNA experimentation, risks to human subjects of experimentation, or irreversible environmental damage, do not apply here. A computed tomographic scan is presumably safe and painless because it is relatively noninvasive. The radiation of the X-rays in the test is within limits that are presently considered safe. At present about 60 percent of the patients who receive computed tomographic scans, however, are subjected to the injection of contrast materials into the bloodstream in order to improve the quality of the data.[11] The risk related to the injection of contrast materials is minimal compared to the alternative of other invasive modalities such as angiography.[12]

The purchase cost of a computed tomographic scanner averages about a half million dollars, with a range of $300,000 to $700,000.[13] Current demand for this equipment exceeds the ambitious production schedules of the manufacturers. The total number of computed tomography installations in the United States increased from less than 100 machines in June 1975 to more than 500 two years later.[14] New medical technologies employing expensive equipment are known to increase health care costs considerably. If half of the hospitals in the United States were to operate a computed tomographic scanner, charges of over two billion dollars per year would be generated.[15]

The sudden proliferation of computed tomographic equipment attests to its high quality of brain imaging and reflects optimism about its effectiveness. Computed tomographic scans of the whole body promise to enhance the armamentarium of diagnostic capabilities even further. Even though research has not unequivocally established the efficacy of computed tomography, particularly of the whole body, this equipment continues to be purchased at a rapid pace.

With the exception of free-standing ambulatory clinics, physicians' offices, and Veterans Administration hospitals, health care providers who plan to purchase a computed tomographic scanner are required to seek the approval of designated health planning agencies. Responsible review of projects is especially difficult for technologically innovative equipment. For such equipment, due to its very nature, extensive clinical experience is lacking. Unless criteria have been established for the review of innovative equipment and services, planning boards at the local level are severely limited in their ability to conduct an evaluation of a proposal. Sponsors of projects are usually knowledgeable, articulate, and enthusiastic about their proposed new service. Without a clear-cut reason to reject the sponsor's avowed aim to do good by reducing suffering and illness, a planning board is likely to approve the project.

Computed tomography is a case of an innovative technology that quickly found acceptance in the medical community. In order to offer the best quality of health care, many hospitals suddenly felt the need to purchase scanners. Planning agencies confronted applications for the approval to purchase equipment which few people could objectively assess. Competition developed among institutions in the same community to be the first to install the latest technological achievement. In some instances unsuccessful sponsors found physicians who provided this new service by purchasing their own equipment without the requirement to secure approval through the health planning process. In other instances hospitals encountered direct competition from enterprising physicians who made large capital investments to purchase similar equipment for their offices or clinics. By the time criteria could be established on the basis of clinical research data, some sections of the country had more computed tomographic scanners than was needed to serve the health service area.

IV

Many modern Americans have been impressed with the triumphs of modern technology and are confident that the solutions to our major human problems are technological. News of successive man-made miracles in the twentieth century has encouraged ever higher hopes for continued technological progress. Fuchs observes that medical training has instilled in physicians a commitment to offer the best care available.[16] This technological imperative assumes that the physician should do everything technically possible for the benefit of the individual patient.[17]

New technologies are commonly adopted by manufacturers in order to reduce their per unit production costs. In the health care delivery system, the adoption of innovations frequently has the effect of increasing costs in order to achieve a higher quality of service. If the progressive increases in cost are paid directly by health care consumers, the benefits of expensive technologies will eventually remain available to fewer and fewer persons who can afford quality care. If access to high-quality care is extended to all, then the material resources available to serve other human needs will be threatened.

Hiatt warns of the imminent danger of expanding the demands on available medical resources to the point where the finite limit of our economic capacity is exceeded.[18] He agrees with Hardin[19] that pursuing the laissez-faire conception of freedom advocated by John Stuart Mill[20] a century ago is economically and morally inadequate today. The rising expectations of individuals cannot be permitted to impose a tyranny on society at large. According to Mechanic,

> As people have learned to have higher and more unrealistic expectations of medicine, demands for care for a wide variety of conditions, both major and minor, have accelerated. No nation that follows a sane public policy would facilitate the fulfillment of all perceptions of need that a demanding public might be willing to make.[21]

The uncritical acceptance of the technological imperative by both providers and consumers to do everything possible for the individual patient may indirectly limit the opportunity to improve the health status of the population as a whole. While advances in technological capacity are a necessary condition for a high standard of living, technological successes may become a factor in creating formidable problems which thereby result in unforeseen technological challenges. Even in an affluent society, both human and material resources are limited. Faced with finite resources, needless or wasteful expenditures create injustices. Preoccupation with the most sophisticated means of treating illness may inadvertently minimize the urgency to improve the effectiveness of preventive care and health maintenance.

Decisions regarding the adoption and diffusion of technological innovations in medical care involve the allocation of scarce resources. The emergence of a public ethic could guide societal decisions on such issues of distributive justice. A public ethic could elucidate public policy decisions on the basis of ethical principles.[22] A few preliminary principles are suggested below. In an economy of scarcity, the needless wasting of human and material resources on nonefficacious technology creates injustices. The unnecessary duplication of technologies which are likely to be underutilized ought to be avoided. Whenever innovative technology requiring the extensive support from "the commons" of society is evaluated for potential adoption and diffusion, the burden of proof rests on the innovator. The proposed technology must be demonstrably superior to existing alternatives currently available.

It is appropriate for developing medical technologies to be subjected to the scrutiny of both empirical research and community values. Empirical data do not establish sufficient conditions for making value judgments, but are necessary conditions for responsible decision-making. Ideally, public policy decisions should involve the participation of the community whose lives are significantly affected by the consequence of the policy implementation. In cases of experimental technologies authorized for public use, the ethical guidelines for the use of human subjects in experimentation are applicable. For example, providers using experimental technologies have an obligation to inform their publics, especially affected patients, of the potential risks and benefits.

Experimental medical technologies that are of low risk to the patient pose subtle problems, since utilization rates may not reflect need.[23] Low-risk diagnostic tests do not require the caution surrounding those that are relatively risky. In the past some medical technologies which became common practice were eventually proven to be of little value, such as gastric freezing machines cited by Hiatt.[24] Fineberg emphasizes that "diagnostic accuracy alone is not a sufficient measure of the clinical value of a test."[25] Adequate justification for diagnostic testing depends on its probable therapeutic impact, the improvement of the patient's health, and a higher health status for the community. Rushmer notes that "the availability of technology to elicit information with minimal or negligible risk tends to assure its excess utilization in the current period of high public expectation, third-party payments, and threats of litigation."[26] Proposed deviations from the current practices are to be encouraged if benefit is bestowed, especially on the least advantaged, medically and economically.

Projected utilization rates may be problematic for innovations whose technical capability and appropriateness have not been established through extensive clinical experience. Underutilization of a new service may create excessive costs for that service which results in an unwarranted burden to either payers of the charges or other departments within the health care institution. Potential overutilization of marginal services is also a subtle danger that merits examination, especially in cases where the referring physician, exempt from regulation, is the owner of the equipment. The present practice of denying reimbursement for unapproved services by third-party payers, who pay two-thirds of the American health care bill, is justified. A study of computed tomography recently released by the Institute of Medicine included a recommendation that

> a procedure be developed at the federal level to identify and evaluate costly innovations before their widespread introduction into the medical marketplace. . . . As long as providers bear no risk for poor investment decisions, as is the case now with the cost-based reimbursement, they will continue to adopt new techniques prematurely.[27]

The recent diffusion of computed tomography is an example of commitment being made to invest in an innovation prior to adequate research, especially of its effect on the health care delivery system and the health status of the population. The National Health Planning and Resources Development Act charged the Secretary of the Department of Health, Education and Welfare to develop National Guidelines for Health Planning within 18 months after enactment of the Act. The first parts of the guidelines were issued 33 months after the effective date of the Act.[28] Those initial regulations did include a standard for computed tomography, at a time after the widespread diffusion of the technology had already occurred. One of the duties of the National Council on Health Planning is to make recommendations concerning "evaluation of the implications of new medical technology for the organization, delivery, and equitable distribution of health care services."[29] Earlier technical memoranda issued by the Department of Health, Education and Welfare[30] were of limited value to planners at the local level, who were struggling with decisions about the adoption of new technologies.

The financial feasibility of introducing an innovative medical technology ordinarily depends in part on its approval for cost-based reimbursement by third-party payers, especially the Federal government and Blue Cross/Blue Shield. Estimates based on early experience with computed tomography indicate that the annual profit on the capital investment ranges from 12 to 65 percent.[31] The charges for scans are commonly based on a depreciation schedule of five years for the capital investment.[32] This relatively short-term depreciation suggests that purchases for developing technologies are vulnerable to technological obsolescence. The prospect of technological obsolescence is one reason caution should be exercised in permitting the widespread diffusion of an innovation.

Inasmuch as the majority of health research is Federally funded, the government is in a position to place a higher priority on research to evaluate the impact of new technologies on the health care delivery system. In an analysis of national health planning goals, Wennberg concluded that "the development of a national policy for the control and evaluation of new medical technology should be a planning goal; in the long run, the control of cost, the assurance of quality, the interpretation of equity and the resolution of the issue of iatrogenesis depend on its accomplishment."[33] The goal of medical research is to improve the health status of the population. This cannot be accomplished fully unless factors are identified that contribute to improved health, and national resources are appropriately allocated to accomplish the goal. In the past, studies of the health care system seldom considered noneconomic factors which are also significant in assessing the impact of technologies on our social and ethical values.[34]

The technological imperative demands that the highest available quality of care be delivered. Acceptance of the right to health care and equal access by everyone to high-quality care in an economy with finite resources is incompatible with the stated national goal of cost containment. As Green notes, "The 'best possible' health care is an unreachable goal whose pursuit can absorb all the resources of even the richest society."[35] As a nation we may not be willing to pay for both the best available health care and equal access to such care by all. In the allocation of limited resources, such as medical equipment and personnel, until the time sufficient experience and research are accumulated, the widespread diffusion of new technologies must be resisted, lest the proliferation of expensive equipment occurs before the empirical data are evaluated and our public values clarified. Unless the proportion of the gross national product devoted to health care is increased, the available resources for health care must be carefully planned. The latter could also involve reassessment of established modalities, especially those whose aggregate cost is large or the risk relatively high. New technologies may be proven to be cost-effective through reducing or eliminating the days of hospitalization, avoiding unnecessary surgery, and replacing procedures whose elimination results in reducing cost, morbidity, and patient risk.

In establishing guidelines for new medical technologies, local health planners depend on research that can be best conducted either as clinical trials at selected centers or through coordination of clinical data from a state or national base. In order to study medical delivery problems of broad interest to its counties, Sweden established the Swedish Planning and Rationalization Institute.[36] The Institute shares its findings with county councils who make decisions about equipment purchases. The Illinois Health Facilities Planning Board recently established "Standards and Criteria for Review of Applications for Permit for Technologically Innovative Equipment."[37] A procedure is specified to identify expensive technologically innovative equipment whose usefulness is not yet fully established. An appointed committee is authorized to study and make recommendations concerning standards for that equipment to the state board. Committee recommendations adopted by the state board are binding on providers and planning agencies. The first product of this process in Illinois was the adoption of "Guidelines for CT Scanners."[38] While guidelines for CT scanners were developed after the proliferation of this equipment had probably already exceeded the subsequently adopted standards, this mechanism for evaluating technological innovation may be a model for others to follow as new medical technologies emerge in the future.

According to Rawls, all individuals desire three primary goods: self-respect, liberty and opportunity, and income and wealth.[39] Justice requires that any deviation from an equal distribution of these goods be clearly justified. Rawls justifies unequal shares of goods whenever an unequal distribution will result in a benefit for those who are least favored.[40] According to his conception of fairness, Rawls asserts that "the higher expectations of those better situated are just if and only if they work as part of a scheme which improves the expectations of the least advantaged members of society."[41] A modification and extension of this principle could be elaborated for the health care delivery system. Without the means to maintain personal health, an individual may be indirectly deprived of one or more of the primary social goods. Individuals, therefore, will need a certain minimal level of care in order to achieve happiness. The assurance of equal access to care by all is, however, a principle distinctly separate from the determination of any specific payment mechanism. The least advantaged persons are those who have the greatest need for medical care in order to minimize morbidity and avoid premature death. Any changes in the delivery of health care that involve major expenditures can be justified if they are beneficial, especially to the least advantaged. In this way, the introduction of an expensive technological innovation is acceptable if it actually contributes to improving the health of patients. Benefits accruing primarily to providers of health care services, such as greater scientific knowledge, improving one's competitive position in the field, or increasing one's income, are not sufficient reasons for the investment in technologically innovative equipment. At the same time, increased expenditures for health care ought not to result in denying other members of society the opportunity to acquire a minimal level of primary goods. It is also assumed here that consuming a significantly greater amount of social resources to achieve the same result (including a comparable level of morbidity and risk of mortality) than a lesser expenditure of resources could achieve is unjust.

The introduction and diffusion of innovative technology which increases the cost of health care to society cannot be justified unless benefits to consumers are demonstrated, especially for the least advantaged. The burden of proof for planning approval should rest on the party recommending the widespread adoption of an innovative medical technology. The health planning process may be viewed as a referendum on the values of the community relating to health care delivery. A sound basis for the evaluation of new technologies is frequently not available to the local decision-makers. It is imperative that timely biomedical research be undertaken to assess the utilization patterns and rates as well as the ultimate effect on the health status resulting from the availability of new technologies.

NOTES

1. U.S. Congress, House, Representative William Lyman of Mass., 4th Cong., 1st sess., May 11, 1796, *Annals of Congress*, p. 1348.

2. American Hospital Association, *Policy Statement on Provision of Health Services* (Chicago: American Hospital Association, 1971), p. 1.

3. U.S. Congress, *National Health Planning and Resources Development Act of 1974*, P.L. 93-641, Sec. 2, 88 Stat. 2226, 42 U.S.C. 300k (1975). [Although periodic changes are made in these federal regulations, it was decided to leave the documentation as provided when the paper was given in 1977. Changes have not affected the argument of the paper.—*Ed.*]

4. Federal Register 4002–4032 (January 21, 1977).

5. Michael S. Koleda, Carol Burke, and Jane Sisk Willems, *The Federal Health Dollar: 1969–1976* (Washington, D.C.: National Planning Association, 1977), p. 27.

6. *Ibid.*, p. 28.

7. Office of Technology Assessment, U.S. Congress, *Development of Medical Technology: Opportunities for Assessment* (Washington, D.C.: U.S. Government Printing Office, 1976), p. 4.

8. O.T.A., *Development of Medical Technology*, p. 25.

9. *Ibid.*, p. 27. Cf. John K. Iglehart, "The Cost and Regulation of Medical Technology: Future Policy Directions," *Milbank Memorial Fund Quarterly*, 55:1 (Winter, 1977), 27.

10. Richard Gordon, Gabor T. Herman, and Steven A. Johnson, "Image Reconstruction from Projections," *Scientific American* 233:4 (October, 1975), 56.

11. Mokhtar H. Gado, "Contrast Enhancement," in American College of Radiology, *Proceedings of the Conference on Computed Tomography in Radiology* (St. Louis, Mo., 1976), p. 102.

12. Office of Technology Assessment, U.S. Congress, *Policy Implications of the Computed Tomography (CT or CAT) Scanner* (Washington, D.C.: U.S. Government Printing Office, Prepublication Draft, 1977), p. 57.

13. American Hospital Association, *CT Scanners: A Technical Report* (Chicago: American Hospital Association, 1977), p. 43.

14. O.T.A., *Policy Implications of the CT Scanner*, pp. 68–73.

15. 3,159 units times $650,000 = $2,053,350,000. Cf. A.H.A., *CT Scanners*, p. 10, and H. David Banta and Joshua R. Sanes, "How the CAT Got Out of the Bag" (background paper for *Conference on Health Care Technology and Quality of Care*, Boston University Health Policy Center, Boston, Mass., Nov. 19–20, 1976), pp. 18–19.

16. Victor R. Fuchs, "The Growing Demand for Medical Care," *New England Journal of Medicine*, 279:4 (July 25, 1968), 192.

17. Victor R. Fuchs, *Who Shall Live? Health, Economics, and Social Choice* (New York: Basic Books, 1974), p. 60.

18. Howard H. Hiatt, "Protecting the Medical Commons: Who Is Responsible?" *New England Journal of Medicine* 293:5 (July 31, 1975), 235–241.

19. Garrett Hardin, "The Tragedy of the Commons," *Science* 162 (December 13, 1968), 1243–1248.

20. John Stuart Mill, *On Liberty*, 1859.

21. David Mechanic, "The Growth of Medical Technology and Bureaucracy: Implications for Medical Care," *Milbank Memorial Fund Quarterly*, 55:1 (Winter, 1977), 63.

22. Albert R. Jonsen and Lewis H. Butler, "Public Ethics and Policy Making," *Hastings Center Report* 5:4 (August, 1975), 22.

23. Louise B. Russell, "Making Rational Decisions about Medical Technology" (paper presented at a meeting of the National Commission on the Cost of Medical Care, Chicago, Nov. 23, 1976), p. 21.

24. Howard H. Hiatt, "Too Much Medical Technology?" *Wall Street Journal,* June 24, 1976. Cf. Stuart H. Shapiro and Stanley M. Wyman, "CAT Fever," *New England Journal of Medicine,* 294:17 (April 22, 1976), 954–956.

25. Harvey V. Fineberg, "Computerized Tomography: Dilemma of Health Care Technology," *Pediatrics* 59:2 (February, 1977), 148.

26. Robert F. Rushmer, "The Technologic Imperative," *American Journal of Roentgenology* 127 (1976), 356.

27. Institute of Medicine, *Computed Tomographic Scanning: A Policy Statement* (Washington, D.C.: National Academy of Sciences, 1977), p. 43.

28. 42 Federal Register 48505 (September 23, 1977).

29. U.S. Congress, P.L. 93-641, Sec. 1503.

30. U.S. Department of Health, Education and Welfare, Public Health Service, Bureau of Health Resources Development, Health Resources Administration, *Technical Assistance Memorandum No. 16,* "Computerized Axial Tomograph Scanners," and *No. 33,* "Additional Information on Computerized Axial Tomography (Now Called 'Computed Tomography' or CT)" (Rockville, Md.: Health Resources Administration, 1974). Arthur D. Little, Inc., "A Health Planning Document: Computerized Tomographic Scanning Systems" (draft of paper prepared for the Health Services Administration, U.S. Department of Health, Education and Welfare, Cambridge, Mass., 1975, and reissued 1977). Harry P. Cain, II, *Memo to HSA Presidents and Executive Directors, SHPDA Directors, and SHCC Chairmen,* from Director, U.S. Bureau of Health Planning and Resources Development, Health Resources Administration, Department of Health, Education and Welfare (Rockville, Md.: Health Resources Administration, Nov. 22, 1976).

31. O.T.A., *Policy Implications of the CT Scanner,* p. 134.

32. A.H.A., *CT Scanners,* p. 43.

33. John E. Wennberg, "National Health Planning Goals" (report of a seminar sponsored by the Health Resources Administration, Department of Health, Education and Welfare, at the Center for the Analysis of Health Practices, Harvard School of Public Health, Boston, March, 1976), p. 32.

34. Clyde J. Behney, "Studies of the Implications of New Medical Technologies" (a staff paper for the National Council of Health Planning and Development, Office of Planning, Evaluation, and Legislation; Rockville, Md.: Health Resources Administration, Aug., 1976), p. 39.

35. Ronald M. Green, "Health Care and Justice in Contract Theory Perspective," in *Ethics and Health Policy,* ed. Robert M. Veatch and Roy Branson (Cambridge, Mass.: Ballinger, 1976), p. 119.

36. Egon Jonsson and Lars-Ake Marke, "CAT Scanners: The Swedish Experience," *HCM Review* (Spring, 1977), 38.

37. Illinois, Rule No. 9, "Standards for Review of Applications for Permit for Technologically Innovative Equipment" (Springfield, Ill.: State of Illinois Health Facilities Planning Board, Aug. 6, 1976).

38. Illinois, "Guidelines for CT Scanners," (Springfield, Ill.: State of Illinois Health Facilities Planning Board, Sept. 2, 1977).

39. John Rawls, *A Theory of Justice* (Cambridge, Mass.: Harvard University Press, 1971), pp. 62, 396.

40. Rawls, *Theory of Justice,* p. 303.

41. *Ibid.,* p. 75.

CULTURAL ISSUES IN MEDICAL RESEARCH

David C. Thomasma

The impetus for this paper stems from my membership on a Patient Participation Committee reviewing research on human beings and from several publications of the Hastings Center. The principal article was "Human Experimentation—The Ethical Questions Persist."[1] And persist they do. Medical research is beset with serious ethical problems.

In this paper I will explore questions raised by research on human beings. Some of these questions are ethical; others involve the broader context of ethics: our cultural value assumptions. The scope of the paper deals only with medical research. In other words, the paper does not deal with the delivery of health care or the training of health professionals.

Moreover, I will narrow my focus on conditions for the occurrence of ethical problems. These conditions constitute an area of "cultural assumptions" to be clarified later.

First I will raise some ethical issues; second, explore the cultural scientific myth leading to these issues; third, argue that scientific assumptions are being displaced by a value myth in our culture; fourth, offer a note of caution about value principles; and finally, offer some considerations leading to improvement of the ethical understanding of scientific research.

Ethical Issues

Although the cases included in the *Hastings Center Report* represent a small fraction of experiments done more responsibly on human beings, they nevertheless highlight several layers of questions, ethical, legal, and cultural, found in all such research.

The cases mentioned include production of abnormal heartbeats in patients, clinical dehydration of patients whose kidneys had been removed, artificially causing asthma in patients prone to the same to test the efficacy of cromolyn sodium (one response in a child was severe). Other cases involved behavior modification using drugs on mental patients, 18-gauge needle dual-site bone marrow samples taken from children (7–12 years) with suggestion of hematologic disease (physical and psychological problems resulted), behavior control of schizophrenics through unmitigated electric shock or withdrawal of food (the purpose was to make people work), the use of placebos resulting in 48 unwanted pregnancies (6 ectopic) in Latin American women, and pain-tolerance tests of 41,119 persons enrolled in a group health plan.

The response to these research experiments ranges from shocked dismay to qualified acceptance. All of the experiments involve high risk without much benefit to the subjects, inadequate consent; for some, a sense of brutality, and for others, a desire for solutions to problems. Humanists would claim the experiments were insensitive to the value of a person; scientists would cite the search for objective facts benefiting society.

The *Report* itself concluded that the following areas needed further exploration: (1) grave risks to subjects; (2) risk to incarcerated or incompetent subjects; (3) rights of subjects in research with placebo groups; (4) responsibility of the institution for harm to subjects; (5) and the pervasiveness of human experimentation.

Many of these areas have been developed more fully since the article appeared in 1973. Some of them have been partially answered by the new HEW guidelines on research.[2] The guidelines control risk-benefit ratios, informed consent, experiments on pregnant women, the right to withdraw and still be treated. In general the guidelines protect the legal rights of subjects. Although this protection is necessary, the ethical questions still remain to be answered. After all, what is legal is not necessarily moral.

In addition to the legal and ethical questions, it seems to me experimentation also raises more profound questions, ones that perhaps can never be solved with a legal fiat. It is to these questions this paper is principally addressed. They involve cultural assumptions.

Our society has not yet coped with the questions of value raised in experimentation, indeed, in the appearance of new technology. As Daniel Callahan argues, we seem to lack the cultural wisdom about what to do with ethical problems.[3]

At issue is the rupture between the moral order (what is right to do) and the scientific-technological order (what works) in our culture.[4] For medical research this rupture takes the following form: a difference between a person-oriented health care and a "benefit-to-mankind" one. In the former, the human person is regarded as supreme. In the latter, human persons may be rightly used in experiments for the benefits of others.

Cultural Assumptions: The Scientific Myth

Assumptions such as the very validity of human experimentation, that all human beings have rights, that the goal of a political system is individual freedom, that science is objective or that truth lies in objectivity, are rarely called into question. These cultural assumptions are values we, as a people, generally take for granted. As such, they are part of our cultural heritage.

The first point I want to raise, therefore, is a cultural one. As a civilization we are beginning to question technological progress. We are doing this out of a set of values not found in technology itself. Examples of this process of questioning abound—ecology curbing industrial growth, political and economic concerns vying with the private automobile, personal values overriding a technical ability to prolong life, and so on.

I would summarize this questioning process as a post-industrialization search for the quality of life.[5] Instead of asking "Can we?" we are asking "Ought we?" Asking "Ought we?" leads to questions of value, and thus to ethics.

If this process is the case, then our culture is moving through a fundamental revolution in outlook. "Can we?" questions are the inherent source of new technology. The goal of technology, its value, is design-implementation. The success of technology is judged on the value: how well does it work? Other values of technological progress are utility, conformity, punctuality, order.

I prefer to describe the process now taking place as a replacement of the "scientific myth" inherited from the days of Newton and reinforced by Descartes by a "value myth." By "myth" I mean the system of value-assumptions embodied in a culture by which meaning is given to everyday tasks and institutions are organized. Myth in this sense is not the "untrue," but what is "most true" in a culture. Myth embodies the ultimate values of a culture.[6]

In proposing that a value-myth is displacing the scientific myth, I am not suggesting that the latter embodies no values. In fact, my next step will be to sort out some of these values. Instead I am arguing that the purpose of the scientific myth is objectivity and problem-solving. This task-orientation necessarily abstracts from individual persons and their values. The focus is not on values, but truth.

The emerging value myth, on the other hand, focuses on personal values. Its object is care, not abstracted truths. The clash between these two myths can be seen by tracing the history of theoretical and scientific medicine over and against clinical practice. This clash is the battleground on which ethical issues in medical research are raised.

After considerable reflection I have distilled from the scientific myth at least three values which I perceive differ from similar components in the emerging value myth.[7] Both myths, by being myths, offer meaning to our tasks. They differ in fundamental values.

The three values of the scientific myth are: analysis, irreducible parts, and a search for objective certitude. Taking this view, to solve a problem means to analyze it into parts. What is valued is a power of analysis, sharp and clear distinctions between parts of the problem. The process of this problem-solving is governed by canons of objectivity and measurement. In other words, objective certitude will solve the problem.

I will offer three examples from the many which have led me to this distillation of values. The first is the atomic theory. I would argue that the atomic theory gradually evolved out of at least the three values mentioned above. A chemical change is a problem. To "solve" it, one analyzes the process (abstracting from all other related processes) into irreducible parts (at first, atoms, then molecules, now subatomic particles). In conducting this analysis, abstraction from personal beliefs and assumptions, feelings and pressures, would constitute the *sine qua non* of gaining an objective report on the process.

The second example comes from the health care system. In order to treat disease under the scientific myth, one had to isolate. Isolation of the diseased became a necessity, moreover, in times of plague or epidemic. Today, isolation often seems to be the real rather than the professed goal of the health care system. One is isolated from work, family, social relationships. Additionally one is then isolated within the hospital from other patients in private rooms, or even isolated from a human environment in an intensive care unit. In a number of cases, patients thus isolated have become psychotic.

In addition to physical isolation, the disease process is isolated from the "container" of the disease. And the parts of the disease are further broken down, assigned to specialists on referral. The end result is often a cure of the disease to the detriment of a person. "Cancer in a 14-year-old girl" is preferred to "a girl, a person, with cancer."

I suggest this isolation is an institutional result of the scientific myth, of analyzing parts to gain objectivity. The person-in-society becomes a part, a unit, a disease, then a part of a disease. No one can quibble with the effectiveness of this approach. One can quibble, however, with its sensitivity to human values and the needs and aspirations of each person. In fact, one can ask if it represents the aims of medicine at all.

My third example follows from the second. Those committed to scientific research on human beings often operate out of the scientific myth. Any challenge, even the HEW guidelines, is frequently viewed as a challenge to scientific progress: breaking down components of disease systems so that understanding these can benefit others. It is easy for a researcher to slip into thinking a person is a subject with a disease, which, after all, has no rights and is the legitimate and beneficial object of scientific study.

The anger some researchers manifest at "do-gooders" and government infringement is a signal that what is really being questioned is not scientific research but the scientific myth itself. Since this myth gives meaning to their lives, they are understandably threatened by the vaguer, less objective discussion of values, rights, risk-benefit ratios, and informed consent. The pressure to consider the subject a person with values, a social being interacting with others, directly contradicts the values of analysis into parts and objectivity.

The Value Myth

The scientific myth is scientific methodology raised to the level of cultural structures. In other words, our culture has adopted or extrapolated science into all areas of day-to-day life. It has done this because science has been successful. Not surprisingly, then, the value myth I discern displacing the scientific myth does not reject science. Instead it appears as the next stage of scientific development: synthesizing what has been analyzed. In its synthesis attempt, human values are more centrally located than in the scientific myth.

The value myth presently emerging has three complementary components, three values opposite those of the scientific myth. These are: instead of analysis, synthesis; instead of irreducible parts, open systems or wholes; instead of objective certitude, vaguer subjectivity and values.

Both the value myth and the scientific myth share the same purpose. That purpose is to solve problems. But the value myth seems to contextualize. In this view, there are no irreducible parts. Everything is linked to everything else. "Parts" are really open systems.[8]

Just as I cited three examples of the scientific myth, I will offer three of the value myth as well.

Gödel's theorem in mathematics and Heisenberg's Uncertainty Principle in physics, each in its own way, contribute to this emerging myth. Gödel's theorem, in essence, states that there can be no absolutely deductive mathematics. Even if we start with X we must have an interpretant Y. In other words, one's subjective interpretation enters even the most formal logic. The Uncertainty Principle deals with the theoretical inability to tell simultaneously whether an electron is a wave or a particle. The observer's very experiment would alter the reality. These discoveries have the effect of discrediting the notion that there are completely "objective" facts divorced from observers' interpretations.

Carried into the broader sphere of daily life, a transportation study seeking to solve a mass-transit problem would have to take into effect the private automobile (with the human value of individual freedom to go where one wishes), the cost of Arab oil, the already existing road economy (including the automobile working force), land-use patterns, existing rights-of-way, and ultimately, the values of Americans in general, including international relationships in the future. In short, transportation is an open system involving values, not an irreducible part unrelated to the whole.

My third example is again from medical research. A disease is not an isolated entity. It is related to organ systems in the body other than its primary location, and it has social and value consequences for the subject, the researcher, and the health care system itself. Cancer not only attacks the body, it ruptures social relationships and forces the subject to reexamine his or her values. Thus when a researcher wants to study hypertension and offers free room and board to the subjects as their only benefit (i.e., free hospitalization), those inclined to the value myth will question such things as the economic status of the subjects. Most people attracted by the offer would be poor, perhaps undernourished. When offered three days of comfortable living, are they being coerced? Is it appropriate for researchers to offer such a benefit? Is health care a benefit or a right for these persons?

A Note of Caution

It is my perception that the ethical issues surrounding research arise in the clash of the two myths described. Researchers are not "unethical"; rather, the values upon which their ethics rests are different from those that proceed from the perspective of the value myth. The value myth deals with the kind of relationships in which interpersonal values emerge as part of the decision to act in a certain way. The scientific myth eschews personal values.

A note of caution is necessary, however. The reason that the value myth tends to "muddy the waters" is that principles of value are inherently ambiguous and incomplete. They are nondeterminative in relation to "the right thing to do." In other words, values are not prescriptive obligations to act in a certain way. An ethics of values is an ethics of what one ought to do but might not, depending on the situation.

Several observations are worthy of note. Principles of value are pluralistic in a pluralistic society. In addition, when two or more persons agree on a value principle, they can still arrive at different conclusions about what to do. This difference stems from three sources at least. The first is the nonprescriptive nature of a value principle. The second is an "instinct," a second-nature reaction built up by different experience. This "instinct" can be formulated as a metaphysical view about the nature of the world and human society. It is the realm of the scientific and value myths previously discussed. A third level of divergence occurs in the juncture between behavior and thought. We do not always act out our own values.

It is clear that the solution to ethical violations in medical research cannot rest on a concern for value principles alone. Once these are introduced through affiliation with the value myth, considerable dialogue must continue. This dialogue will involve conflicts of "metaphysical instincts," conclusions, and behavior.

Thus, to raise the ethical issue about proposed research is to challenge values, assumptions about the world not easily defended, and the behavior of medical research in general. Legal prescripts, like those of HEW, represent steppingstones in an endlessly bubbling stream of values.

Application to Medical Research

If the argument so far presented is valid, then ethical issues in medical research arise out of a clash of values embodied in what I have called the scientific myth and the value myth. Succinctly put, the scientific myth would aid an investigator in prioritizing values by reinforcing the need for objective, theoretical knowledge. By theoretical knowledge I mean what Aristotle meant, a body of knowledge concerning what is, sought for its own sake. St. Thomas adds that speculative sciences "must be things that cannot be made or done by us, so that our knowledge of them cannot be directed to activity as an end."[9]

It should be obvious that medical research should not aim at theoretical knowledge, although in its nontherapeutic form it may seem to. Yet even nontherapeutic research is aimed at practical knowledge, for its goal is to discover some knowledge which may be directed toward an activity, namely, some manipulation of the human body. In other words, all medical research, whether therapeutic or nontherapeutic, should be aimed at a therapeutic imperative. The reason is that medicine itself has such an aim.

Two conclusions flow from this line of reasoning. The first is that the scientific myth is an inappropriate set of cultural assumptions for medical research. It has entered the design of experiments in a nonreflective way because medicine has been taught as a science, since the Flexner report (he himself was a scientist, and not a physician). This observation is not intended to demean the scientific character of modern medicine. Instead it is meant to reinforce this character, but in the form of a practical science.

The second conclusion may be more difficult to uphold. If the scientific myth is an inappropriate value-prioritizer when applied to medical research, then can it be that medical research diverted, however remotely, from its eventual therapeutic aim is *de facto* unethical? If the aim of medical research is some cognitive art of manipulating the human body, then are experiments aimed at the theoretical alone unethical?

It seems to me they are. The judgment on the research cited at the beginning of this article was universally negative. Not only philosophers, but clinicians too abhorred their insensitivity to human life, which should be a hallmark of all facets of medicine. Subsequent guidelines developed by HEW have done much to remove this kind of research from the medical scene in the United States. But an important question remains: *Why* are those experiments judged unethical?

Our first temptation is to cite a deontological check on utilitarianism. Put another way, this argument would run as follows: nontherapeutic research relies on a form of utilitarian ethics for its justification. It presupposes that the greatest good for the greatest number justifies putting some individuals at risk. These experiments erred either in lack of adequate consent or in inherent scientific design, such that the experiments violated the right to self-determination or did not produce valid scientific results to justify the risk taken by subjects. Medicine must do no harm.

This line of reasoning, however, neglects two important factors: first, it fails to recognize the validity of nontherapeutic research which may fail to support a hypothesis about disease; and secondly, it assumes that modern medicine can gain ethical support from utilitarianism and/or deontological ethics. This assumption frequently leads to a simplistic bifurcation of medicine into clinical and research, the former governed by an ethic of not harming and the latter governed by an ethic of harming for the greater good.

A better line of argumentation will conclude this paper. The reason the experiments in question are judged inadequate is that they implicitly rest on a scientific myth inappropriate to the nature of medicine as a practical science. Knowledge gained in the experiments was not reintegrated, not applied synergistically to subjects. In other words the experiments failed to be medical experiments, aimed at therapeutic evaluation.

Prioritizing values in medicine rests on the value myth, a myth more appropriate to a practical science. Hence an important distinction: scientific research, including biological research, is that aimed at theoretical knowledge. Scientific research on human subjects may, in fact, be justified by appeal to the common good and may be supported by a form of utilitarian ethics with adequate consent. However, scientific research masquerading as medical research is always unethical, even with adequate consent, because it lacks therapeutic evaluation in its design.

Medical research, resting on the value myth, may therefore be either therapeutic or nontherapeutic. In both cases the judgment of the ethical nature of the experiment is guided not on the basis of objective knowledge gained, but of potential therapeutic effect. Although nontherapeutic does not offer therapeutic benefits to subjects, it is designed to prove or disprove a hypothesis having direct bearing on eventual "medical knowledge." That is, its aim is holistic and evaluative of practical effects on the lives of human beings.

In this view, all medical research is ethical, including nontherapeutic failures, because it rests on a value myth which more adequately reflects the nature of the medical enterprise.

Finally, no artificial bifurcation of modern medicine into two opposing camps appears in this argument. Neither utilitarianism nor deontological ethics can adequately deal with the full range of present-day medical endeavors. A new ethic is needed and might be drafted in light of the value myth proposed in this paper.

Enhancing the "quality of life" would be articulated, not in quantitative units, but against the backdrop of an open system of interlocking relationships, relationships in which values occur. Altering human life would be seen as a holistic responsibility not solved by analysis into parts in terms of the scientific myth. However great the knowledge gained in medical research, it could never be gained ethically at the expense of the quality of life understood in terms of the value myth.

Although more development of this line of reasoning is necessary, some initial conclusions can be reached.

1. Medical research is judged ethical by appeal to implicit or explicit prioritization of values.
2. Values are prioritized according to certain assumptions of a cultural myth.
3. The scientific myth inadequately represents the nature of medical research. Accordingly, research which neglects therapeutic intent in medicine is unethical.
4. Medical research rests on a value myth which stresses integration, wholes, and personal values.
5. Therapeutic and nontherapeutic medical research is ethical provided the proper consent is obtained.
6. Norms such as risk/benefit and adequate consent reflect the moral nature of the therapeutic intent of medical research.
7. Medicine should not be divided into clinical and research. All medicine is therapeutic experimentation.
8. Deontological and utilitarian ethical systems are rarely capable of analyzing the whole of modern medicine.
9. A new ethic is needed which will be developed from the value myth cited in this paper.

NOTES

1. *Hastings Center Report,* 3 (June, 1973), 1–3.

2. I have heard some humanists argue that the guidelines do not stop research. Instead, they offer more possibilities while assuaging objections of humanists by their inclusion on committees reviewing experimentation. But this view assumes all researchers are moral dolts, not borne out in my experience. Point 5 of the report, however, is rarely discussed.

3. *The Tyranny of Survival* (New York: Macmillan, 1974). Bertrand Russell once remarked in this regard: "If a scientific civilization is to be a good civilization, it is necessary that increase in knowledge be accompanied by an increase in wisdom. I mean by wisdom, a right conception of the ends of life. This is something which science in itself does not provide" (as quoted in "Medicine's Technical Feats Forcing New Emphasis on Study of Ethics," *American Medical News,* Dec. 2, 1974, p. 17).

4. This rupture is explored in relation to dying by Eric Cassell, M.D., "Dying in a Technological Society," *Hastings Center Studies,* No. 2 (May, 1974), 31–36.

5. "Post-industrial" is not an original phrase. A collection of essays on cultural critics by William Kuhns is entitled, *The Post-Industrial Prophets: Interpretations of Technology* (New York: Harper Colophon Books, 1973).

6. Analogous to "myth" is the German, *Weltanschauung.* For those unfamiliar with this usage, I would point out that it is common in cultural and religious studies. Thus, American Studies, in seeking what is unique about America, would point to the "myth" of rugged individualism, the frontier mentality, pragmatism, and the "new Eden." These myths are discovered by analyzing the behavior of our society.

7. In this sense, I agree with those who argue that there is no value-free science. An important distinction must be drawn here between the components of the scientific myth and of science itself. The components of the myth are value assumptions. The components of science, its methodology and practice, abstract from personal values. But this abstraction is itself a value of the myth (what gives meaning to the scientific enterprise). In other words, the scientific myth is value-laden, not scientific methodology. Failure to make this distinction causes all sorts of chaos. Those who argue that science itself is value-laden seem to their opponents to absurdly state that objectivity is not part of real scientific methodology. Those who counter by arguing science is not value-laden fail to see that "abstraction from values" is itself a value.

Intellectual chaos is not the only result of failure to distinguish science from its myth. Institutional and personal chaos also result, as is apparently happening at Dutch universities. For an enlightening article on this very point, see Henry L. Mason, "Reflections on the Politicized University: II. Triparity and Tripolarity in the Netherlands," *AAUP Bulletin,* 60, No. 4 (1974), 383–400.

8. The most popularized expression of open systems can be found in the thought of Buckminster Fuller (see, for example, his *Ideas and Integrities* [New York: Collier Books, 1969], pp. 65–66; 178; 239–240). In his view the entire world is synergetic process. By "synergy" Fuller means the behavior of wholes unpredicted by the behavior of individual parts. The phenomenal structural support of his geodesic dome cannot be predicted by analyzing the weaker support of any individual part.

Cybernetics is another source of open systems approaches. It has been discovered that the behavior of whole social systems works best when individual parts are *not* functioning at their optimum. Perhaps the best example of this insight applied has been unearthed by corporation consultants. Puzzled by efficiency reports that analyzed a department (Public Relations, let us say) but did not improve the function of the whole corporation, consultants found that the whole corporation functioned differently than did the sum of its parts. (See Ludwig von Bertalanffy, *General System Theory* [New York: Braziller, 1968]; *Robots, Men and Minds* [New York: Braziller, 1968].)

9. *In Boethius' de Trinitate,* translated by A. Mauer, *The Division and Methods of the Sciences* (Toronto: Pontifical Institute, 1963), q.V, a.1, reply, p. 7.

SYMPOSIUM: REVIEWS OF RECENT BOOKS IN PHILOSOPHY OF TECHNOLOGY AND MEDICINE, WITH AUTHORS' REPLIES

REVIEWS OF BERNARD GENDRON, *TECHNOLOGY AND THE HUMAN CONDITION*

I

Paul T. Durbin

I should like to begin by praising Gendron. In *Technology and the Human Condition,* with respect to the problem of controlling technology, he offers a radical solution, but a particularly reasonable one. Though not a radical myself, I admire many that I have known. I particularly appreciate their perceptiveness, long before Watergate brought things to the consciousness of the general public, with respect to the civil liberties threats to our society in the sixties. (Such radicals as are still around and still speaking out would, of course, maintain that those liberties are still threatened.) I also appreciate their perceptiveness on problems of technology.

Research in Philosophy & Technology, Volume 3, 1980, pages 77–130.
Copyright © 1980 by JAI Press Inc.
All rights of reproduction in any form reserved.
ISBN: 0-89232-102-4

What I would like to do here is divide my treatment of Gendron's book into three parts. In the first, I outline why I think he deserves praise for his reasonableness in defending an unpopular thesis. Second, I will summarize some of what I see as the main points in his argument. Then, finally, in the third part of my review, I turn to an argument against Gendron in the name of what one might call a "techno-liberalism," an optimistic but non-Utopian (in Gendron's sense) liberal scheme for the democratic control of technology.

Gendron's Mode of Argumentation

The carefulness of Gendron's argumentation—*Technology and the Human Condition* is really one long argument—does not show up in terms of the now customary style of American and British analytic philosophers, with elaborate logical apparatus, imaginative counterarguments to probe the conceptual rigor or consistency of statements, and the like. Rather, for Gendron, "reasonableness" seems to mean persuasive rationality in the public discussion of an issue—here the control of technology. What he offers is particularly congenial to me because it is so close to the sort of Aristotelian-style "dialectic" that I so often defend. Gendron assumes that the point (or points) at issue is not yet definitively decidable, and asks, What are the best reasonably arguable opinions around? If they are not satisfactorily reasonable to him as presented by their leading proponents, he redoes the arguments to make them more persuasive—exactly the opposite of the approach of many philosophers today, who would rather attack the inconsistencies in an argument than attempt to find out how the same conclusion could be more agreeably arrived at. Gendron then examines the consequences of the various reasonably defended opinions—his candidates for the understanding (and control) of technology he calls Utopianism, Dystopianism, and Socialism—and compares them with one another in terms of their potential.

Two examples may suffice to illustrate how this procedure works.

For one example, consider Gendron's treatment of Jacques Ellul's by now all too well known thesis about *technique* as out of all control. Gendron renders this "runaway technology," treats it under Dystopianism, and says this at the beginning of his treatment:

> To the sophisticated and academically trained reader, the thesis of technological fetishism may appear to be wholly bizarre and implausible, the product of a confused mind. It may appear untestable or not worthy of being tested. . . .
>
> This response, I think, is unfair. Technological fetishism can be formulated in a way which is clear and testable, though it usually is not (p. 149).

Though my point here is to emphasize Gendron's approach to getting at the reasonableness or unreasonableness of an argument, and not the argument itself, I may be pardoned a brief digression to indicate how Gendron attempts to render "technological fetishism" credible (before showing why he thinks it is wrong). If we can find cases, he says, where as a society we know further technological growth will be self-defeating, yet this in no way deters us from pursuing it, we have a clearly testable meaning for technological fetishism. Gendron then gives his examples: "If indeed the imperatives of technology commit us to a spiraling arms race and increasing ecological breakdown, and if we nevertheless pursue technological growth, then it is prima facie evidence that we have made a fetish of technology" (p. 156). Gendron does not in fact think this is the explanation, for instance, of the "warfare state," but it shows how far he is willing to go in trying to understand how a reasonable person could maintain the technological fetish thesis.

A second example. At the end of his treatment of John Kenneth Galbraith on the "technostructure," Gendron summarizes his argument:

> We asked what basis Utopians could have for believing that the postindustrial revolution in technology will lead to the elimination of scarcity. No Utopian I know of has given a systematic defense of this claim. What I did was to reconstruct a defense seemingly implied in the works of some important Utopians (p. 60).

Again I am interested in the procedure but—since it is important and will not be treated elsewhere here—I will repeat Gendron's summary of the argument:

> It goes like this: to set the stage for the elimination of scarcity, postindustrial technology must put an end to those trends (such as toward poverty, pollution, and compulsive consumption) that counteract the scarcity-inducing effects of growing technology. It does this by replacing industrial technology and eliminating the capitalist corporate system, . . . bring[ing] about the rise of a technocratic corporate system, and with it a technocratic government system which will ultimately eliminate poverty, . . . pollution, and . . . compulsive consumption (p. 60).

An Outline of Gendron's Overall Argument

In the introduction to his defense of a socialist view of technological promise (in Part III), Gendron summarizes points of agreement and disagreement with Utopians and Dystopians (pp. 189–190). I summarize these views in Figure 1.

In chapter 11 Gendron claims to be arguing against the "technocracy thesis" of both Utopians and Dystopians. In fact, the major burden of the argument is anti-Utopian. The argument is in two steps.

Step one argues, against the technocracy thesis, that capitalists, not "manager-technicians (experts)," in fact still control the corporations. The arguments in favor of the technocracy thesis—that if expert recommendations were frequently overridden by stockholder-owners there

Figure 1.

Gendron's summary of socialist agreements and disagreements with utopians and dystopians (*Technology and the Human Condition* [1977], pp. 189–190)

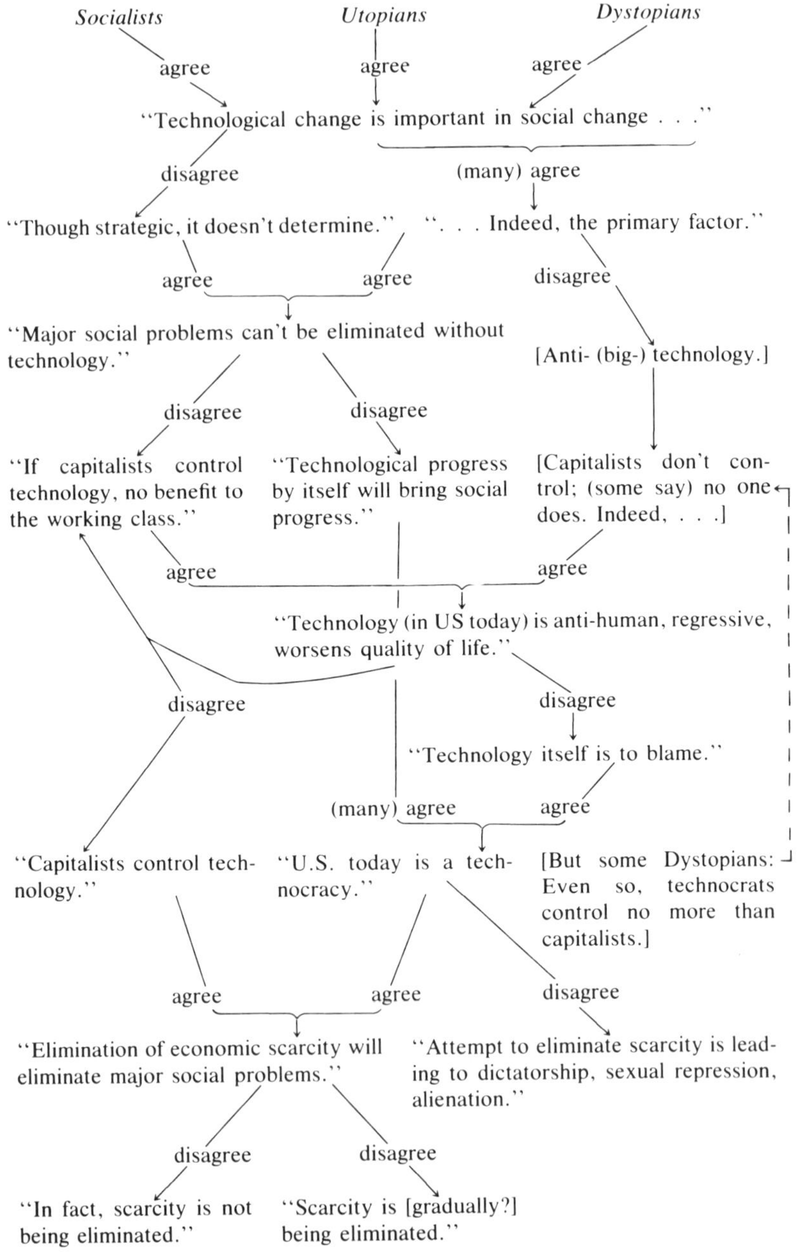

would be "wholesale havoc" (this is called a "theoretical" argument), and that financiers no longer have the economic wherewithal to control modern corporations (called a "more empirical" argument)—are fairly easily brushed aside. Familiar material is brought forward to show (a) that very small concerted voting blocs are sufficient to control even the largest corporations, and (b) that banks and other financial institutions, especially when acting in concert in a small number of major economic "interest groups" scattered across the United States, exercise enormous control over the biggest corporations, many of which are heavily into technology.

In step two, Gendron argues that corporate capitalists exercise effective control over the Federal government. He is fairly sophisticated here, arguing not that there is direct control but that, "If . . . public government does all this [protect the 'private power system' from excessive regulation and consumer demands], then the public power system, though it theoretically has authority over the private power system, is in effect its instrument" (p. 205). His arguments (again) are fairly standard, that corporations have privileged access to the Federal government (e.g., by way of lobbying), and that they, and their capitalist owners, share disproportionately in the largess of the Federal government (especially through tax breaks). (Gendron does not make as much as he might, at this point, of fat contracts to big corporations, often for sophisticated technological gadgetry.)

If at this point one is tempted to ask, What is the relevance of all this to technology?, two things need to be recalled. Gendron is arguing against a technocracy thesis. In addition, his view in agreement with the Dystopians is that technology is somehow to blame for many of our current social ills. (The "somehow" is in there to reflect the differing emphases: Gendron would say it is *capitalist* technology, while many Dystopians would say it is technology itself.)

In chapter 12 Gendron turns to the problem of scarcity. Passing over his claim that Dystopians think the very attempt to eliminate scarcity through the use of technology brings with it the evils of dictatorship, sexual repression, and alienation, the chapter is again mostly anti-Utopian. Scarcity is *not* being eliminated, gradually or otherwise, and the culprit is a stubbornly persistent capitalism, not technology.

The final substantive chapter (13) is Gendron's Marx-based positive view of how a "real socialism" (p. 242 and elsewhere) can manage technology beneficently.

I might argue at once that Gendron's vision of the future beneficent management of technology under a "real" socialism is optimistic, utopian, even visionary. He even asks himself the question: "Are we to agree with the cynics that real socialism is impossible because real [worker-controlled] democracy is impossible?" His reply—"We should be careful here not to take too rigid a view of human history. All the evidence is not in yet"—seems strange enough coming from a follower of Marx with the latter's belief in "scientific" laws of history. But there is more: "Furthermore, socialism seems to offer the only hope" (p. 244 for all this). That is it. That—except for the counterarguments against Utopians and Dystopians—is the final and the only reason given for expecting a beneficent management of technology under real socialism, just hope.

This is a little unfair to Gendron, however, since this same criticism can be and often has been leveled against any Marxist-oriented view. It is not peculiar to Gendron. To his credit, Gendron does bring forth some evidence to support his belief. Such promise as he offers us that the program *might* work comes from Maoist China (though later Gendron admits that "China is not strictly speaking a socialist [worker-controlled] society . . . [any more than] the Soviet Union, . . . Eastern Europe, or . . . North Korea" [p. 243]):

> During the Cultural Revolution, the Maoists instituted a number of reforms to correct the excesses of the [Liu] policy of "technics takes command." . . . One of the most dramatic ploys was the institutionalization of "three-in-one" committees. . . . Technicians were never to be left alone; they were to interact with workers and party members at every level of decision making. And technicians could not appeal to their expertise in order to intimidate workers and cadres. . . . [E]ducated workers and cadres would [even] take part in research projects with technicians and scientists. Meanwhile, technicians would be periodically sent out to work in the countryside or in the production lines of factories (p. 242).

A "Liberal/Progressive" Critique of Gendron on Technology

If, rather than focus on the visionary character of the socialist view—something often noted by critics and not peculiar to Gendron—I focus on a peculiar weakness to be found in *Technology and the Human Condition,* it has to be on Gendron's treatment of the political process in technocapitalist America.

Because he brings it up at so many crucial points, I take it that Gendron recognizes *pluralism* as the key concept in the Utopian defense of the technocracy thesis—so crucial to its response to socialism. At several points Gendron summarizes his understanding of pluralism. For example:

> Supporters of the pluralist view have of course argued that the above [the federal government as instrument of the private sector] is not the case. As the story goes, the emerging twentieth-century welfare state has encroached on the economy . . . [and] increasingly succeeded in subjecting the business community to external regulation by representatives of other groups (such as consumers) or by spokesmen of the public interest (p. 205).

At the beginning of the crucial section of his argument, Gendron defines what he understands to be the pluralist view and indicates how he thinks it is opposed to his view:

> Here Socialists run into the very popular pluralist view that no one group controls the federal government. Rather, according to the pluralist view, major political decisions of the federal government are a consequence of pressures from many groups, each competing with the other. This view concedes that capitalists have a good deal of influence over major political decisions; but so do labor unions, medical associations, and educational associations. No group gets its way all the time or loses all the time. The federal government is the principal arena of competition for the various groups and the locus . . . of equilibrium (p. 201).

Gendron is not clearly wrong here, but when I first read this passage, as well as Gendron's treatment of corporate influence on the Federal government through lobbying as well as the rewards received, I had an uneasy feeling that something was wrong, that there was some misplaced emphasis. I still have that feeling. In the remainder of this review, I will try to rationalize my uneasy feeling and attempt to put forward a better account of the pluralist political process than the one Gendron associates with Utopianism.

Gendron's most obvious misplaced emphasis is in focusing almost exclusively on the Federal government, and on presidential and other "media event" politics such as the selection of Cabinet and sub-Cabinet officials. He makes much of the William Domhoff thesis about capitalists as having a stranglehold on the top executive appointive positions, as well as on most Cabinet office advisory commissions and such policy research institutes as the Council for Economic Development and the Brookings Institution. He also recognizes, though he emphasizes it less, the influence of "big money" in lobbying Congress and in campaign financing. But his lack of perception shows up in this: "Labor, on the other hand [i.e., as opposed to the capitalists], does not have the same kind of automatic impact on governments. Other than the important role of helping to determine wages, labor unions have little to do with the everyday decisions involved in running the economy" (p. 204). This seems fishy, to say the least. But so is the almost total ignoring of the myriads of other groups that make up the pluralist mix in contemporary society: the political parties all the way down to the local level, the military (so important, especially in its R & D and procurement components, relative to technology), not just the national education lobbies but educational institutions at all levels, the churches, volunteer associations such as charities and recreational groups (consider, for instance, the NRA), society, the "celebrity" world of entertainment, the elaborate "cops and robbers" system of the courts, the police, and the criminal subculture (many radicals have made much of the importance of the appointment of U.S. attorneys in the maintaining of a quasi-police state), etc., etc., etc. Pluralism is a very complex phenomenon if it has any meaning at all, and Gendron's analysis badly oversimplifies it. None of this necessarily counts against Gendron's thesis that capitalists are in control—they are in obvious control of many of these additional groups in the pluralist mix, but it makes the thesis much more difficult to support.

A better account of pluralism can be found in the following quote. It comes from Amitai Etzioni, whom Gendron would probably classify as a Utopian (he is not, however, on Gendron's basic list, p. 4) even though the quote comes from a review of Alvin Toffler (one of Gendron's explicitly listed Utopians) that is highly critical of the latter for his unrealistic understanding of the political process. Here is Etzioni:

> [Toffler's] conception of political processes is innocent, to say the least. . . .
> As I see it, . . . society is composed not of elites and masses but of a plurality of groups who command *un*equal political power, ranging from forceful groups of manufacturers to weak groups of people on relief. The power of each group is determined by numerous factors, including resources, size of membership, level of education, political consciousness, organization and mobilization for political action. The course society steers—within the constraints set by its environment—is largely shaped by the interplay among these various groups (*Trans-Action*, November-December, 1970; p. 103).

Why do I make so much of this seemingly minor difference in emphasis? Gendron and other critics of contemporary technological society who draw their inspiration from Marx must, if they would be consistent with their mentor, pin their hopes—slim and visionary as they sometimes seem—on a working class awakened to the inherent contradictions in late capitalist society. I am sufficiently radically inclined to find appealing a faith in the militant working class. (Ours currently is certainly not such; in fact, many salaried employees who ought to identify with the working class—engineers, for example, as well as other salaried professionals—are strongly antimilitant.) But such optimism as I have with respect to the democratic control of technology is rooted in a different stratification scheme from Marxist classes. In line with a tradition going back to Aristotle, I pin my hopes on the middle class—though defined in a particular way.

Some sociologists, or at least their popularizers, seem to define socioeconomic classes more or less exclusively in terms of income, status, and such large material possessions as homes and automobiles. (For instance, I happen to have been in Boston when a feature story appeared, "The 13 Steps of Metro Hub's Social Ladder," by Robert Levey, February 22, 1976, in which the major feature of differentiation was income and place of residence; the sociological study on which the article was based, by Richard Coleman and Lee Rainwater, appears to be more balanced.) I prefer the approach of Herbert Gans—for instance, in *The Levittowners* (1967)—who defines "middle class" primarily as a set of attitudes or cultural way of life. According to Gans, the upper middle class

> . . . is already the culture of the college-educated, cosmopolitan population, trained to be interested in and to participate in the larger world. . . . Upper middle class people participate not only in voluntary associations but in the entire community. . . . For this population, community participation is almost a cultural duty. Although upper middle class people are as distrustful of politicians as others, they have both the skills and the status to become involved in government and to fight for what they think is desirable (p. 30).

It is interesting that Gans's comments are made in the context of inhabitants of Levittown (N.J.), which one would not expect to be upper middle class in the usual demographic sense. It is my view that such "middle class" political consciousness exists in all strata of society, naturally in different degrees, and that (as Etzioni already noted) it is one of the principal sources of whatever political muscle a group has.

If we feed this attitudinal definition of "middle class" into the pluralism theory, it comes out quite differently from what Gendron defines. Particularly important would be a distinction between a (reasonable) "statesmanlike" and a (partisan) "politician's" attitude. I believe there are reasonable, statesmanlike individuals in all levels of society, and that interaction between them is largely what generates the vitality of a pluralist system.

Someone else who has noted this is John W. Gardner, the founder of the citizen movement Common Cause:

> Some critics [ask] . . . How . . . could we assemble "a constituency for the public interest" when most of our members are involved in their own special interests? . . .
>
> Perhaps most physicians think primarily in terms of the vested interest of the medical profession. But there is a percentage (to avoid undue optimism, let's say a modest percentage) who see their profession as part of the larger social enterprise. . . .
>
> Similarly, . . . a modest percentage [of businessmen] see that it is a matter of elementary self-interest to work as well for the good of the larger community. . . .
>
> In the same way, we have found in every other field—labor, the professions, the universities, and so on—people capable of seeing how much their own self-interest is caught up in the public interest.

Further on, Gardner says this can be found "in every social class and every segment of society . . . among members of every race, every level, every cultural tradition . . . a housewife here, a union member there, a business entrepreneur, a salesman" (*In Common Cause* [1972], pp. 93–96).

What Gardner professes to find in this is a new (or, really, a revitalized) "constituency for change":

> Then in the 1960s a feeling for citizen action reappeared with extraordinary vigor. It was foreshadowed in the 1950s in the civil rights movement. In the 1960s the students raised the cry of "participatory democracy." Among the poor the phrase was "community action." The peace movement, the conservation movement, the family planning movement emerged as potent elements in our national life (p. 73).

Gardner even makes an analogy with the Founding Fathers: "When this nation began in the 1770s, it had a population of about 3 million, yet it produced at least a dozen statesmen of extraordinary quality. . . . Today, with sixty times greater population, we surely have far more men and women of that caliber" (p. 110).

As I would, Gardner contrasts his vision with what he calls "interest group pluralism":

> There are those who [say] . . . lobbying . . . should be left to the special interests because no group of citizens, no matter how high-minded, can take it upon themselves to say what is in the public interest. This view reflects a lingering belief in so-called interest-group pluralism. According to that doctrine, each special group pursues its own selfish concerns, but taken all together they somehow balance one another out and the public interest is served (p. 80).

Aside from the "hidden hand" aspect of this last quote, it seems to be pretty much this sort of pluralism that Gendron has in mind in his attacks on Utopianism. (The reservation is brought in because Gendron depends so much on Galbraith, who believes not in a "hidden hand" but in expert, technical, managerial planning.) I think Gardner's understanding of pluralism is far better than Gendron's. I believe there is a public-interest constituency for change. And I believe one of the best exemplifications of it is in the so-called "public interest science movement." And if enough scientists and other members of the technical community have not yet joined in that movement, I believe more will—and, in any case, the technical community already contains reasonable, statesmanlike individuals enough to discuss policy alternatives in our technological culture in a reasonable way.

I would not expect my remarks to have convinced Gendron, who, if he is a really convinced radical, is likely to see what I have said as every bit as utopian as the Utopianism he argues against in *Technology and the Human Condition*. What I hope to have done is to enter into the spirit of Gendron's argument, that is, to have raised another reasonable voice in the debate in which he is engaged. In the discussion to follow he may convince me that my optimistic liberal techno-progressivism is not really different from his target Utopianism; but after all, discussion is one thing we have in mind today.

II

Michael Goldman

Technology and the Human Condition is a very nice book, and it is a pleasure to be able to review it. Professor Gendron's strategy in the book is to consider three major views about technology: the Utopian, the Dystopian, and the Socialist, and to subject the major claims and arguments of each to close and usually impeccable scrutiny. Since the Utopian view has been so unfashionable for over a decade it is refreshing to see it so clearly and sympathetically stated. While Professor Gendron ultimately rejects it, it fares surprisingly well under his careful probing. Conversely, the Dystopian position, which has been preeminent at least in intellectual circles since the early days of the Vietnam war, does not fare as well as its popularity among thinking people would lead us to suspect. Professor Gendron ends by favoring the Socialist position, at least insofar as he is able to find a coherent articulation of it. This is not to say that he agrees wholeheartedly with all its elements, but that it is the most promising of the three global positions considered. There is a wealth of material in this book, a good deal of it, not surprisingly, concerned with social, economic, and political issues and their relationship to the role of technology in human affairs. There is no way to discuss all or even most of these issues here. Let me strongly urge those of you who have not read this book to do so. As for myself, I intend to make as much use of it as I can in classes in the next few years. What I will do in this review is discuss several key topics which come up at least once and usually several times in the book. In a few cases I disagree with Professor Gendron's position; in a few others my remarks are intended merely to extend or clarify his position. I hope that in this way the breadth and uniqueness of the book will become apparent to those who have not read it. The issues I will deal with are: the concept of wealth, the concept of poverty, alienation and decentralization, the demands of capitalism and economy-ecology.

Wealth

Throughout history there have been a number of ways of conceptualizing wealth. For my purposes it is useful to distinguish and discuss two of them: the physiocratic notion of wealth and the Marxist-humanist notion. According to the first, wealth can be identified roughly with "the bounty of nature"—specifically land and its unchangeable agricultural yield, but by extension we can also include natural resources. Since land and its yield are, according to this view, finite and since natural resources are also, this idea of wealth implies a fixed (though perhaps large) total available wealth and hence a fixed (though perhaps large) "carrying capacity" for the planet. It also implies that the process of consumption must take place within the constraints of a "zero sum game": the more I consume, necessarily the less there will be for someone else to consume, not merely now but forever into the future. Such a notion of wealth naturally gives rise to great speculation about distributive justice, because my consumption can only be conceptualized as a drain on the world's wealth forever more; hence it must be very clear that I deserve my portion of it. While the physiocrats were active as such only through the eighteenth century, their notion of wealth did not die with them. It persists today especially in the works of Garrett Hardin and in the influential book *Limits to Growth* sponsored by the Club of Rome. The practical consequences of such a notion are apparently well known—conservation, zero economic growth, etc.—but, as I will show later, are not usually carefully investigated. Parson Thomas Malthus, while not exactly a physiocrat, nevertheless came fairly close to this view when he insisted that population growth would outstrip the growth of food resources. But the notion that food resources could grow at all even at arithmetical rates is a bit more sophisticated than what we find in the physiocrats.

The Marxist-humanist concept of wealth is quite different. Here, there is no notion of a fixed resource, no idea of a fixed agricultural yield, no "zero sum game" to contend with. Wealth is potentially unlimited insofar as human creativity is potentially unlimited. The following quote from Engels sums up this idea quite well:

> Where has it been proved that the productivity of the land increases in arithmetical progression? The area of land is limited—that is perfectly true. But the labor power to be employed on this area increases together with the population; and even if we assume that the increase of output associated with this increase of labor is not always proportionate to the latter, there still remains a third element—which the economists, however, never consider as important—namely science, the progress of which is just as limitless and at least as rapid as that of population.[1]

Professor Gendron's quote from Buckminster Fuller expresses exactly the same sentiment:

> Experimentally demonstrated, wealth is: energy compounded with intellect's know-how. Science's Law of Conservation of Energy states that "energy cannot be created or destroyed." The first constituent of wealth—energy—is therefore irreducible. . . . Every time man uses the second constituent of wealth—his know-how—this intellectual resource automatically increases. Energy cannot decrease. Know-how can only increase. It is therefore scientifically clear that wealth which combines energy and intellect can only increase, and that wealth can only increase with use and that wealth increases as fast as it is used.[2]

The notion of wealth here is humanistic because it locates wealth in human activity and human potential (labor and science), not in some fixed, unchanging agglomeration of things (land or resources). According to the physiocratic notion of wealth, a resource always has been and always will be a resource, and hence always has had and always will have a fixed value. According to the humanistic account, a resource is fundamentally a human *creation* and its value is the result of human creative and productive activity. Consider for example petroleum. Up to a point in history it is useless or less than useless junk. It sometimes even dramatically decreases the productivity of land when it spontaneously seeps up through it. When science and the appropriate technology figure out a way to use it (through, for instance, the theory of the internal combustion engine), and when labor creates the material means for its implementation (the engine itself), petroleum *becomes* a resource. Human civilization "runs out of" resources only relative to a technology, never absolutely. Current talk about resource depletion must be understood in this context. All we are "running out of" are resources appropriate to the technology we have now, not to any or all possible technologies. What perplexes me about Professor Gendron's book in this connection is his apparent recognition of the humanistic rejection of the physiocratic notion:

> The Malthusian model, however, suffers from one glaring weakness: it fails to consider the dramatic role of technology in consistently expanding (or providing substitutes for) the supposedly fixed stock of nonrenewable resources, and in consistently increasing the productivity of land and capital. . . . Technology then tends to attenuate, if not eliminate, the scarcity of resources and investment funds (pp. 167–168). . . .

followed just a few pages later by an apparent acceptance of the *Limits to Growth* argument: "There is little question now, given present trends, that key nonrenewable mineral resources (such as lead, mercury, manganese, tin and tungsten) will be severely depleted in the not-too-distant future" (p. 177), and that, therefore, economic growth must be severely curtailed. If he *does* reject the physiocrats, why does he accept limits to growth?

Perhaps, though, this is not Professor Gendron's main argument, for he continues by shifting ground to a new consideration:

> But even if these options [exploitation of lower grade ores, outer space exploration] opened up indefinite opportunities, they would still raise further insurmountable ecological difficulties. Each would require continued increases in the input of energy for every output of mineral resource . . . the use of increasingly more energy inevitably leads to the emission of more pollution, at least in the form of thermal pollution (i.e., waste heat). Now, it may well be the case that thermal efficiency will continue to improve, but this would only be a short-term solution. As long as the use of energy keeps increasing exponentially, the absolute amount of thermal pollution

> will ultimately increase beyond tolerable bounds even after the maximally feasible efficiency is reached (which of course can never be 100 percent). Clearly, then, there is no way to avoid the conclusion that growth in extraction of minerals must come to an end sometime, and most probably within the next century (p. 178).

But surely this argument is weak. For even if we admit the need to reduce to *zero* the increase in energy use—or perhaps even aim to maintain consumption at a lower rate than it now is, thermal pollution must continue. Perhaps we will thermally pollute at a slower rate, but if thermal pollution is a planet-threatening problem, it is not clear to me that we can solve it by putting off for a few years the time when it becomes critical. If thermal pollution is indeed an insurmountable deadly threat, I am tempted to say let's enjoy our last years rather than become neurotic kilowatt counters, austerely preserving our last BTU until disaster strikes. However, there are two considerations which mitigate against this dismal prospect: (1) The use of fossil fuels and direct solar energy are methods which in effect temporally redistribute solar heat which is already on the planet; they do not create any heat which was not already here. This is not to say that such redistributions don't have potentially serious consequences: I have often wondered if a worldwide solar energy economy would not alter our climate by temporally redistributing some (admittedly small) percentage of the solar heat which now is instrumental in producing it. I'd like to see an environmental impact study of this Naderesque project. Nevertheless these consequences are not necessarily bad, and they do not necessarily involve thermal pollution problems. (2) Fission and fusion sources do indeed increase the heat energy on the planet, but there is no reason to assume that more heat is worse than less. The notion of thermal *pollution* depends on the assumption that there is some relatively fixed, eternally desirable level of heat energy for the earth's biosphere. This in turn presupposes that there is some eternally desirable ecological system. But what reason is there to suppose this? The history of the biosphere is characterized by very *many* ecologies, and no doubt the biosphere will continue to display many more. It would require a proof, which I have never seen or heard of, to show that the one we have now is better than any other, or for that matter, that *any one* is necessarily the best. Besides, the time will surely come when most manufacturing, and hence most thermal pollution, will take place in other parts of the solar system. If these considerations seem speculative and "unrealistic," consider that the alternative implies that we are all doomed by thermal pollution no matter what we do, short of social suicide.

There is further evidence that Professor Gendron accepts the physiocratic notion of wealth, and this strikes me as the most serious weakness in the book. For example, he insists that the value of socialism will be principally to slow down the consumption of resources, and hence to solve this aspect of the ecological crisis. This position is predicated on the physiocratic theory. It is peculiar, especially in light of the unambiguous remark of Engels quoted above, that Professor Gendron attributes to socialism the ability, much less the desire, to slow down resource consumption.

Poverty

One of the most important and interesting elements in Professor Gendron's book is his attempt to define poverty in such a way as to capture our intuitions about it and at the same time make an assessment of the real and potential significance of technology for alleviating it. What I want to do in this section is offer an account of poverty at the social level, which is, I believe, coherent with Professor Gendron's account, but which introduces an important element that he only hints at.

It is useful to begin, as he does, with some undefined notion of a basic, non-culture-relative biological need: enough of whatever is necessary to keep the heart beating, metabolism functioning, etc. While the needs of a given individual in this regard may differ from the needs of some other individual, we can nevertheless represent the *totality* of such needs across a given population as L. The question which then arises is, in the world defined by this population what does it take to assure that L will be available? Consider for a moment the situation in which all and *only* those needs are met; in other words a situation in which there is just L, and it is rationally distributed—that is, everyone gets just enough to stay alive. Clearly such a world would rapidly become a nonproductive one. The productive network required to produce L would begin to erode as those who do the physical work to provide the essentials would lack the strength to continue (as their caloric intake would barely sustain life, not productive life). The education needed to bring people to productive intellectual levels would also, by hypothesis, be absent. In short, a world in which everyone consumed only enough to stay alive would quite rapidly become a world in which there was not enough for everyone to stay alive. These considerations allow us to pinpoint a certain consumption level, call it C, which is substantially greater than L, but which nonetheless is the minimum necessary to assure the production of *at least L*. This may seem paradoxical but it is not. It also establishes conceptually the crucial link, not really developed by Professor Gendron, between consumption and production. But, not surprisingly, we cannot stop our analysis with C. Necessary to the maintenance of the social productive level (and hence to the consumption level) is, in the long run, an actual increase in social consumption. To take a simple example: it now takes a much larger energy input to extract a given BTU level of fossil fuels than it did a decade ago; it also takes more sophisticated technology and hence, on the average, (somewhat) more sophisticated engineering talents in the population. In other words, in addition to larger energy consumption in the future, a larger consumption of education (teacher time, books, computer time, etc.) and a larger amount of nonproductive but highly consumptive education time are required *now*. Also, the production of technologies which will only become useful later is necessary now. Thus C is really a fiction, a value derived from conceiving society as a nondynamic, ahistorical set of interactions, almost as a frozen temporal cross-section. When viewed in its actual temporal dimension, we need to add somewhat to C, so as to assure our future against this problem of "diminishing returns." We can call this new quantity of minimum social consumption D, with the understanding that D itself shows a tendency to increase over time. Two interesting corollaries, neither of which will be explored, follow from the analysis so far: (1) the concept of "zero economic growth," where this is taken to mean no increase in social consumption, is in fact a call for a positive *decrease* in the human standard of living, and is therefore a vicious social policy which ought to be resisted with all the moral and political force available to us; (2) we can now more clearly see the mistake of the "zero sum game" concept of consumption which is implicit in the physiocrat's conception of a "fixed" stock of wealth. For while it might have seemed that whatever I consume diminishes the total available for others, we now see that my consumption under many circumstances might be required for a positive increase in total wealth.

Notice now where we are: we have nowhere assumed any extraordinary "compulsive consumption," nor have we even tried to bring a concept of "progress" into our perspective. We have not even taken into consideration the needs of an increasing population, nor have we assumed the legitimacy of any felt need for luxury. Nevertheless, we are forced to conclude that the maintenance of a population at a bare subsistence level is an impossibility, and that life itself demands a substantially higher level of consumption. Surely this is the proper meaning of Marx's observation, quoted by Professor Gendron:

> The worker's natural wants, such as food, clothing, fuel, and housing, vary according to climatic and other physical conditions of his country. . . . On the other hand the number and extent of his socially necessary wants . . . are themselves the product of historical development and depend, therefore, to a great extent on the degree of civilization of a country (p. 217, quote from *Capital,* Vol. 1, pp. 170–171).

Professor Gendron interprets "socially necessary" as "psychologically desired" and hence as, primarily, determined by a subjective reaction to one's culture's general level of consumption. I think, however, that "socially necessary" ought to be understood literally as *necessary* to the social productive needs of a given culture. This would put Marx's theory in Professor Gendron's category of a variable absolute income approach to poverty, where I think it belongs.

We are now in a position to offer a social meaning of poverty: a society will be poor if its consumption level at a given time drops below the value of D for that given time. If we prefer to be on the safe side, we might define a value D′ which is somewhat larger than D, and define poverty as a consumption level below that point. We should keep in mind that D (or D′) takes into account the needs of the foreseeable future: for instance, scientific education, whose fruits will not be manifest for a generation but which we feel confident will be absolutely necessary then, count in D. Thus we can assert that the last decade's unconscionable de-emphasis of scientific and technological education when viewed in light of the obvious need now for sophisticated alternatives to fossil fuels, might well have come close to qualifying us as a culture of poverty, despite appearances to the contrary.

Of course none of this gives us a measure for individual poverty, which can still be abstractly considered as consumption below the subsistence level. But it does suggest that when viewed dynamically through time, individual poverty would be a failure to increase one's consumption level in such a way as to increase one's socially useful talents at a rate at least consistent with the growing social productive needs (assuming that there is no social poverty to begin with).[3] Of course this cannot be the whole of it, since it gives us only a necessary condition for avoiding an individual's falling into poverty *if* one begins without individual poverty. What we need is a nondynamic account to use as a base, and all I can offer here is a very speculative account which I know is incorrect but which perhaps can be used by others as a basis for further thought: persons are poor (in a society which is not poor) if their consumption level is such that they cannot productively contribute to the extent that they are potentially capable of, and if that consumption level can be increased without detracting from the productive levels elsewhere in society. I think this is what Professor Gendron would call a variable absolute characterization, and meets his objection to other kinds of measures.

Alienation and Decentralization

Professor Gendron's discussion of alienation and the related question of decentralization are the aspects of his book that I find least satisfying, not so much because he is wrong but because he seems to operate with two senses of alienation, leaving an ambiguity both in his analysis and in his solution. Let me begin with what seems to me to be a proper account of alienation. In the first place, alienation is fundamentally a class phenomenon: the working class is alienated from its labor and from itself as a class. This means that an individual worker is alienated as a logical consequence of his/her relationship to the means of production, and not as a psychological consequence of that relationship. What this is meant to deny is that the working class is alienated as a class *because* each or most of its members are "alienated" in their work. The many anecdotes, stories, psychological studies, etc., which describe the psychologically disastrous conditions that prevail in many modern industrial activities are only peripherally related to the problem of alienation. This is because what characterizes alienation is that social relationship to the means of production in which the immediate object produced is determined by another class in the interest of that class, and the methods of production are designed to serve those interests. The working class is alienated from itself precisely because the *ultimate* product of its activity is human life, the life of the working class itself, but it produces this only, as it were, accidentally, never self-consciously, never with the *awareness* of what it is doing, much less the *power* to alter what it is doing. I agree with Professor Gendron that technology has almost nothing to do with this phenomenon. A highly technological society no more than a moderately technological one can manifest this class relationship. Neither does the related issue, the extent to which the productive process is subdivided. A highly subdivided process may well be preferred by a working class which is not alienated. (Professor Gendron's reference to the life of medieval craftsmen as less alienated than contemporary factory workers because their work is more varied, and because they had more leisure than the modern factory worker, is seriously mistaken. Medieval craftsmen were only an infinitesimal percentage of the medieval working class, most of whom were agricultural workers who did repetitive work with very little leisure. By supplying the needs of the nobility, who paid them with the appropriated product of the serfs, these "manufacturers" were social parasites.) The psychological "meaninglessness" of work is not a function of how specialized it is but of one's knowledge or belief that the work has no social significance, that the worker has no direct interest in its successful completion, that the worker's interests were not instrumental in the decision to produce this object in this way, etc. Modern "efficiency experts" are sometimes quite skilled in giving workers the illusion of participation in the decision-making process within the confines of capitalist production. Workers who are seduced by these experts are no less alienated than those who are not, even though they may think their work is meaningful.

It is sometimes assumed that the best, if not the only way to overcome alienation is through "worker control" of the means of production. This is true by definition, but it is often badly misinterpreted (usually because "alienation" is misunderstood) so as to imply that decentralized methods of production and control are necessary for worker control. The argument, often implicit, goes something like this:

> The farms, mines, factories, transportation systems, retail outlets, banks, etc., are the means of production. If each of the persons who work in those workplaces were to have some control over that workplace, then, collectively, the working class would control the means of production.

Unfortunately this argument commits the fallacy of composition in a very serious way. The "means of production" is the cooperative, interactive network of farms, mines, factories, transportation systems, etc., not each of these taken separately. The operative "working class" are all the people who work in all of these elements. To suppose that because each element of the means of production is controlled by some subset of the working class, that therefore the entire means of production is controlled by the entire working class, is to commit the fallacy of composition. In fact, for practical purposes, there is no better way for the working class to lose control of the means of production than for small subsets of it to "control" small elements of it. Since the crucial productive questions concern matters which transcend the particulars of a given factory's operation, such "control" is essentially meaningless for the elimination of alienation. At best it can give the illusion of control, as the efficiency expert sometimes can.[4]

Decentralization might be considered a way out of this problem *if* each productive unit could be made self-sufficient. In that case each productive unit would be the means of production for those who come under its jurisdiction, and that particular "working class" would in fact not be alienated. Two problems, insurmountable I believe, mitigate against this. The first is a moral one: the drive toward universality, toward sharing and cooperation on a wider and wider level seems to me to be fundamental to any moral notion of social progress. The isolationism implicit in such decentralization schemes seems morally appalling. The second objection is practical: to support a world of four billion or more people requires an enormous social effort, but this amount of effort can be achieved only through cooperation, not through duplication. To suppose that these four or more billion people could split up into self-sufficient communities of several thousand people is, practically speaking, absurd. So-called "self-sufficient" communities inevitably rely upon the external, world-wide productive network for survival: tools need to be purchased, or at least the steel to produce them needs to be. Energy almost always needs to be externally produced, and even a so-called "decentralized" solar energy economy, assuming it to be possible, requires sophisticated centralized manufacturing processes for solar collectors, conversion units, etc. (Even the "beer can" theory of solar collection requires that someone produce beer cans, which requires that aluminum be produced, which requires mining operations, etc.) A return to a "simpler" form of life, in which there is no need for large amounts of energy or large productive networks, presupposes, in a world of four billion people, that "nature's bounty" is practically infinite, something which neither the physiocrats nor the Marxists-humanists believe. Self-sufficiency is no longer possible, if it ever was, and is surely no longer desirable. Those who call for it almost inevitably have a vision of a much smaller world population, and those who scheme to put it into practice are condemning much of the world's population to death by devolution.

A common fear implicit in the objection to large, centralized productive networks is the fear that these networks will be controlled by "others" in the *interest* of others. Decentralization in theory allows *you,* as an equal, to look squarely in the eye of your fellow controller, thereby feeling reassured of his/her nonthreatening intentions. I suspect that this is what lurks behind Gendron's mistrust of the Soviet Union and Eastern European systems of social organization. Yet what seems crucial to me is not whether or not you personally know the people who are ultimately responsible for decision making, but whether you can have confidence that these persons recognize that your interests are coextensive with their own, and whether or not a system of representative democracy exists which is not corrupted by systematic class antagonisms. The idea that if you don't personally know your fellow controllers, if you don't live near or with them, then you have to fear anything that they do, is reminiscent of the "human nature" arguments which Professor Gendron so brilliantly demolishes: if you can't see them don't trust them. Let us hope Professor Gendron is right in supposing that this is *not* human nature but a consequence of a class society, and, we hope, the elimination of classes will eliminate the peculiar remnant of mistrust which decentralists apparently think will be justified even in a classless society. Perhaps I am naïvely trusting of my fellow human beings, but I welcome the efficiency that centralization in a noncapitalist society can produce, especially as this means the availability of more sophisticated technologies. As long as everyone in a position to make a difference recognizes the essential interdependence of human interests, and as long as there is a decent system of representative democracy, I also welcome the moral advance that such centralization would represent.

The reason I mention this at such length is that Professor Gendron expresses admiration for the decentralization and regional self-sufficiency policies of the People's Republic of China. Rather than reflecting a move toward a more rational, more moral use of technology, such schemes seem to suffer all the objections to local control mentioned above, and there is reason to believe they are now being reversed.[5]

The Demands of Capitalism

The following is a paraphrase of an argument which I have heard repeatedly over the past five or so years:

> Under capitalism, continued economic growth is required for the maintenance of profit. But profit is exploitative in that it takes from working persons what is rightfully theirs. Therefore economic growth is wrong.

Fortunately, Professor Gendron does not use this argument against economic growth, since this argument is hopelessly fallacious.[6] He does, however, make use of a key premise in that argument in a somewhat different context. His argument, paraphrased, seems to be this:

> Economic growth is ecologically disastrous. Under capitalism continued economic growth is required for the maintenance of profit, and what capitalism requires, capitalism gets. Therefore, the elimination of ecological disaster requires the elimination of capitalism.

This argument is perfectly valid, but it has two false premises: the one it shares with the former, invalid, argument, i.e., the assumption that capitalism requires continued economic growth, and the assumption that economic growth is ecologically disastrous. In this section I will consider the first of these premises. In the next section I will consider the second.

It is not thoroughly clear to me why the assumption that capitalism requires continued economic growth has come to be so widely, almost universally held, especially among academicians, when there are so many evident counter-examples. During the Great Depression, for instance, major capitalists—those that Professor Gendron identifies as being the most powerful, the most able to exert their will on the political and economic life of this nation and the world—did not suffer at all; if anything, they got richer as the value of their holdings increased. Perhaps the assumption is based on the "common sense," though false, belief that capitalists get rich only by selling increased amounts of "stuff" which they appropriate from wage earners who are in turn paid less than the selling price of what they have produced. This is no doubt one way that capitalists get rich, but it is certainly not the only way. An alternative method, one seen during depressions, is for capitalists to appropriate wealth from the working class by lowering its standard of living. This method does not require economic growth at all. Professor Gendron seems to recognize this as at least a theoretical possibility when he says that under conditions of no growth in wealth there must either be depression or greater impoverishment of the working class (or, I might add, both), leading to class antagonisms (pp. 181 ff.). This seems good enough for him as a refutation of the possibility of no growth under capitalism. But depressions, as I have said, do not particularly hurt capitalists, or at least those who count, and class antagonisms can be mitigated by psychological manipulation (the current zero-growth, pro-austerity movement seems to be such a process[7]) or if necessary by overt fascistic moves, such as in Chile, Brazil or Argentina. Of course capitalism must have economic growth to survive *indefinitely,* but so too does any economic system. Furthermore, capitalists count wealth in general and profit in particular in terms of the "accounted" or "book" value of their holdings (for instance, a slum building which has no, or very little real value may nevertheless have a high accounted value, and it is this value that capitalists seek to maximize). As real material holdings increase (and as resources are "used up" in their creation) so too might accounted value. But accounted value can increase without the increase of material holdings: for instance, through speculation on a fixed amount of a commodity, through the devaluation of some currency or the upward valuation of some other, etc., none of which implies any economic growth in an ecologically threatening sense at all. There are other, more complex but more systematic ways for this to happen, but I do not wish to discuss economic theory here; my observations are meant merely to suggest that economic growth may not at all be necessary for capitalists. Perhaps the best way to conceptualize the point is to distinguish between "economic growth" as the growth of accounted values and "economic growth" as the growth of material values. The former has no necessary connection with resource usage or the environment, and the two forms of "growth" are only occasionally, and never necessarily, correlated.

The world-wide call by capitalists, especially bankers and their governments, for "austerity"—a call heard in Britain, Argentina, Italy, France, New York City, and many, many other places—all testify to their current desire to hinder, rather than encourage, material economic growth for the purpose of "growth" of accounted value.

Although I find Professor Gendron's reasoning unsound, his conclusion is nonetheless correct, though for quite the opposite reason: socialism is necessary for the elimination of ecological disaster precisely because capitalism, committed as it is to growth of accounted value, *cannot* continue to guarantee material economic growth, and only such growth can avoid ecological catastrophe. In other words, Gendron's first premise is also incorrect. This is what I wish to consider in the final section.

Ecology-Economy

Professor Gendron identifies economic growth with growth in the gross national product. While this certainly is one of the common measures, it has the disadvantage of using the accounted values of things, and therefore gives us a very distorted reflection of what we want to measure: consumption of resources. Indeed, GNP can even increase when resource consumption does not. A better measure of the sort of growth Professor Gendron is concerned with is per capita energy consumption. This is a measure which he would agree measures potential ecological disaster, so I don't think I am distorting his position by using it. As far as I can tell, the only positive argument offered against continued economic growth is the already mentioned thermal pollution argument, which we have seen is seriously weak, and, occasionally, the exhaustion of nonrenewable resources argument, which is seriously undermined by the Marxist-humanist conception of the relativity of resources.

On the other hand, Gendron does favor technological growth, by which he means greater information about and control over nature. He believes that such growth, when not in the service of capitalism, can minimize ecological disaster and may even be necessary for doing that (pp. 175 ff.). But Professor Gendron also believes that technological growth is possible without economic growth (p. 179). This, I think, is a profound mistake. If we use the United States as a model, it is easy to be seduced by the argument that a reorganization of energy consumption, perhaps even a reduction of it, could take place without hindering technological growth. Yet to be so seduced would be a mistake for two reasons: (1) The immediate cutback would only be temporary, as a result of "cutting the fat," the pure waste from our current consumption. For reasons to be discussed below, consumption would necessarily begin to climb again. (2) There is no fat to be cut from most of the world's consumption. Even if we chose to "maintain" consumption at a relatively low level, there is no morally acceptable way to keep it as low as it is now, and no practical way to do so in a context of technological growth.

I have already suggested two reasons for thinking that technological growth requires increased energy consumption, but let me reiterate: (1) The problem of diminishing returns requires greater output merely to maintain our current level of control. And when it is no longer possible to do that with existing resources, we require a shift to radically new technologies, which demand enormous energy expenditures over and above what is needed for maintenance. Thus technological growth, if it means maintaining a certain level of control over nature, requires economic growth. Indeed, as I suggested above, merely staying alive as a civilization requires this. (2) A growing control over nature requires (a) growing knowledge and (b) the material means to implement this knowledge. But (a) means more and more education, which implies not merely greater consumption of educational resources, but, more importantly, more and more time which is not socially necessary labor time. But such time can only be purchased at the expense of labor-saving, energy-intensive technologies, thereby requiring economic growth (technology alone will not suffice, of course, but neither is it superfluous). Finally, (b) is necessary because knowledge alone does not bring control. Control must be achieved by the material implementation of scientific and technological innovation. This requires increased energy consumption also. Therefore, I conclude that to the extent that technological growth is necessary and desirable, so too is economic growth.

Before I end let me say something a bit more positive about the relationship of economic growth to environmental health. The issue of nuclear power development is before us. While he does not confront that issue, we can infer from his remarks that Professor Gendron is undecided about it. On the one hand, he must recognize that reserves of fossil fuels are limited, if not as limited as the oil companies and the government report, nevertheless dangerously low. He also finds that "the technical obstacles to full conversion to solar energy are mind boggling" (p. 178). This suggests nuclear power as the most plausible alternative. On the other hand, this alternative—indeed any alternative which maintains anything like current world-wide consumption—involves economic growth, as capital expenditures are required for plants and delivery systems. But Professor Gendron does not favor economic growth. This highlights the dilemma of his position. If we choose not to grow, then we condemn ourselves, in the words of an appalling recent television show, to "freeze in the dark." If we choose to grow, we must, he says, pollute. The only way out I think is to choose growth. The first choice implies certain death, at least for an enormous percentage of the world's population. While strictly speaking this is not an "ecological" disaster in the usual sense, it is in any sense morally equivalent to one. (And if we count human beings as part of the ecological system it certainly is an ecological disaster.) The devastation indirectly wrought on the environment by such massive "diebacks" (in Garrett Hardin's words) would be staggering. There is no doubt that growth will affect our ecological system also: all will not remain the same and those who long for a return to "simpler" times will have to go on longing. Will the change be for the better? My answer is highly speculative, but the evidence of biological history suggests that it will be: the biosphere is characterized by growth, by differentiation, by ever increasing complexity. Human intervention, while not entirely benign, has nevertheless increased the earth's carrying capacity: the fact of four billion people, and the known potential for billions more, cannot be ignored. What we can be sure of is that if we choose not to grow, the results will be devastating.

NOTES

1. From *Outline of a Critique of Political Economy,* reprinted in *Marx and Engels on the Population Bomb,* ed. by Ronald Meeks, Ramparts Press.

2. P. 17, quoted from Fuller, *Utopia or Oblivion: The Prospects for Humanity* (New York: Bantam), p. 288.

3. It is revealing to notice that whenever one sees formal or informal discussions of economic well being emanating from the Soviet Union, figures are given or at least something is said about how many workers are "improving their talents" through training for more sophisticated jobs, night schools, paid leaves of absence, etc. What is significant is that this is viewed in the context of social wealth, not individual ambition.

4. A plan suggested for New York City recently is to divide the city into about 50 locally "controlled" areas, each responsible for its own educational, sanitation, fire, etc., services. The city would provide some (but not as much as before) revenues for these communities. The illusion of local control here masks the fact that the city will continue to pay its enormous bank debts at the expense of services, while each community will seek its special advantage at the expense of another, similarly situated community instead of at the expense of the banks. Not surprisingly, this suggestion was made by bankers and those who have close ties to them.

5. The situation in China is by no means clear, but the death of Mao seems to be leading to the death of Maoism, which is being masked as an attack on the "Gang of Four." We should see a move away from regional self-sufficiency, a re-emphasis on theoretical education, and moves toward further industrialization, if I am correct about this.

6. The argument, reconstructed, looks something like this:

1. If there is capitalism, then if there is profit, then there is economic growth.
2. There ought not be profit.

C. There ought not be economic growth

This is abysmal, unsalvageable logic. The best one could do would be to insist that economic growth is *sufficient* for profit under capitalism (something which is very unlikely), and get:

1′. If there is capitalism, then if there is economic growth, then there is profit.
2′. There ought not be profit.

C′. If there is capitalism then there ought not be economic growth.

This conclusion follows only if we ignore the modalities, but even then all it says is that economic growth under capitalism is wrong; it says nothing about its desirability under another system such as socialism.

7. Even a cursory examination of the main support found for this movement suggests that the major funding agencies are almost always capitalist foundations, especially the Ford and Rockefeller Foundations. If growth is indeed so important to capitalists why are they funding antigrowth movements?

REPLY: GROWTH, POWER, AND THE IMPERATIVES OF TECHNOLOGY

Bernard Gendron

I appreciate the pains that Professors Durbin and Goldman have taken in writing their supportive and helpful reviews. Both direct their criticisms primarily at the sections of my book dealing with the Socialist view of technology. These criticisms are both stimulating and penetrating; I hope to be able to do justice to them. Goldman concentrates on my analysis of economic growth, and Durbin on my analysis of power. I will deal first with Goldman's criticisms, and then with Durbin's.

I

The issue of economic growth runs through my book. Goldman is quite correct in singling it out for special attention. The Utopians, Dystopians, and Socialists all have something important to say about this topic; and the differences in their views on economic growth reflect the differences between their views on the social role of technology. Both Utopians and traditional Socialists think that indefinite economic growth is a good for humankind, and indeed one of the primary sources of human progress; in contrast, Dystopians believe that the pursuit of indefinite economic growth will lead either to ecological disaster or to some politically undesirable steady state (e.g., some technocratic dictatorship). Both Utopians and Dystopians claim that ultimately the modern tendency toward sustained economic growth is due to the imperatives of sustained technological growth. The Utopians of course praise technology for this imputed causal role, and the Dystopians blame it. Socialists, though not denying the importance of technology for economic growth, put more stress on the social relations peculiar to capitalism (and those peculiar to socialism) in order to account for it. And they emphasize that economic growth within socialism is qualitatively and quantitatively superior to that within capitalism.

In my book, I avoided these traditional alternatives, and argued for some form of long-run, no-growth socialism. I concluded that economic growth must come to an end within the next few centuries. However, I also maintained, with equal emphasis, that economic growth remains for the near future an indispensable tool for successfully coping with the world-wide problems of poverty and deprivation. I made use of these conclusions about economic growth to add new weight to the argument that capitalism is not desirable in the short run and not viable in the long run. In the short run, capitalist economic growth will do little to alleviate economic scarcity even in its more extreme forms (for it is always in the process of creating new forms of scarcity to replace the old). In the long run, capitalism is compulsively committed to economic growth, and will not be able to bring it safely to a halt when ecological conditions require that. Socialism, I maintained, suffers from neither of these short-term and long-term growth liabilities.

Since the publication of my book, I've moved away somewhat from my (long-run) no-growth posture. The case for (long-run) no-growth is simply not as strong as I once thought it was. Professor Goldman is quite correct (and persuasive) in pointing this out. However, I have not gone over to the other camp either (as Goldman thinks I should). Essentially, I am agnostic about the whole debate over growth: I don't think that one can make a definite argument either for indefinite growth or (long-run) no-growth. The social sciences, in their present state, lack both the theoretical and factual base to do this. So I think Goldman's arguments for indefinite growth are as weak as, and indeed weaker than, the normal arguments for no-growth. Furthermore, I think that most of his criticisms of the arguments for no-growth fail to hit the mark; indeed, these arguments, though not decisive, have much more plausibility than he gives them credit for. Let me explain this somewhat.

II

In my book, I developed my views on economic growth within the broad context of an analysis of the Dystopian perspective on technology. First, I supported the Dystopian belief that indefinite economic growth in the long run will wreak ecological disaster. Next, I criticized the Dystopians for imputing the impetus for this indefinite economic growth to the imperatives of technology rather than to the compulsions of capitalism. Goldman takes me to task for both these endeavors. He thinks my argument(s) against indefinite economic growth are based on a faulty conception of natural resources, and on a faulty assumption about the growth of thermal pollution. Further, he contends against me, both that technological growth requires economic growth and that capitalism does not. He further contends that the maintenance of an above-subsistence standard of living presupposes indefinite economic growth. So, for Goldman, indefinite economic growth, far from being a scourge, is necessary both to avoid increasing poverty and to further human progress.

Goldman claims that my no-growth position is based on a physiocratic conception of natural resources. Indeed, he wonders why I would be at all inclined to take a no-growth position if I did not accept the physiocratic conception. But here Goldman is wrong. I emphatically reject (as Goldman does) the physiocratic conception; and nowhere do I (unconsciously) flirt with it. Similarly, I reject the humanist conception which Goldman suggests any good Marxist should accept. Each of these conceptions is equally one-sided. The "physiocrats" are clearly wrong in thinking that the stock of extractable natural resources is fixed, for all practical purposes, and that the size of that stock at any given time is a function exclusively of the physical structure of the human environment. For human ingenuity and human practical know-how—that is, human technology—also have a role in determining what is to count as a natural resource and what not. *Everything else being equal,* the stock of extractable natural resources will grow as human technology grows. But not everything else is equal, as we shall see below.

The "humanists" then are quite correct in accentuating the historical variability of the stock of natural resources and its relativity to human technology. But they are clearly wrong in locating natural wealth exclusively in "human activity and human potential," or in saying that its value is the result exclusively of "human creative and productive activity." But this is the conception that Goldman seems to be imputing to them.[1]

In reality, it takes two to tango. The materials of our natural environment are useless to us without technology; conversely, our technology is useless to us without the appropriate materials. As Goldman put it so nicely, petroleum up to a point in history "is useless or less than useless junk." But, similarly, nineteenth-century industrial technology, in a physical environment altogether bereft of oil, coal, and iron, would appear as little more than ornamental, exotic knowledge.

The only adequate conception of natural resources is a "mixed" one, which stresses both the importance of the physical environment and human technology. Whether some item, at a given time, is to count as a natural resource depends on its physical structure, its immediate physical environment, the state of technology, and its absorbability into the human environment. The last of these four conditions says this: it is not enough for something to be extractable by the current technology to count as a natural resource; it must be extractable in a way compatible with the carrying capacity of the human environment.

Unlike the physiocratic or humanist conception, the "mixed" conception of natural resources does not commit us a priori either to an indefinite growth or a no-growth position. This point requires some explanation. At any given time, the stock of known reserves of natural resources is being added to and subtracted from. There is an outflow of resources from this reserve base due to extraction, and an inflow due to new technologies or the continued successful application of old technologies. We extract old resources and discover new ones.

In a growing economy, the outflow of resources out of the stock of known reserves will increase continually as extraction increases. If the total stock is minimally to remain stable, then there must be a continually increasing inflow of resources, due to new discoveries or technologies, to compensate for the increasing outflows. We have good reason for thinking that the world economy can grow indefinitely only if we have good reason for thinking that the inflow of new reserves can grow indefinitely. At this point in time, we have no conclusive reasons for believing either that future technology will, or will not, be able to support indefinitely increasing inflows into the resource reserve base. There is no doubt that technology can keep this inflow going; but there is doubt that it can keep it growing.

However, conclusiveness aside, a good case can be made that the carrying capacity of our environment ultimately will not allow indefinite growth in the inflow of resource reserves. In my book, I argued for this as follows. Increasing extraction of resources involves increasing energy expenditure. In turn, increasing energy expenditure (in the long run) entails increasing thermal pollution. Assuming ultimate limits on the amount of thermal pollution absorbable by the human environment, we must assume ultimate limits on the growth of the inflow of extractable resources into the reserve base.

Goldman thinks this argument is inadmissibly weak. I admit that I did not present it as persuasively as I could have, and certainly not as persuasively as others (e.g., Robert Heilbroner) have. Nonetheless, I think that each of Goldman's objections misses the mark.

First, he says,

> For even if we admit the need to reduce to *zero* the increase in energy use . . . thermal pollution must continue. Perhaps we will thermally pollute at a slower rate, but if thermal pollution is a planet threatening problem, it is not clear to me that we can solve it by putting off for a few years the time when it becomes critical.[2]

Here Goldman is confused. He is treating thermal pollution as a stock which accumulates over the years. But thermal pollution is a net flow rather than a stock. Let me explain. Money can exist both as a stock and a flow. When hoarded it is a stock, and when in circulation, a flow. Ultimately, all heat flows into our atmosphere from the sun, and most of it is dissipated, i.e., flows back, into space. Some of the heat is stored in living or dead organisms, in the latter case producing coal, oil, and other fossil fuels; these fossil fuels represent our long-term stock of heat. As a stock, heat does not pollute the environment. Nor does it merely as a gross flow. The gross flow of heat is the total amount of heat coming into our environment from the sun, in a particular time interval, plus that produced here on earth by human or nonhuman sources (e.g., the burning of fossil fuels). The net flow is the gross flow minus the heat which is dissipated into outer space or otherwise stored by objects in our environment. Heat pollution grows with growth in the net flow of heat.

What I asserted was not that heat pollution will need come to a halt, or equivalently, that the net heat flow on our planet should be reduced to zero (as Goldman seems to think). I said rather that the growth of heat pollution, or of net heat flow, will need be reduced to zero. This eventuality would be quite compatible with some growth in gross heat flow (so long as the rate of dissipation also grew) as well as in the stock of stored heat. To treat my argument for zero growth in thermal pollution as, in effect, an argument for zero production of thermal pollution (as Goldman seems to do) can only make sense if one construes thermal pollution as some indestructible or immovable stock which fails to grow only when it is no longer produced.

Goldman's two other objections to my no-growth argument are equally inadmissible. He maintains, for example, that "the use of fossil fuels and direct solar energy . . . in effect temporally redistribute solar heat which is already on the planet; they do not create any heat which is not already here."[3] With respect to fossil fuels, Goldman is dead wrong. He is again confusing stocks and flows of heat energy. It is true that the burning of fossil fuels does not add to our stock of heat energy, since it only dissipates what was already stored there anyway. But it surely adds to the net flow of heat, and hence to thermal pollution. The use of direct solar energy does merely redistribute the inflows of solar heat. But there are definite upper limits to the use of solar energy, and continued growth in energy demand would ultimately run against those limits.

> Present emission of energy is about 1/15,000 of the absorbed solar flux. But if the present rate of growth continued for 250 years emissions would reach 100% of the absorbed solar flux. The resulting increase of the earth's temperature would be about 50°—a condition totally unsuitable for human habitation.[4]

In the long run, increasing energy demand cannot be met simply by redistributing the solar heat flow on this planet; other heat (say, from nuclear fission or fusion) must be added to that flow.

Goldman further argues that

> The notion of *thermal pollution* depends on the assumption that there is some relatively fixed, eternally desirable level of heat energy for the earth's biosphere.[5]

It may well be the case that at different times (probably at quite distant intervals) the biosphere can sustain different levels of net heat flow. This is not something that no-growth theorists need deny. All they need deny is that the tolerable level of heat absorption can increase indefinitely, to keep up with the indefinite increases in net heat flow which in the long run follow upon sustained increases in energy demand.

III

I had argued that capitalism, and not technology, is responsible for the compulsion toward economic growth in Western societies. Goldman takes the contrary position, namely, that technology, and not capitalism per se, requires technological growth. Indeed, he goes so far as to say that economic growth is required for the normal sustenance of any economy.

My position on capitalism and growth is that the health of capitalism (though not necessarily its survival) requires economic growth. If capitalist economies were to stop growing, I maintain, they would be faced, on the economic front, with frequent (if not constant) recessions and depressions, and on the political front, with intensifying class antagonisms. Goldman, at least for the sake of the argument, concedes these two consequences, but denies that they are seriously detrimental to the health of the capitalist system (or equivalently to the welfare of the capitalist class). About the first consequence, he says

> During the Great Depression, for instance, major capitalists . . . did not suffer at all; if anything they got richer as the value of their holdings increased.[6]

And of the second

> . . . and class antagonisms can be mitigated by psychological manipulation (the current zero-growth, pro-austerity movement seems to be such a process), or if necessary by overt fascistic moves, such as in Chile, Brazil, or Argentina.[7]

First, with respect to depressions, one must distinguish (which unhappily I did not do explicitly in my book) between classical (short-term) depressions and Keynesian (long-term) depressions. The classical kind is a corrective device for restoring the balance of the market system, and in that sense is healthy for capitalism; the depression comes to an end once it has made the appropriate corrections. Even then, of course, the capitalist class suffers somewhat insofar as its income (if it has not absolutely decreased) is not as high as it would have been had the period of boom prevailed. Indeed, the evidence (though somewhat controversial) seems to be that the overall capitalist income declines both absolutely and relatively during a classical depression. The capitalists' proportionate share of total income actually goes down.

Keynesian recessions or depressions are not corrective devices for restoring the capitalist market to normalcy. Rather, they result from chronic dysfunctions of the modern capitalist market, such as excessive and relatively constant disinclinations to invest. In these conditions, the economy remains in a state of quasi-recession or depression, with upswings which never restore the full employment of productive capacities. For the past ten years, we have been experiencing a mild version of Keynesian recession.

My argument comes down to this. The cessation of growth would lead to serious Keynesian depressions. And no one can deny that these are bad for the health of the capitalist economy.

Goldman, I think, really overestimates the capacities of capitalism to keep class antagonisms under control in bad times. (No doubt, in good times, as in the long boom following World War II, manipulation comes fairly easily.) There have been times when class conflict has brought capitalist societies to the verge of collapse, as in the period immediately following World War I. And, during the depression, the American capitalist system was forced to make substantial concessions to the working class in order to maintain some semblance of class harmony. What we need to note here especially is that such concessions could only be made realistically within the framework of a (long-term) growth economy.

It seems easy to overestimate the stability of fascist regimes and their ability to suppress class conflict. At any rate, the relatively successful fascist regimes, far from being austerity regimes, have brought much of their social peace through economic growth. The real question is: how successful would capitalist states be in bridling class conflict in the complete absence of economic growth? In no-growth, the irritants of class conflict would be on the rise, and the resources for dealing with it on the decline. Capitalism would need have recourse to more desperate measures. We don't know whether they would succeed in the long run.

IV

Goldman's evaluation of economic growth is the extreme opposite of mine. He argues that growth is necessary (paradoxically) even merely to maintain a subsistence economy. I find these arguments unpersuasive—though, I should add, his analysis of the necessary conditions for subsistence is otherwise quite original and provocative, and certainly helpful to me.

His first reason for believing that "an actual increase in social consumption" is necessary "to the maintenance of the social productive level" is that "it now takes a much larger energy input to extract a given BTU level of fossil fuels than it did a decade ago."[8] It is rather surprising to see that Goldman, given his enthusiasm for the humanist conception of wealth, subscribes here and elsewhere to the Malthusian-Ricardian principle of diminishing returns, in order to make his case for economic growth.[9] For if this principle held absolutely (which I deny), then the time indeed would come when growth would have to stop (and even that regression would set in): the steadily increasing ratio of BTU inputs per unit output of energy source would finally make it too costly to produce energy in sufficient quantities to support economic growth. But there is no reason why the ratio of BTU inputs per unit output of energy source must inevitably increase during every time interval. The discovery of easily accessible reserves of a conventional energy source (such as past discoveries of large oil reserves) or the substitution of a new source (e.g., nuclear fusion) for a conventional one (e.g., oil) through technological advances, will oftentimes lead to a reduction of BTU units per unit output of that energy source. Technology can and oftentimes does neutralize the principle of diminishing returns.

Furthermore, increasing BTU inputs per unit output, where it does occur, may well be neutralized by decreasing labor inputs or "hard" capital for each unit of energy source produced. Put otherwise, decreasing returns to BTU inputs may be accompanied by increasing returns to labor and, say, mineral inputs. This means that real production costs of energy sources do not necessarily rise as BTU inputs rise. Hence, the economy does not need to grow.

Goldman's second reason for postulating economic growth as a necessary condition for maintaining a subsistence economy is that even the maintenance of a constant standard of living requires technological growth, which in turn, he argues, requires economic growth.[10] The history of pre-industrial societies does not square with his hypothesis: centuries of minimal change in the standard of living have been accompanied by technological and economic stagnation.

But suppose that technology must grow. Does it follow that social consumption must also? Goldman's affirmative answer is in part based on the principle of diminishing returns to energy inputs, with which I have already dealt. But then he goes on to argue that technological growth, insofar as it involves growing knowledge, must require "more and more education"; and that the material implementations of technological knowledge, becoming increasingly capital intensive, require increasing energy consumption (per unit of output, I assume).[11] On both counts, Goldman seems wrong. Growing knowledge requires improved education; but this does not mean longer or more education. As knowledge grows, the means of transmitting it improve; what was known forty years ago is more easily transmitted now than it was then. But improved education does not necessarily cost more than the more deficient education of the past. Hence it does not necessarily require growth.

Nor is it obvious that the implementations of new technologies are necessarily more capital-intensive than the past. But, more importantly, even if they were, this would have no implications for economic growth. That a technology is capital-intensive entails, not that it requires more physical capital (e.g., energy) per unit of output, but rather that it requires more physical capital per input of labor. Now it is possible, at the same time, both to be increasing the capital-intensivity of capital and to be decreasing the labor and capital inputs per unit of product; in such a situation, increasing capital intensivity simply means that the capital inputs are not decreasing as quickly as the labor inputs per unit of output. In such a situation, clearly, economic growth is not required.

In short, Goldman's a priori arguments for economic growth, though quite provocative, simply don't work. Finding out why they don't work, however, teaches us much about the cluster of concepts associated with economic growth, and thus proves to be a very worthwhile exercise. And it has forced me to say things which I should have said but didn't say or say clearly in my book. For this I am very grateful to Professor Goldman.

There are other sections of Goldman's review which are quite challenging and which deserve comment, but which I won't touch on to ensure unity and brevity. On the most fundamental points, our views do converge; that is why it has been important for us to clarify our differences in such detail.

V

Professor Durbin's review is considerably shorter, and more expository, than Goldman's. My response to him will correspondingly be briefer.

Durbin has described my argumentative strategy quite well,[12] certainly better than I have anywhere. Indeed, the paucity of my own remarks about my method has caused confusion in the minds of some readers. One reviewer bemoans the fact, for example, that my book is insufficiently philosophical and excessively empirical. It is true that I spend little time directly discussing social and political norms. Nonetheless, what I am doing in that book is normative political philosophy.

Most people think that the major problem facing normative political philosophy is adjudicating the basic norms which divide the major political camps (e.g., Marxist, liberal capitalist) from each other. I think that this problem has been greatly exaggerated. Indeed, within the framework of philosophical argumentation, the different camps can agree on ultimate values, such as freedom, democracy, and efficiency. Then the issue boils down to which type of sociopolitical system comes closest to achieving these ideals. On this interpretation, the major points of unresolved difference between the camps are empirical. It can easily be borne out, I think, that the traditional debate between socialism and capitalism has indeed been empirical in this sense. And so is the debate between Utopians, Dystopians, and Marxists, on the social progressiveness or regressiveness of technology.

Most people also think that if the primary normative issues in normative political philosophy are resolved, then the remaining work in that discipline can proceed apace without much trouble and fanfare. But this is seriously mistaken. I would argue that the central empirical issues of normative political theory are much more difficult to resolve than the attendant normative issues. The former are not the sort which can easily be settled by appeal to ordinary experience or the more sophisticated repertoire of the behavioral and social sciences. Normative political theory has to confront highly global and foundational empirical issues, which have not yet been resolved by science and are not about to be in the foreseeable future. The following is a sampling of such global empirical issues which I had to confront in my book. Is the human proclivity for aggression innate or learned? Are technocrats now in control of corporations and governments? If not, who are? To what extent does technology cause alienated work? Is technology running out of control? Can democracy withstand technology? Is genuine workers' control of industry feasible?

We have no noncontroversial answers from the social sciences to these questions. And we may never succeed in getting such answers from them. These issues, though straightforwardly empirical, seem to be too hard (and too hot) to handle for the behavioral and social sciences, as we know them. Yet, for the purposes of social action, we need to come to some provisional answer. Such is our quandary. The major objective of my book was to demonstrate ways of rationally resolving in a tentative manner those global empirical issues which at this time (and perhaps always) admit of no "scientific" resolution.

VI

Professor Goldman and I engaged really in an in-house debate within the socialist camp. Professor Durbin comes at me from the liberal capitalist camp. Our issues of contention are correspondingly different.

Durbin says that my defense of socialism, as the only framework for a beneficent use of technology, is "optimistic, utopian, and even visionary"; and further that my reasons for preferring socialism are based on mere "hope."[13] This I think is a misinterpretation, though Durbin correctly intuits that I am in disagreement with the usual versions of "scientific socialism."

There is a certain amount of repressed pessimism (or nonoptimism) in my book which reveals itself in the paragraph quoted by Durbin. I make it clear that we are in a terrible mess. I point at a way out, but vaguely and with not much confidence (certainly with less in the later drafts than in the earlier). There are severe limits, I think, to rational speculation about future alternatives. With the "scientific socialists," I think there is little or no scientific (or rational) basis for speculating about the fine structure of future emancipatory societies. But, unlike the "scientific socialists" I do not think we can show that the present contradictions of capitalism somehow presage the rise of a future emancipatory socialist society. There are some "socialist" societies today, and presumably their rise has had some connection to the contradictions of capitalism. But these are not what I have called "real socialist" societies, that is, societies in which the working class is firmly in control of political and economic matters. Real socialist societies don't exist now, and we have no strong reasons for thinking that they will in the foreseeable future. Thus my nonoptimism.

Nonetheless, my book is a defense of real (i.e., worker-controlled) socialism. In its boldest form what I am saying is: the only way to avert technological disaster is to transform the present capitalist and "pseudo-socialist" societies (e.g., the Soviet Union) into real socialist societies. Consider our options. On the one hand, the probability of successfully instituting real socialism is very low; but I argued, the probability of averting technological disaster, and using technology in an emancipatory way, is very high in real socialism. On the other hand, the probability of maintaining the existence of capitalism (and pseudo-socialist) societies is very high; but the probability that these societies will use technology in an emancipatory and disaster-averting way is very low. We are left then with two options whose chances for success are both low: either attempt to patch up capitalism or pseudo-socialism so that they avert technological disaster, or attempt to bring about a real socialist revolution. There is strong evidence—some of which I have marshaled in my book—which says that neither capitalism nor pseudo-socialism can be sufficiently patched up to avert technological disaster or technological repression. That is, the chances here are almost nil. On the other hand, we have little notion of how high the odds are against implementing real socialism; we know they are high, but not how high. In effect, we are left with the choice between an option which has been persuasively argued to be well-nigh impossible and an option which, for all we know, may well be possible. Clearly, under these descriptions, it is more rational to opt for the former rather than the latter. It is for these reasons that I say that working toward real socialism, however uncertain and risky it may be, represents our only hope, or our highest hope, of averting technological disaster and of converting technology into an agent of emancipation.

VII

It is consonant with his liberal philosophy that Durbin should take the strongest exception with my criticisms of pluralism and my defense of the view that the capitalist class is the ruling class of American society. He does this in two ways, first by raising objections to my treatment of the standard pluralist theory, and then by outlining an alternative pluralist theory which he thinks is immune to such criticisms.

I devote only six pages to pluralism in my book. The reason for this is quite straightforward. The evidence against pluralism and for the capitalist ruling class theory is quite familiar and massive. What I have found lacking, however, is an argumentative strategy for maximizing the import of this data. So what I decided to do in the brief section on pluralism was to sketch the bare outlines of such a strategy. There are important facts which I don't give but whose sources I reference. For this reason, I don't think that Durbin is altogether justified in saying that my analysis of pluralism "badly oversimplifies it," because it almost totally ignores "the myriads of other groups that make up the pluralist mix in contemporary society," such as "the political parties all the way down to the local level, the military . . . , not just the national education lobbies but educational institutions at all levels, the churches, volunteer associations . . . , the celebrity world of entertainment," and so on and so forth.[14] First, the few arguments that I did present show the capitalist class to have such overwhelming access to both private and public government, and to benefit so overwhelmingly from the outputs of government, that it did not seem worthwhile within the space limitations to speculate on the size of the small pieces of the action that these other groups get. Second, one can easily argue on the basis of the evidence I put forth that some of these groups, such as the military and political parties, insofar as they are part of the government, are controlled by the capitalist class. Finally, I did reference some key works, such as Grant McConnell's *American Power and American Democracy* and the books by Nader's associates, that detail the relative political impotence of the groups not controlled by the capitalist class.

VIII

Durbin is correct in pointing out that I only attack directly one of a number of pluralist theories. But I think that my evidence and my arguments bear indirectly on the other theories also.

We can distinguish between empirical and normative pluralism, and between interest group and public interest pluralism. Pluralism is empirical if it claims to describe the contemporary social and political reality; it is normative if it claims to articulate the ideal toward which all social and political structures should strive. Interest-group pluralism is usually cast as an empirical theory: according to it, government decisions reflect the state of equilibrium resulting from the collisions of the many interest groups. In contrast, public interest pluralism is usually cast as a normative theory: ideally, it says, government decisions should be made in the public interest, by the public-interested members of the various special-interest groups.

When discussing pluralism in my book, I was mainly interested in claims about power realities in contemporary capitalist societies. So I focused my critical attention on empirical special-interest pluralism. Durbin, on the other hand, puts forth a normative public-interest brand of pluralism. Does the latter theory escape the arguments I presented against the former? Not altogether. For one thing, by proposing pluralism as an ideal, Durbin implicitly concedes my arguments against pluralism as a description of reality, and perhaps also my arguments for the capitalist ruling class theory. But if power in capitalist societies is not pluralistic now, then what basis is there for thinking that it can plausibly be pluralistic in the future, especially in the "public interest" sense? Can public-interested members from the different special-interest groups (including the capitalist class) wrest power from the special-interested capitalist class?

According to my arguments, such a power transformation is highly improbable. These arguments were meant to show not merely that capitalist society is not pluralistic at this time, but also that capitalist society will never be pluralistic. Part of the basis for this is inductive. The evidence indicates that political decision-making in capitalist societies has never been pluralistic; capitalist societies have always had a capitalist ruling class. This entails (inductively) that political power in capitalist societies will never be pluralistic.

Further, in addition to detailing the present and past control over the state by the capitalist class in capitalist society, I have (albeit too briefly) tried to show how the institutional structures of capitalist society inevitably and overwhelmingly bias the state toward the interests of the capitalist class. It is not merely a matter of capitalist intrusions into the state. It is also a matter of the inertial tendencies of the capitalist state and the capitalist corporate system. For example, in virtue of their size and interconnections, the major corporations inevitably "play a major role in settling many key social questions," such as, "the allocation of resources, . . . the distribution of income, the extent of economic development and backwardness of many communities, . . . the direction of research, the shape of school curricula, and the direction of population flow. . . ."[15] Also, consider the fact that

> in their everyday decisions and actions, corporations automatically put pressure on the federal government, even in the absence of such pressures as lobbying, financial support of candidates, and media campaigns. No government can succeed in operating successfully in a crisis-free atmosphere when it does not have the confidence of business. The reason is obvious. When governments lose the confidence of business, this puts a damper on investment and leads to some form of recession, depression, or other economic crisis. Even socialist-oriented governments in liberal capitalist democracies have gone out of their way to put business at ease. . . .[16]

In effect, the obstacles to the achievement of pluralism are structural as well as attitudinal. Any viable normative pluralist theory must show how these can be overcome.

Finally, I doubt that there is a common (public) interest that underlies all the sectional interests, and especially that there is a common interest that unifies the capitalist and working classes. Durbin is right in distinguishing between "statesmen" and "partisans" in each interest group. But statesmanship is not to be identified with action in the public interest. In each group, some have a narrower view, and others a broader view, of the group interests and the overlap of these with other interests. The former are partisans and the latter "statesmen." But the broad view of "statesmen" in one interest group will not coincide with the broad view of "statesmen" in another. The "statesman" still reflects the bias of his group, however broadly this is conceived. Bebe Rebozo, I take it, is a capitalist partisan, whereas Bernard Baruch was a capitalist "statesman." But Baruch's view of the cosmos, however sweeping it may have been, was shaped by capitalist lenses. Though public-interested in his intent, the business "statesman" Baruch most probably conceived of the public interest in quite different terms than, say, the labor "statesman" Eugene Debs, the multi-versity "statesman" Clark Kerr, or for that matter the Maoist "statesman" Chou En-lai.

IX

The stimulating reviews of Goldman and Durbin have forced me to develop and clarify my thoughts. My answers to them are not always as confident as they appear on paper. Indeed, I suspect that in some places I may have modified the ideas of my book somewhat in order to deal with their objections. But this is all for the good. Reviews of this caliber prevent one from merely shelving one's ideas once the book is completed and safely published. At any rate, I am more aware now of the more subtle gaps and incoherencies of my book. If this is humbling it is also enlivening.

It should be kept in mind that the Goldman and Durbin reviews focus on a fairly narrow region of the book, namely, that which bounds my analysis of the socialist perspective on technology. The section on the

Dystopian perspective is equally controversial and has been subject already to a number of criticisms mainly from the loosely bound antitechnology camp. My analysis of the Utopian perspective has raised few hackles, perhaps because few people are willing to speak out for that perspective at this time.

NOTES

1. Goldman, p. 89.
2. *Ibid.*, p. 90.
3. *Ibid.*
4. Robert U. Ayres and Allen V. Kneese, *Economic and Ecological Effects of the Stationary State*, Resources for the Future, Reprint #99, December 1972, p. 16; quoted in Robert Heilbroner, *An Inquiry into the Human Prospect* (Norton: New York, 1974), p. 51.
5. Goldman, p. 90.
6. *Ibid.*, p. 97.
7. *Ibid.*, p. 98.
8. *Ibid.*, p. 91.
9. See also p. 100.
10. *Ibid.*, p. 92.
11. *Ibid.*, p. 100.
12. Durbin, pp. 77–79.
13. *Ibid.*, pp. 81–82.
14. *Ibid.*, pp. 83–84.
15. *Technology and the Human Condition*, p. 202.
16. *Ibid.*, pp. 203–204.

REVIEW OF LANGDON WINNER'S *AUTONOMOUS TECHNOLOGY*

Stanley R. Carpenter

At a time when courses about technology are burgeoning on college campuses, when government money is being channeled into technology assessments, and when philosophers such as ourselves are trying to discern the outlines of a philosophy of technology, Langdon Winner's book *Autonomous Technology* provides a note of seriousness and substance. One is tempted to use the term "exhaustive" in describing the range of ideas and authors which Winner is not only aware of but which he also handles with confidence and clarity. A serious examination of this book should dispel once and for all the facile conclusion that technology as a social phenomenon is conceptually barren as compared to science, as some philosophers of science would seem to suggest. A second myth which is bound to go is the belief that technology is ethically neutral, a mere means to preselected and clearly normative ends. Those of us who have tried to reflect on our modes of life in a technological society, tried as it were to get our bearings, have encountered that sense of bewilderment at the curious combination of material comfort and "unfreedom" which exist nowadays in uneasy tension. Winner addresses this experience and goes a significant way toward explaining it.

He sees clearly what the majority of current technology assessment personnel miss, namely, that technology is a political phenomenon. This is a much stronger claim than the widely accepted admission that many impacts of technology are political. Rather, it is the claim that large-scale technologies which dominate the industrialized countries determine the locus of political power, define the life styles of citizens, and ultimately determine the consciousness itself of the human actors. Such a position, Winner acknowledges (p. 251), turns out to be a theory of culture itself. In formulating his argument he appears to have been most deeply influenced by Ellul. He does in fact provide the best explication of Ellul which I have found to date. One is much less likely to dismiss this obscure and frustrating thinker if he reads Winner first. There is also a considerable debt to Marx reflected in Winner's development of the concept of the technological order and its definitive impact on human needs and aspirations. In both cases, however, he departs in substantial ways from these men.

The concept of autonomy is the core of Winner's argument. Its explication leads him, as it did Hegel, Marx and Nietzsche, to its paradoxical embodiment in the slave/master relationship (p. 187). In appearances the slave is least autonomous, the master the most free. In reality the reverse occurs as the lord comes to depend on the slave. The autonomy of the slave is expressed in his power to subvert the wishes of the lord and a reverse domination sets in. Applied to the technological order the slave, technology, created to liberate man from economic necessity, ends up by dominating its would-be masters. The autonomy which persons might have hoped for is snatched away and assumed by a technological milieu of practices, institutions, goals which appears to have a life of its own, clearly beyond human control. To make sense of this process, to, in fact, bring ourselves to see it at all, we need to acquire a selective focus by asking "What governs?" rather than "Who governs" (p. 173). What, Winner inquires, are the conditions, constraints, necessities, requirements, imperatives governing the advanced technological societies independent of those who may happen to wear the trappings of power? To begin to answer this question he develops a theory of technological politics which subjects the entire structure of the technological order to critical scrutiny (p. 226).

His approach is to look at the literature of technology both past and present. The examination ranges from the past technocratic visions of Bacon, Saint-Simon, Veblen, to the images of a technology out of control drawn by Carlyle, Dickens, Hawthorne, Thoreau, and several others. He offers a critical appraisal of the modernization and industrialization literature exemplified by Rostow and deals with efforts to trace technological excesses to fatal flaws in our intellectual history as proposed by Weber, White, Passmore. System theorists such as Bertalanffy, Forrester, and Simon are discussed. Many other authors are cited and placed in context. Special attention is given to Price's *The Scientific Estate,* and Galbraith's *The New Industrial State* and to Marcuse. As mentioned, Ellul and Marx come in for the most thorough examination. *Winner's journey through the literature has in fact provided us with the best single orientation to technology studies of which I am aware.*

This, of course, is not his main objective. By discovering premises and extending the arguments which he uncovers in his search he aims to provide the beginnings of an original theory of technological politics. His concept of "reverse adaptation" will provide a novel approach to such a theory (p. 252).

To lay the groundwork for his theory Winner is forced to deal with and reject a number of attempts to explain the sources of technological dynamism. In some cases he argues for the basic incorrectness of a particular theory while in other cases the fault of the theory may lie in its not going far enough. For example, when taken *in toto* his argument is a radical rejection of the technology-as-tool, technology-as-neutral-instrument, or technology-as-mere-means mode of understanding. All such formulations, so simple and seductive in their appeal, fail to appreciate matters of structure and modes of operation which modern technological systems require for their survival. The idea of the clever artisan perceiving his needs and meeting them in inventive ways is virtually pure romantic myth.

Winner also rejects fatal flaw theories of technology. Here he finds Passmore's location of technological dynamism in the utilitarian orientation toward nature of classical Greece more convincing than Lynn White, Jr.'s widely read argument that our ecological crisis stems from the Hebrew desacralization of nature and metaphysical dualism. In the end, however, he claims the historical facts fail to support either view. Weber's location of the dynamic of capitalism in Calvinist thought is also mentioned, though not treated in detail.

Space does not permit a detailed examination of Marx and Ellul, whom he does not seek to discredit (though he seems to carry on an interesting love-hate relationship with Ellul throughout the book). His main quarrel with Ellul centers on the latter's claim that advanced technology tends toward centralization in a supremely powerful technology-oriented state. Winner's own theory may be seen as a corrective to this centralist argument (p. 252).

Disagreement with Marx seems primarily a matter of emphasis. Here

Winner contends that technology, despite its undeniable importance in Marx's theory of social change, does not receive the detailed articulation that is required. In fact, he notes, both Marxist and liberal positions frame their analyses in terms of the labor theory of value, capital accumulation, etc., leaving the varieties of technological forms relatively underdeveloped. His main complaint with the neo-Marxists of the Frankfurt school—Adorno, Horkheimer, Marcuse, Habermas, and others—is their failure to see the similarities between the technological orders of first and second world countries.

A final group of analyses end up, Winner believes, as elitist theories of the political order, and while he does not deny their accuracy he faults them for underestimating the impersonal forces of the technological order which bear on all of us and which are beyond the conscious direction of any elite group. Here he includes Price, Galbraith, and coming from a different direction, Marcuse. He is distressed that Price and Galbraith both describe new patterns of political enfranchisement in American society—Price's order of estates and Galbraith's "technostructure"—yet neither finds it particularly distasteful that his picture has no place for the average citizen, the sovereign consumer, the small stockholder, or the home-grown politician (p. 170). Winner's message is clear at this point. Such theories have a ring of truth about them but they are in fact morally repugnant.

In the last analysis Marcuse comes back to the concept of ruling-class interest as shaping technical reason and thus the technological order (p. 264). Winner therefore locates him, somewhat uneasily perhaps, with other proponents of elitist theories.

By chapter five Winner has concluded his comprehensive survey of the literature. Elitist theories, which answer the question "Who governs?" are laid aside, though nearby, and the question "What governs?" is tackled head-on. Winner's reply contains several basic theses. 1) Rather than looking at scientific and technical elites, we must examine the notion of a technological order; 2) in the technological order no persons or groups rule in the true sense; 3) structures and processes carry more weight than elites, 4) because technology is embodied within such an order it must not be regarded as neutral but as a potent political force (p. 173).

Winner does not weigh us down with jargon, nor does he opt for catchy phrases in lieu of a coherent theory. What he gives us are two central concepts, "the technological imperative" and "reverse adaptation" and nine subsidiary ideas with which he proposes to describe the essential features of the technological order. Included among the subsidiary ideas are artificiality, extension, rationality, size and concentration, division, complex interconnection, dependence and interdependence, the center, and apraxia. Each in its way describes one characteristic of the technological order, with apraxia requiring special mention. The fact of complex interconnection between major sectors of the technological society, communications, manufacturing, transportation, etc., makes the possibility of catastrophic breakdown, or apraxia, a real one. To keep the system going becomes a moral imperative contributing to reverse adaptation in which means replace ends as goals of action (p. 185).

The concept of the technological imperative goes to the heart of the matter. The kinds of decisions required by the technological order itself, rather than consciously selected ends directed either toward human purposes or special interests, turn out to dominate. Providing for basic needs, food, clothing, shelter, mobility, etc., is replaced by technologically specific objectives—more oil, increased electric power, wider roads, higher buildings, etc. Such redefinitions of original human demands in terms of large-scale, sophisticated, and intermeshed technologies result in man, the ostensible master, acquiring a pathological dependence on technology, the supposed slave. Winner pushes the analyses deeper along Marxist lines. As the technological order respecifies objectives, human expectations themselves change to match the respecification. Perceptions, judgments, feelings, desires are reshaped under the lordship of the order itself. Ways of speaking, thinking, behaving along with patterns of work the individual must tolerate are transformed. A mode of life is defined and so ultimately is human consciousness itself. There is an inexorable quality to the process as more highly developed rational-artificial structures overwhelm and replace less well-developed life forms whether they be customs, craft technologies, community identities, local institutions or the like (p. 212). The "slave" imposes increasingly stringent discipline on its "human" masters. The bondage of economic necessity is replaced by the domination of technical means.

With regard to his second core concept, reverse adaptation, Winner is no doubt correct in claiming it to be novel to the politics of technology literature (I must say, however, that it helped me to make sense of some of Ellul's own characterizations of *la technique* which had heretofore defied comprehension). The description of reverse adaptation is indeed significant. It goes far toward explaining the slightly mad quality of life one becomes conditioned to accept, when, for example, one passively accepts tomatoes bred for picking rather than tasting or when one is willing to put out good money for armor-plated oranges that must be believed to be eaten. As he makes clear, reverse adaptation does not suggest that technological means are totally divorced from human wants, needs, cultural standards, political decisions, but rather that ". . . technical systems become severed from the ends originally set for them and, in effect, reprogram themselves and their environments to suit the special conditions of their own operation'' (p. 227). The major obstacle to ''seeing'' reverse adaptation at work is the seductively appealing tool-use concept of technology, what Winner calls the straight line notion of means-ends relationship. Man has a need, he envisions the goal that will meet the need, then devises means to achieve it. This notion, however, is simplistic to a fault. It overlooks the complex, interconnected and large-scale nature of major technological systems and obscures the fact that the tool users of the system are nearly always distinct from the consumers who merely utilize the end product essentially in total ignorance of the production and distribution processes which made it possible.

Winner describes the genesis of reverse adaptation in the following way (p. 259). An originally accepted purpose such as communication or transportation gets pursued by specific means. But the means require their own means, resources, and social power such as government subsidies, right-of-way legislation, foreign policies, regulations, etc. As a result a situation is created in which actions and decisions are far removed from original purposes. (One recalls the army official who defended the obliteration of a Vietnam village as necessary to make it safe from communism.)

There are two forms of reverse adaptation. The first occurs when means *specify* ends: mobility becomes having a car; communication means having a telephone; or food is translated into a refrigerator, stove, and supermarket. In a second sense, however, means can *become* ends. The ever-present threat of apraxia to the technological system consumes the attention of the technocratic actors. Planning demands a predictable and orderly market. The government is involved as further insurance. The shudder that recently passed through the airline industry at the possibility that fares might be deregulated is a case in point. It is not uncommon, as Winner notes (p. 248) for ''the System'' to create a scapegoat to mobilize support for its expansion. Foreign powers, crime in the streets, ecologists, pollution controls come readily to mind. To mix the metaphor, energy companies and the Department of Defense are not averse to crying ''wolf'' using proprietary data to describe its size. The point is to save ''the System,'' and ends which are disparate and possibly incompatible with this goal are simply junked (p. 241).

But does Winner's concept of ''the System'' require a centralized technocratic hierarchy ultimately embodied in the one superior center, the state? Here Winner disagrees with Ellul and with elitist theories in general. His theory is more subtle. Reverse adaptation makes such centralization unnecessary. Technological politics shape the expectations and perceived needs of technocrat and citizen alike. Each becomes an unwitting actor in the mode of life which the technological imperative defines. And each in his way contributes to its survival. To paraphrase Marx's famous quote, the technological order is not determined by consciousness, but consciousness by the technological order.

While the heart of Winner's argument is in chapters five and six, his final two chapters flesh it out. His discussion of the inherent complexity of the technological order helps explain how human actors both inside and outside the technocratic establishment are not in control. Clearly technological politics are not democratic politics. Traditional categories of political systems, however, don't fit. Oligopoly is out, since the role of technological elites is not decisive. Fascist designations are not accurate, since the power of the technological order is not centralized in the state. What we appear to have is an emergent order, a novel political system, made possible by large-scale technological systems. I find Winner's characterization of this order both compelling and important.

What disagreements I have with the book are mostly small points and matters of emphasis. The first may be a small point, though I am not sure. Virtually all of Winner's analysis of the technological order rings true with one exception. It does not appear to possess the necessary properties for calling it "autonomous." Technology is out of control. It is plainly out of human control but it may also be out of any control whatever. For technology to be autonomous would require some core, some nerve center, some mechanism for self-direction and correction. Winner has rejected, correctly, I believe, the idea of centralization in an all-powerful state. Similarly, he does not consider the role of technocratic elites as decisive. Rather than autonomous or self-directed technology, what we may have is an ensemble of actors each of whom has been deluded and anesthetized by the technological milieu that itself possesses a certain logic but which for the most part is a kind of onrushing, poorly coordinated muddle. A tyranny of small decisions rather than the autonomous direction of a technological "organism" may account for its momentum. In a word, I believe it could be argued that the technological order has no *telos*.

The possibility seems real that the technological society may be more anarchistic than coordinated. This is certainly the legacy that liberalism intended. Ferkiss has made the case against liberalism in *The Future of Technological Civilization*[1] and I am inclined to think that an analysis of the elements of liberalism—laissez-faire capitalism, market economics, and utilitarian ethics—can go a long way toward explaining the roots of the technological order which, as Winner so perceptively describes it, has outgrown its origins and transcended its original intentions. Nor do I think that finding the legitimation for the present order in Lockean liberalism conflicts with the fact that socialist as well as nonsocialist industrialized countries share the same kinds of problems and suffer together from technological domination. Winner notes Lenin's deliberate choice of the "technological maintenance state" for the short run, over ideology-based revolutionary goals (p. 267). And the "technology maintenance state" is liberalism wrapped in a red flag. The technological order is of a piece in the first and second worlds because it has sprung from the same sapling.

In this regard I mention a matter of emphasis which I might have wished different. Winner is inclined to place less importance on the role of economic theory than I might have expected. I would want to argue that much of the dysteleology of the industrial state can be traced to an economic theory which legitimated a sharp bifurcation between internal and external costs and knowingly sacrificed the poor, the environment, earth's resources, and the future for short-term profit. Economists knew the wind they were sowing, at least the more perceptive ones. Keynes's poignant admission comes to mind. "For at least another hundred years we must pretend to ourselves that fair is foul and foul is fair; for foul is useful and fair is not. Avarice and usury and precaution must be our gods for a little longer still."[2] And having sown the wind the economic whirlwind we now reap is the technological order itself.

To summarize, while I have difficulty with calling the existing order "autonomous" and while I would think liberalism is more of a root cause of our present powerlessness than Winner would, I find his characterization of the technological order both brilliant and sensitive.

A few minor points. I think the connection between invention and innovation is less automatic than Winner suggests (p. 73). It is true that there is a fairly predictable sequence from invention to innovation in the large corporations. But these inventions tend to be fairly unsurprising, more evolutionary than surprising. As solar energy gains in popularity inventors are finding time after time that they are reinventing rather than creating afresh. Most inventions fail to mount the innovation track because they threaten the control and inherent conservatism of the technostructure itself.

I also believe the appropriate technology movement will prove more significant for the future than Winner indicates. I find it a more positive tool for subverting the existing order than his "methodological Luddism" though I am not completely sure what the phrase really means. In fact, one could argue as Harman has persuasively done in a recent book, *An Incomplete Guide to the Future,*[3] that the old order is simply not working, that its vital signs are in fact weak. Stavrianos, too, reminds us in his slightly Pollyannish book, *The Promise of the Coming Dark Age,*[4] that the collapse of Rome dismantled vast bureaucracies that had swelled to support the empire. One can imagine more than one person criticizing bureaucratic unresponsiveness before the fall. If we are in fact reaching limits in our present technologies I should hope that some soft technologies are around to provide positive options. I can imagine here that Winner would remind me of Mumford's reluctant conclusion that polytechnics will lose to megatechnics every time.

Winner's extensive notes at the end of the volume are worth the price of the book alone, though it is admittedly steep. They can even entertain. At one point in his crucial chapter five he states, "It is not possible for me to acknowledge everything I have learned or been brought to think about through my readings of Ellul, Marcuse, Giedion, Mumford, Habermas, and others. If I have not seen farther, it is because there are giants standing on my shoulders" (p. 356). Winner has borne the weight well and seen farther than most.

NOTES

1. Victor Ferkiss, *The Future of Technological Civilization* (New York: Braziller, 1974).
2. *Essays in Persuasion* (New York: W. W. Norton & Company, 1963), pp. 371, 372.
3. Willis W. Harman, *An Incomplete Guide to the Future* (San Francisco: San Francisco Book Co., Inc., 1976).
4. L. S. Stavrianos, *The Promise of the Coming Dark Age* (San Francisco: W. H. Freeman and Co., 1976).

A RESPONSE TO STAN CARPENTER'S REVIEW OF *AUTONOMOUS TECHNOLOGY*

Langdon Winner

Professor Carpenter is correct in seeing that the central focus of *Autonomous Technology* concerns the political character of technology. In my approach, the philosophy of technology becomes the political philosophy of modern artifice. My entrance to this subject came through the study of classical political theory. As I studied that tradition, it seemed to me that after a certain point most political philosophers in the modern age had failed to address adequately the conditions under which the recurring questions of politics now find their context. During the past two centuries the Western world has been reconstructed in many of its basic institutional forms. In this radically reconstituted world, the basic themes of political theory—order, justice, freedom, membership, community and others—must be seen in the light of whole new ranges of facts and possibilities. Yet all too often political philosophers have been content to assume in effect that the terms and understandings of the seventeenth or eighteenth century still apply to our situation without modification.

In much the way that a painter sometimes begins with a generating moment—a shape, a color, an idea—and works outward from there to produce a larger work, I began with a generating theme: the notion that technology has gotten out of control. Over the last century and a half that theme has found its way into modern novels, theater, poetry, motion pictures and, finally, political and social theory itself. The first part of my inquiry was to locate, catalogue, and analyze as many important manifestations of that idea as I could find. I was assisted in my work by the fact that when I would tell people that I was writing on technology-out-of-control, they would tell me of their experiences and conclusions on the matter. Listening precedes thinking. I tried to listen and found that my theme was a commonly reported observation or complaint among intellectuals and ordinary people as well. The bus stop was as useful as the library as a place for my research and reflections. After a time the logical structures of the various reports and complaints began to take shape. Thus, the book provides arguments and historical examples which attempt to clarify issues of technological evolution, drift, determinism, the technological imperative and other questions of that kind. There are perhaps a couple of dozen separate-but-subthemes about "technology-out-of-control" in the book. Most of these issues are not in fact unique to the technological age. They are, instead, manifestations of the paradoxes of human action, paradoxes which re-occur throughout history as humans have tried to shape their world.

After sorting through a number of these issues, the book begins its discussion of a more systematic perspective: the theory of technological politics. Professor Carpenter is right in observing that this is the book's true center of gravity. In twentieth century writing about technology and politics one finds that important thinkers of widely different ideological pedigrees—liberal, conservative, Marxist, and anarchist—have begun to share many basic understandings about the character of modern technological society. Despite the obvious differences that separate Ellul, Marcuse, Mumford, Goodman, and Giedion, they also share a good amount of common ground. *Autonomous Technology* tries to spell out the categories and logic of a theory which takes technology as the primary structuring phenomenon of the modern age. Many social scientists, by contrast, now agree that technology is the source of many unintended consequences and therefore requires legislation. Theorists of technological politics go beyond this to assert that technology itself is legislation, that it is a constitutional phenomenon in the sense that ostensibly technological structures and processes are responsible for much of the shape of modern society. I try to argue in agreement with this perspective that technologies are forms of life. As regards large-scale, complex sociotechnical systems, we do not use technologies so much as live them. And while we still talk about these systems in the traditional language of a tool/use understanding, many of the terms in this language no longer apply. Our "tools" have no handles.

My explication of the theory of technological politics does not constitute an argument that all other conceptions of politics should be discarded. I offer it as a perspective on dilemmas which orthodox liberalism, Marxism, conservatism do not recognize adequately enough.

Professor Carpenter points out that I do not pay sufficient attention to liberal economic theory in accounting for the rise of our technological order. I must confess that here I followed the maxim of John Birks Gillespie—known to us less as a philosopher than as America's premier jazz trumpeter—that "It's not what you put in that matters; it's what you leave out." The categories and analyses of neoclassical economics—whatever their truth may be—represent working ideology of the technological maintenance state. One approach to the inadequacies of this ideology is to proceed as Marx did—to explore and explode it from within. I chose another path—that of interpreting technological phenomena in the terms of classical political thought. In helping us understand why we have the technologies that we have, the categories of economics are rigid, narrow, and fundamentally apolitical. I see the kinds of descriptions and explanations that I have given as a partial corrective to the blindness of economic theory. In this sense the book bears a certain kinship with historian of technology Arnold Pacey's recent work, *The Maze of Ingenuity*. While not denying the importance of economic determinants, Pacey sets out to write a history of technology sensitive to other kinds of human motives—religious and social idealism in particular.

Professor Carpenter points out that I have not shown that the technological order is "autonomous." That is true. It was never my intention to do so. The only social theorist who argues an extreme position of that kind is Jacques Ellul. For him, the technical ensemble is a self-determining entity which has a life of its own. That position, it seems to me, is not defensible. One can always find examples of human will, choice and action which refute Ellul's claims. At the same time, however, in various specific instances reports of impersonal agency in the technological order do make sense as symptoms too unsettling to ignore. One such symptom appears as the moral excuse that in the context of large, complex systems one is not responsible. "I was merely a cog in the machine." "I just take orders." Such excuses arise often enough in modern moral discourse to merit our attention. It does seem to me futile to attempt to reawaken the sense of ethical responsibility in scientists, engineers, planners and yet ignore the fact that such persons work in sociotechnical structures which increasingly mock the notion of individual, responsible moral agency.

Professor Carpenter observes that I am not sufficiently enthusiastic about "appropriate technology." Perhaps that is true. In the last chapter of *Autonomous Technology* I point to a diverse collection of people who have begun to realize that technologies are not neutral, that they are forms of life and that choices of technology amount to choices about the kind of society you are going to have. Many of them—including my students—are now at work devising the hardware of an "appropriate" or "alternative" technology. I am hopeful that they will find what they are looking for, hopeful but skeptical. For what does it mean to propose alternative technologies? Are there technical systems which do not have the annoying properties which the theory of technological politics describes and yet which could sustain the world's population at an acceptable level of material well-being? Much of the work I am involved in at present consists of a critical look at contemporary and historical attempts to propose structural alternatives to the large-scale, centralized, complex, high-energy systems that characterized modern technics. From Robert Owen to the New Alchemists, many difficulties both theoretical and practical have vexed the search for an emancipated society founded upon alternative technical means. The underlying question, however, is a perfectly good one: After the criticism of technology, then what?

PHILOSOPHY AND MEDICINE: SOME REFLECTIONS ON A CRITICAL ASSESSMENT BY CAROLINE WHITBECK[1]

H. Tristram Engelhardt, Jr.

Professor Caroline Whitbeck has, for the most part, pointed out accurately and in a helpful fashion both the vices and the virtues of the Philosophy and Medicine series. I will as a consequence not join issue with any of the particular points she has raised, though I am not in perfect accord with them. I will rather address the more general issue of the purpose and function of the series and how it has and should bear upon the areas of research which have become styled "the philosophy of medicine." After all, the individual volumes of the series, and in fact the series itself, is of general interest in terms of what it promises for our understanding of arguments in philosophy and medicine. In this brief response I will (1) provide a brief history and prospectus of the series in order to indicate what Professor Spicker and I, the co-editors of the series, take the field to be, (2) address the question of whether there is a special field of "philosophy of medicine," and (3) consider whether the recognition in scholarly periodicals, series of books and departments of philosophy of such a field runs the risk of cutting the field off from the general ambience of the biomedical sciences which has been its origin and its inspiration.

Though the review touches upon the first two volumes of the Philosophy and Medicine series, a third volume, *Philosophical Medical Ethics: Its*

Nature and Significance, and a fourth volume, *Mental Health: Philosophical Perspectives,* will have appeared by fall 1977. A fifth volume, *Mental Illness: Law and Public Policy,* a sixth volume *Clinical Judgment,* and a seventh volume, *Organism, Medicine and Man,* are already edited and on their way to the press. As the titles of these forthcoming volumes indicate, we have seen the series as a means of exploring the heterogeneous dimensions of the biomedical sciences and arts. The Philosophy and Medicine series and the Symposia on Philosophy and Medicine, which have generated the majority of the volumes, have been meant as introductions to various areas of the field of the philosophy of medicine. Volumes in the future, though, will likely be increasingly focused on more restricted areas, and will presuppose some prior acquaintance with the field. The sixth volume on clinical judgment which constitutes the proceedings of the Fifth Symposium is an example. Further, that volume is likely to be followed in the future by other symposia and volumes addressed to particular issues that fall under the general rubric of "Clinical Judgment." So, after laying out a general map of the field of the philosophy of medicine, we hope to address particular areas with increasing order and sophistication.

Having alluded to the field of the philosophy of medicine, I should provide at least a sketchy justification of the term. First and foremost, there are heuristic grounds to speak of "the philosophy of medicine." Medicine and the biomedical sciences are a salient and crucial element of our culture. No matter how conceptually heterogeneous medicine and the biomedical sciences may be, they cohere in a sort of social collage. That is, a number of conceptual issues cluster together because the social practices to which they refer are joined. Moreover, they can be treated together with mutual benefit. At least Professor Spicker and I, in forming this series, had to presume (and I believe the facts have sustained the presumption) that it is useful to explore the meaning of clinical judgment together with exploring the meaning of models of explanation in medicine, and talk about special issues of explanation, such as concepts of cerebral localization (the mind and brain problem in medicine) and analyses of the legitimacy and place of references to mental states in the development of explanatory and therapeutic models in medicine, etc. In exploring how explanations of states of affairs and modes of therapy are constructed, we suspected that discussion of moral issues would be helpful as well. So, even if the philosophy of medicine represents simply the pursuit of traditional problems in moral philosophy, the philosophy of science, the philosophy of biology, and the philosophy of mind, but not in medicine, there are heuristic grounds for treating them conjointly.

In particular, our view has been that issues in bioethics could be adequately understood only against the background of a more basic understanding of the nature and significance of the biomedical sciences. If it is the case, as many have argued, that concepts of health and disease are in significant measure evaluative concepts, one would expect to find medical-ethical questions disguised as explanatory questions. Or at least one should find manifestations of evaluational issues, disputes concerning what is proper or good for humans expressed in terms of concepts of health and disease. Our goal has been to broaden the focus of philosophy of medicine beyond those of bioethics, as well as to stress the interdependence of conceptual issues in medicine. As a result, though we have committed one volume to issues in bioethics, we have been most concerned to address issues concerning models of explanation in medicine, problems of the philosophy of mind in medicine, questions in philosophical psychology in medicine, and problems in the philosophy of technology in medicine.

We have also been very concerned that discussions of issues in the philosophy of medicine achieve a credible academic quality without becoming professionalized in a narrow sense and cut off from their roots in medicine itself. In the development of individual symposia and volumes we have, therefore, striven to have an ample representation from physicians, biomedical scientists, and other health professionals. One of the great dangers for the field is that it may cut itself off from medicine and the biomedical sciences. It was, after all, medicine's actual problems that stimulated the growth of this field and offered this field of philosophy an exciting practical importance. It would be disingenuous to forget about the obligation to help medicine, if not to solve its problems, then at least to be clearer about the nature of its problems. As periodicals and scholars develop in this field they should keep contact with medicine.

On the other hand, we have been concerned that a tradition of careful scholarship develop in the field. We have therefore sought contributions from philosophers likely to address the conceptual issues in medicine with rigor. We would hope that in the future, fields of interest and scholarship would develop within the general domain of the philosophy of medicine. To some extent this has already occurred with respect to analyses of the concept of health and disease, investigations of the nature of clinical judgment, not to mention the field of bioethics. We believe that there are a number of such clusters of issues which can be explored with both theoretical and practical benefit. In particular, the appearance of journals concerned with issues in the philosophy of medicine other than bioethics, such as the *Journal of Medicine and Philosophy* of the University of Chicago Press and *Metamed* of Burg Verlag, Trecklenburg, West Germany, signal an important maturation of the philosophy of medicine.

In closing, then, we acknowledge that the initial volumes of the Philosophy and Medicine series are introductory. The field itself is *in statu nascendi.*

We are grateful for the authors, readers and attendants at the symposia that have helped us midwife this field. We are in particular debt to Caroline Whitbeck for her attentive review of the first two books in this series. Given the significance of the problems of understanding the nature of the concepts of health and disease, the character of goals of therapy, the place for references to mental states and activities in medical explanations and treatment, as well as the criteria used in resolving bioethical issues in medicine, we believe that an authentic stimulus exists to maintain the growth of this field. In any event, there are conceptual quandaries with sufficient content to insure that future volumes will not simply be introductory.

NOTE

1. See *British Journal for the Philosophy of Science* 50 (1979): 399–406; the review was presented originally at the 1977 Delaware Philosophy & Technology conference.

PART II:
OTHER CONFERENCE PAPERS, 1978–1979

THE METAPHYSICS OF PAOLO SOLERI

Henryk Skolimowski

In our present circumstances, environmentalists and ecologists are often philosophers of our times. They are rethinking, or at least trying to, our methodological and our metaphysical heritage *de novo*. One of them is Paolo Soleri. He is, of course, best known for his architecture, which is pregnant with philosophical and theological implications. One could equally well write about the metaphysics of his architecture as about the architecture of his metaphysics. The latter is the subject of my presentation. I will attempt to show that his metaphysics is uniquely rooted in and defined through his methodology, which is tantamount to arcology.

Arcology in the simplest rendering stands for very compact three-dimensional cities—cities that go up rather than sprawl around—in which architecture and ecology are blended together to the degree that they represent a new ecological and social, and ultimately metaphysical reality. The metaphysics of arcology lies in its aspiration to become not only a

Research in Philosophy & Technology, Volume 3, 1980, pages 133–152.
Copyright © 1980 by JAI Press Inc.
All rights of reproduction in any form reserved.
ISBN: 0-89232-102-4

new form of urban habitat but also a vehicle of furthering evolution toward increasing spiritualization of the world.

Martin Heidegger maintained that in order to create a new metaphysics it is not sufficient to write a book on metaphysics; first, one has to re-experience and reconceptualize the world around. Thus metaphysics becomes much more than a new linguistic arrangement of the old conceptual pieces. It requires remolding the furniture of the cosmos and reconceiving the ways we interact with it. It is in this sense that we can consider the idea of arcology a form of metaphysics.

1. The Aestheto-Genetic Imperative

Soleri is above all an architect. Though this is an obvious statement, I have often wondered whether his true vocation is not that of the metaphysician or theologian. Be that as it may, let us look at arcology vis-à-vis the traditions of twentieth-century architecture.

The dominant ideology of architecture of the first part of the twentieth century was that of the Bauhaus, a militant *and* extremely talented group, inspired by the image of technology triumphant. "Form follows function" was their slogan. It still is in some quarters, but held without any conviction—a mere shadow of the memory of the past. "Form follows function" simply meant that you must do what technology invents. You must follow all the technical inventions and improvements because this is *progress*. So intoxicated were we with the image of technology triumphant that the slogan ("Form follows function") was seen as one of those indubitable Cartesian truths. Even the greatest minds succumbed to its perilous grip and created architecture shaped by the perverse influence of technological rationality. I use the phrase "perverse influence" quite advisedly. For "Form follows function" was but an endorsement of the technological imperative: whatever can be done, must be done; whatever is technologically possible must be actualized. This is a distortion and perversion of real Reason, which should inform us not how to be enslaved to technology, but how to use it imaginatively to human purposes and in order to actualize human ends.

The technological euphoria of the late 1920s and the 1930s so dazzled and mesmerized even the great minds that both Buckminster Fuller and Le Corbousier announced that "House is a machine for living," a very telling metaphor. The case of Le Corbousier is the strangest of all. He was a real poet of great sensitivity and imagination. Had he lived today he would have been among the most outspoken of ecologists, crusading for an end to the ecological degradation and urging us to live in a symbiotic relationship with nature. I can almost hear his vibrant and pronounced utterances on the subject. As it happened, he was one of the culprits. His considerable influence helped to spread and legitimize the reign of the spurious linear rationality in architecture. The result was the geometrically planned city, rational in its linearity, but otherwise sterile. Another result was the high-rise building, a supreme technological achievement, but in many ways a social disaster. The ideology of "Form follows function" is now dead. If we are to encapsulate our vision in one crisp sentence, then our motto should be "Form follows culture," as I have tried to argue in my essay, "Space, Quality of Life, and Architecture."[1]

The second major imperative of our time is the energy imperative, formulated in the 1950s by Buckminster Fuller. He saw the folly of following blindly the dictates of the technological imperative and insisted that our design activities should be guided by an intricate calculus of energy, which is the ultimate commodity we must reckon with. In a sense, Fuller starts where the Bauhaus leaves off. Technological inventions and innovations are well and good. But we must look into them more searchingly and see how they resolve themselves into the ultimate calculus of the universe—energy.

There is a metaphysical scheme behind Fuller's energetic imperative. He saw the cosmos as one comprehensive technology working toward greater and greater efficiency. He summarized this tendency as the principle of ephemeralization: doing more with less. The ultimate logical limit of ephemeralization is: doing everything with nothing. Fuller actually did not push his principle of ephemeralization to its logical limits.

Fuller is truly a Renaissance man, perhaps a man of genius. But his energetic imperative is an expression of a grand folly. He thought, at least throughout the 1960s, that we can make everybody "a success on earth"; that is, by providing everybody with energy slaves on a scale unprecedented in human history; on a scale, actually, which was to match the consumption of energy in the United States, which he calculated as the equivalent of 400 to 500 energy slaves, supposedly working for every citizen. Fuller envisioned vast and fanciful schemes according to which we could hook up all the energy networks of the world into one giant global one. Fuller hoped that we could make not only 4 percent of the earth's population a success on earth but perhaps 90 or 95 percent. Fuller's imperative is actually a very refined version of the technological imperative; it also is an expression of the American dream, for Fuller is a typical Homo Americanus.

With the energy crisis looming larger and larger on our horizons, with the increasing awareness that the way to the future is via the reduction of our energy consumption and of our *psychological dependence* on energy slaves (see Ivan Illich, *Energy and Equity,* 1975), and with the increasing awareness that frugality with dignity is much more preferable than abundance amidst spiritual squalor, Fuller's energetic imperative has lost its attraction and indeed its sense.

Both the Bauhaus and Bucky Fuller represented the highest attainments of the technological civilization, but at the same time were terribly constrained by the narrowness of vision of this civilization. It is as if the god of technical virtuosity gave its prowess to some people, but at the same time blinded them to human concerns.

Paolo Soleri starts where Buckminster Fuller leaves off. Soleri proposes the third imperative, the aestheto-genetic imperative. The energetic imperative is not enough. Man is more than an energy system. We must start with a human imperative. The business of architecture, of city planning and any human habitat is to provide a life-enhancing environment.

There are two components in the idea of the aestheto-genetic imperative. The first component is related to our understanding of evolution and the genesis of life, and also of our place in the chain of being. The life forms called human beings are exceedingly complex and intricate. The story of life in us, which also explains the marvelous capacities we possess, in spite of our rather limited physical container, is the story of an ever-increasing complexity, which is managed in exquisite hierarchical patterns.

The three basic concepts that govern the ascent of life, also conscious life on the human level, according to Soleri, are: complexity, miniaturization, and hierarchy. Life is complex, for life occurs in the thick of complexity. Health and radiance of life forms are in proportion to the diversity of the environment in which they reside. Now, the attribute of complexity which is an essential characteristic of life, evolving into ever more mature and capable forms, should be sharply separated from mere complication which is an attribute of ever expanding bureaucracies and of high-power technology, and which is life-suppressing.

Complexity must go hand in hand with the process of miniaturization, for otherwise it would become unmanageable for organisms. The process of imploding, another way of describing miniaturization, is making things internally so exquisitely related that little space is required by an organ or an organism to perform exquisitely and sometimes almost miraculously. Miniaturization must be governed by hierarchies, those supreme orders that make other orders fall into their respective places. For Teilhard, morphology recapitulates spirit, spirit being the supreme hierarchy that governs all other hierarchies.

Complexity, miniaturization, and hierarchy are seemingly physical attributes of organisms in evolution. But they are also the attributes that create radiant forms of human consciousness. And they are also the ones that make human habitats alive and flourishing. They are therefore an essential aspect of our design activities, whatever the nature of the design. To design well is to govern complexities by means of exquisite miniaturization in the hierarchical universe, which is continually self-transcending, and within which we are the conscious agents of further transcendence.

The second component of the aestheto-genetic imperative is related to our understanding of the phenomenon of the aesthetic and the phenomenon of residual anguish. For Soleri, "the aesthetic defines the compassionately beautiful, and as compassion, the child of reflectivity, is the charisma of humaneness, the aesthetic is peculiarly man's creation, the compassionate is the stuff that consciousness is elaborating throughout the universe (by man, in the case of the earth) or more precisely, it is the hyper-structural character such stuff coordinates itself in. In this sense, the aesthetic is the form of the compassionate, that beautiful which is willed" (*The Genesis of Leisure,* 1970).

About anguish Teilhard de Chardin writes in *The Phenomenon of Man,* "Conscious or not, anguish—a fundamental anguish of being—despite our smiles, strikes in the depths of all our hearts and it is the undertone of all our conversations. This does not mean that its cause is clearly recognized—far from it. Something threatens us, something is more than ever lacking but without our being able to say exactly what."[2]

Those two aspects of human existence—the aesthetic which is compassionately beautiful (and we must remember that the compassionate for Soleri is "the stuff that consciousness is elaborating through the universe, by man in the case of earth"), and anguish, which the aesthetic attempts to overcome—must be woven into the aestheto-genetic imperative which is an essential part of our design strategies for evolving man.

All things are so connected together for Soleri that any subject may crop up in any context of his writings. So while discussing the genesis of leisure, he gives a succinct summary of the idea of the aestheto-genetic imperative—as related to everything else! The twelve points that follow are again peculiarly revealing of Soleri's language, in which the descriptive constantly intermingles with the normative. These points reveal another of Soleri's propensities: making his point so succinct and his thought so dense that it verges on the obscure. On close reading, however, one realizes that the ideas are not obscure at all.

1. Life makes the cosmos vectoral.
2. The two infinites (space, time) of the entropic universe are nesting in themselves a third infant infinity: complexity (the self-creating god).
3. Complexity has its arrowhead in self-reflecting consicousness: mankind.
4. Physiological evolution is transferred by man to the world of aesthetic-compassion (neo-nature).
5. Such a world is not only post-statistical as life has been from its inception, but also post-rational with man seeking a man-made universe.

6. In the process of this humanization of cosmos, man produced the *functional* so as to achieve the *aesthetic*, gets tangled in the *extravagant* and out of indifference emulates the *statistical*.

7. The functional is because it functions (obsolescence is its second nature).

8. The aesthetic functions (i.e., is meaningful) because it is.

9. The extravagant is the pseudo-functional as well as the pseudo-aesthetic. Ultimately a counter-life proposition.

10. The functional can and will be delegated to the mechanically controlled world (cybernetic).

11. If the aesthetic will be suffocated by the extravagant with which the functional tries to humanize, embellish itself, then it stands to reason that a mathematical world, rational and heartless, will take over.

The self-creating god will wither away. The selectively rational (robot-like) will rejoin the statistical-rational proto-life. That is, man will be an aimless, expendable, unjustifiable, if explainable, phenomenon.

12. Leisure, becoming a central parameter of life, will transfigure the world or debase it and with it man, its originator (*Genesis of Leisure*, 1970).

The aestheto-genetic imperative is another articulation of Soleri's methodology of complexity. The first impression is that the aestheto-genetic imperative does not really belong to the sphere of methodology but rather to the sphere of ontology or axiology, as it articulates peculiarly human traits. We should remember, however, that methodology at one point becomes the city, arcology itself. And the ultimate definition of arcology is the love bondage. Given the concept of the city as the love bondage, the aestheto-genetic imperative merges with it perfectly; indeed, gives it a distinctive coloring and ethos, informing us that material satisfaction is not enough and never will be enough, that anguish is an inherent part of the human condition, that the aesthetic is a way of partially overcoming anguish, and that further transcendence is a necessary condition of the continuation of life as love bondage.

The aestheto-genetic imperative represents an ingenious way of tying things together: the physical aspects of evolution—complexity, miniaturization, hierarchy; with transphysical aspects of evolution—man's consciousness, the aesthetic, residual anguish. The aestheto-genetic imperative is a prism which enables us to see in a unified way the double nature of things, physical and spiritual, without splitting them and mutilating them in the process. The aestheto-genetic imperative is a bridge between matter and spirit, a peculiarly Solerian concept.

The aestheto-genetic imperative is also a vehicle and a partial expression of eschatology: the anguish of being human is an inseparable part of perfecting ourselves in the process of arriving at more etherealized stages at which spirit will increasingly assert itself over our somatic existence. The anguish of being human is a necessary part of our spiritual odyssey.

2. The Socio-Genesis

In his funeral oration of 429 B.C., Pericles proclaimed, "A man who takes no part in public business some call a quiet man: we Athenians call him useless. Speech we do not regard as a hindrance to action, but as a necessary preliminary: other people are made bold by ignorance, timid by calculation: we can calculate, and still be audacious. Also, we are generous, not out of expediency, but from confidence. In fact, our polis is an education to all Greece."[3]

The spirit of the polis inspired and, in a sense, compelled the Athenians to excel themselves, both as individual beings and as social beings. Great individuals though the Greeks were, sometimes with vastly enlarged egos, their achievements of the classical period are inconceivable without the unique organization of the polis which channeled and fostered their energies, visions, aspirations and dreams. This historical instance alone should demonstrate that the propitious organization of the city can be a great forming and inspiring influence.

Arcology attempts to recapitulate the best features of the ancient polis, as the place in which spirit is fostered and human energies channeled to the benefit of humanity, society, and individuals. Arcology does not attempt to mimic the polis or recapitulate any of its particular features; the polis was a unique phenomenon created in unique circumstances and probably never to be repeated. We can retrieve from it, however, the spirit of excellence. And we can be inspired by the realization that great things are possible even for small groups of people, if and when they are united by a common bond in the pursuit of a vision encircling the whole of humanity.

In arcology, too, people who take no part in public business, and who do not actively participate in the multifarious affairs of the community, are considered useless. The social ideal or the social ideology has not yet been worked out for any of the arcologies, including Arcosanti. It would be only too easy to impose one of the existing utopian schemes on the habitat which may turn out to be altogether different from anything we have experienced in our social history. So such a temptation must be resisted. The success or failure of the idea of arcology, however, will very much depend on the social order it will create. Because so much is at stake in this realm, and because of the inordinate difficulties we confront when we attempt to design a new social order, the builders of the first arcology, called Arcosanti, have been cautious and slow in this venture.

The social genesis will have to work itself out in the actual living creative practice.

Paolo Soleri himself has been conspicuously reticent about the sociopolitical aspects of arcology. He once said: "My reticence [regarding the social conditions which will be congruent with the overall ideal of arcology] is the result of my ignorance . . . maybe it is my blessing that I do not know about these things. . . . It is a precarious approach. Life is a precarious process." As if in afterthought, he added, "It may be that it is my background that makes me unwilling to make a political commitment, to talk about the political aspects of arcology. I was raised in the fascist regime in Italy, and we felt we were not responsible at all for the political conditions; that politics was out."

I do not wish to give the impression, however, that the social order of arcology is all but *tabula rasa*. Some implications and consequences are obvious; and some are already implemented at Arcosanti. Arcology is basically a *sharing* community, which means that rampant, egoistic, narcissistic individualism is out; also rampant exploitation to assert your individual prowess is out; also some of the "freedoms" of rugged individualism are out, such as the freedom to exploit, to plunder, to get rich fast by whatever means and at whomsoever's expense.

In the era of ecological cooperation with nature, sharing immediately implies recycling and frugality. "Frugality with dignity," the term coined by Steve Slaby, will have to become the order of the day. This form of frugality has nothing to do with abnegation and deprivation. It is a positive assertion of a new philosophy of life, partly born of the realization that the material overabundance and overconsumption not only produce bad livers, heart attacks, and other diseases of civilization, not only result in ecological degradation, but also bring about human degradation as they produce slovenly, slothful, loveless, indolent, egotistical, uninteresting human beings.

Frugality is thus not only a principle of a truly rational economic management, not only an expression of care for natural resources and for the well-being of ecological habitats, but also a principle of the inner discipline which enables the individual to perform optimally. True performance is based on the maximal use of our inner resources, and not on the maximal consumption of material things, which in the long run only enfeebles our body, mind, will and imagination.

Expressed in aesthetic terms, frugality is a precondition of inner beauty. Frugality at this level of analysis is a metaphysical principle which unites the economic, the psychological, the aesthetic, the ecological and the evolutionary, and makes them a part of an alternative pattern for living.

The three concepts alone—sharing, frugality, and recycling—when analyzed with sufficient depth, already suggest an alternative social paradigm. However, one needs much more than sharing, frugality and recycling in order to create an alternative society and an alternative life style. Many communes and communities fell apart, without leaving a trace of their existence, although they preached and practiced the three principles. One needs a larger philosophy which makes sense of these principles and which can weave them creatively into the texture of our individual lives and into the fabric of the social life. For Soleri this larger philosophy resides in his conception of evolution and particularly his methodology of complexity.

Frugality with dignity also means frugality in the service of excellence. One of the main divisions in our times is between those, on the one hand, who think of alternative social orders in minimal terms, and for whom decentralization, frugality, sharing, and the rest are instruments for mere survival, who, in other words, entertain a rather shrunken ideal of society, and often conceive alternatives in the image of early agrarian societies; and those, on the other hand, who think of alternative social orders in maximal terms, who think about frugality, recycling and the rest not as a retreat into the past, but a positive transcendence to a future of an unimaginable human promise and spiritual wealth. Soleri belongs to the latter category, to the maximalists for whom survival is not good enough: in order to exist as human beings we must not be satisfied with mere subsistence, we have to exist plus; we have to go beyond where we are now; we have to transcend; we have to entertain social orders far beyond anything that has ever been created in human history.

The difference between the minimalists and the maximalists in envisaging new social orders is quite deep and often appears to be an unbridgeable chasm. The minimalists seem to be "opting out" in a philosophy of mediocrity, while the maximalists aspire to a philosophy of excellence. Soleri himself has been particularly adamant in maintaining that we must not accept a philosophy of "dimness," for the sake of some temporary economic gains. His uncompromising stand for excellence is the heritage of Nietzsche by whom Soleri was profoundly influenced in his youth. Nietzsche thought of life as the pursuit of excellence. This view is, of course, confirmed by our perceptive reading of evolution. From the standpoint of unfolding evolution, the pursuit of life, toward more and more evolved and accomplished forms, is the pursuit of excellence. Soleri's methodology of complexity is not only congruent with the idea of life as the pursuit of excellence; it is indeed a codification of this idea, and represents spelling it out in fine details.

For this reason, Soleri translates "small is beautiful" into "small might be mediocre," for the movement which is spreading under the inspiration of the slogan "small is beautiful" seems to be rather conservative and mediocre in ultimate qualities, the quality of spiritual life, and aims fundamentally at the same goal that the consumptive society is after, but only with less waste. This judgment may be harsh, but some truths are harsh.

For the same reason, Soleri has often been at odds with environmentalists and conservationists, who want to preserve and preserve and preserve without having a clear vision what it is for, and who often actually look nostalgically back to the past as the only viable paradigm.

The minimalists have often accused Soleri and other visionary maximalists of "elitism," of not caring enough for the opinions of ordinary people. The assumption here is, of course, that ordinary people, the populace at large, has more common sense and more fundamental wisdom than all those visionaries. This view stems from Rousseau and Tolstoy (not very ordinary people themselves!), who firmly believed, or at any rate passionately argued, that there is an irredeemable wisdom residing in the folk which we must obey. Rousseau thus writes in *Social Contract:* "When we see among the happiest people in the world bands of peasants regulating the affairs of state under an oak tree, and always acting wisely, can we help feeling a certain contempt for the refinements of other nations, which employ so much skill and mystery to make themselves at once illustrious and wretched?"[4] A very beautiful and *sentimental* statement, but so far from the truth of the present condition of ordinary people in the Western industrial societies (who are hopelessly corrupted by material affluence, hooked on their energy-consumptive gadgets and unable to resist the continuous stroking of their egos) that this statement now reads like a farce. For a variety of complex historical reasons, ordinary people in industrial democracies are nowadays the citadel of conservatism and backwardness. If it were left to them to decide things, they would certainly continue buying those monstrous cruisers of motor cars, they would overheat their houses in winter time and keep the air conditioning going in summer time, and would keep on with the consumptive orgy of "The ball of life happens only once and to hell with the rest of the world."

No viable social philosophy can be designed by following the instinct and wisdom of the masses in industrial societies. The consciousness of these masses has been so radically conditioned by the consumptive ethos that they are not only a shadow but a forceful expression of the ideology of consumerism. The populists and egalitarians should realize that nowadays egalitarianism is not an expression of the protest of the many against the oppression by the few, but an endorsement of the brutal quantitative ethos which has been slowly destroying our civilization during the last couple of centuries. If we translate "elitism" into "qualitism" and "egalitarianism" into "quantitism" then we shall see at once the root of our problem. For "quantitism" (egalitarianism married to consumption) is but a subtle social translation of the quantitative, impersonal, objective idiom of our entire civilization. "Elitism," on the other hand, is in our times a retreat into qualitative forms, by which human life is governed and through which alone it can be developed further.

By so saying, I do not wish to endorse any form, and all forms, of elitism, particularly those which accrue privileges and advantages for the few at the expense of and while oppressing the many. I do wish to point out, however, that irresponsible and slapdash use and overuse of the term "elitism," with regard to anything that is different and based on quality (as if there were this ultimate wisdom residing in the masses), is a bogus philosophy. The sooner we realize that, the better. For then we shall realize too that a new social order will have to be based on the ideal of quality, the ideal which the masses will first be very reluctant to endorse, but which ultimately will make "the masses" more and more articulate human individuals.

This, of course, still leaves the main question unresolved: what about the social order of arcologies? Specifically, what shall we do about the ideal of democracy? Shall we throw it away altogether as it seems to be unworkable nowadays and totally corrupted by the instrumental rationality of technology and its imperatives of efficiency? We can do better than that. But we must not constrain our vision to the next five or ten years and we must not be hopelessly tied to things as they are. Let us remember what Henry D. Thoreau has said about democracy: "The progress from an absolute to a limited monarchy, from a limited monarchy to a democracy, is a progress toward a true respect for the individual. . . . Is a democracy, such as we know it, the last improvement possible in government? Is it not possible to take a step further towards reorganizing and organizing the rights of man?" (*Civil Disobedience*). Let us also be mindful of Walt Whitman's words: "We have frequently printed the word democracy. Yet I cannot too often repeat that it is a word the real gist of which still sleeps quite unawakened, not withstanding the resonance and the many angry tempests out of which its syllables have come, from pen or tongue. It is a great word whose history, I suppose, remains unwritten because that history has yet to be enacted."[5]

For Soleri, city is an organism of a thousand minds. If and when it does become such an organism, many new options, unimaginable now, may become visible to us. No doubt this will require an augmentation and further maturation of our consciousness. But life is a continuous process of the augmentation of consciousness.

In summary, the socio-genesis or the social order in the making in arcology is still *in statu nascendi,* only implicitly there. Given the idea of evolution, however, as perpetuating itself; given the idea of excellence as the propelling force of change, toward new and more evolved forms; given the slow but inevitable process of integration and development of consciousness; given the constraints of our planet in terms of physical resources; and given the imperative of frugality, sharing, and recycling, some compelling consequences follow. Future generations will be able to put these consequences in coherent and viable structures.

3. The Methodology of Complexity

The whole idea of arcology may be conceived as a methodological device. And so too can be conceived the idea of the city, right from its inception. But this kind of methodology, which arcology represents, must not be mistaken for particular methodologies, as developed in specialized sciences. Methodologies of specialized sciences are those of *simplicity*. Soleri's methodology is that of *complexity*. Since life is not simple, and since architecture and human habitats must nourish and nurture life, architecture and the city must be subtle devices for organizing complexity.

Soleri's methodology of complexity may be looked at as something peculiar to one man and his idea of arcology. Thus viewed, it appears a bit "far out" (which is to say "unreasonable" and "unjustifiable"). On the other hand, it can be looked at as part of a new methodological tradition in the making. Viewed in this way, it appears as an instrument of great conceptual power, actually anticipating many new methodologies of complexity, which have emerged in recent years. It is of course the latter approach to Soleri that is more promising and illuminating than the former.

In recent years, rudiments of a methodology of complexity which attempted to free itself from the straitjacket of the methodology of simplicity, were created by Ludwig von Bertalanffy—General Systems Theory; Herbert Simon—the architecture of complexity; and more recently by West Churchman—the idea of philosophy of complexity.

Still prior to these attempts was Warren Weaver's pioneering paper (1948), "Science and Complexity," in which he argued that we can distinguish three kinds of problems:

1) problems of simplicity;
2) problems of disorganized complexity; and
3) problems of organized complexity.

He further argued, rather compellingly, that science, which he identified with physical science, "has succeeded in solving a bewildering number of relatively easy problems, whereas the hard problems, and the ones which perhaps promise most for man's future, lie ahead." Problems that are hard and which promise most for man's future are, according to Weaver, problems of organized complexity which involve simultaneously "a sizable number of factors which are interrelated into one organic whole."[6]

A couple of years later (1950, 1951), von Bertalanffy proposed the distinction between open and closed systems, which was the foundation of his General Systems Theory.[7] Whatever shortcomings of Bertalanffy's theory we can find by laboring over its alleged imprecision and its formal inadequacies, it was a genuine effort to work out an alternative methodology, in which the properties of life can be analyzed as life, and not analyzed out in terms of physics and chemistry. Ironically, recent developments of General Systems Theory, which have attempted to make it formally more respectable, have succeeded in making it an adjunct to specialized theories and methodologies, which General Systems Theory deliberately and consciously wanted to overcome and transcend.

Herbert Simon's idea of "the architecture of complexity" was originally published in 1962, and subsequently republished in 1969, in a book rather unfortunately entitled *The Sciences of the Artificial*. It must be noted that Simon is not concerned with architecture per se, as he is an artificial intelligence expert, but rather with complex problems the architecture of which he attempts to reconstruct. He argues that "by a complex system I mean one made up of a large number of parts that interact in a nonsimple way. In such systems, the whole is more than the sum of its parts, not in an ultimate metaphysical sense, but in the important pragmatic sense that, given the properties of the parts and the laws of their interaction, it is not a trivial matter to infer the properties of the whole." He further argues that "complexity frequently takes the form of hierarchy. . . . If there are important systems in the world that are complex without being hierarchic they may to a considerable extent escape our observation and our understanding. Analysis of their behavior would involve such detailed knowledge and calculation of the interactions of their elementary parts that it would be beyond our capacities of memory or computation."[8] In another context he maintains, "Because organisms are not energetically closed systems, there is no way to deduce the direction, much less the rate of evolution from classical thermodynamic considerations."

All these variations on the theme of methodology of complexity (as well as many similar ones undertaken by David Bohm, Karl Deutsch, Ahron Katchalsky, John Platt and others) are important steps in overcoming the deficiencies of methodology of simplicity. But they stop at the level of the biological organism. They do not seem to dare to enter the distinctly human world. They clearly shy away from a commitment to a normative methodology. Soleri, on the other hand, consciously and unapologetically concerns himself with the grandeur of human life. There is only one step that separates Soleri from other proponents of methodology of complexity, but it is an important one: the recognition of the normative character of methodology and complexity when it is concerned with man. In brief, what Bertalanffy, Simon, and others attempt to accomplish in their partial ways, and in reference to biology, Soleri attempts to do on a grander scale: in reference to human life as developed in distinctly human habitats: past, present and future. Soleri's methodology is the bridge that ties together his philosophy of man and his philosophy of nature. The city (arcology) is an inherent part of this methodology and indeed the most articulate part.

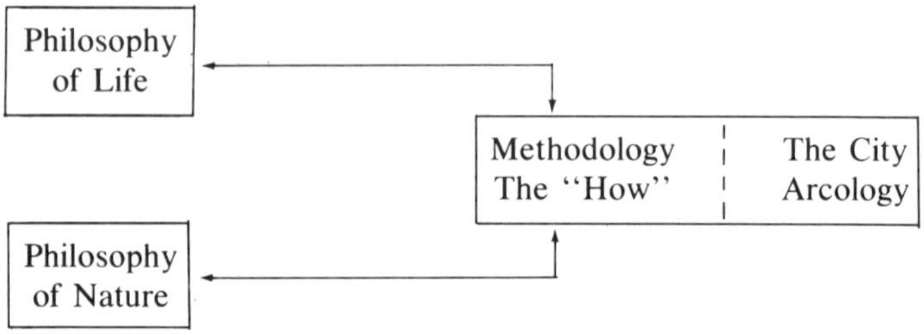

Now, let us be clear: *all* methodologies are normative. Each methodology judges and attests what is right and what is wrong, what is good and what is bad *in the light of its own specific criteria.* Methodology of physical and behavioral sciences insists on objectivity, depersonalization of the observer, quantification of results, operationalization of language. All these characteristics: objectification, depersonalization, quantification, operationalization are *specific values* of this methodology. These values are *postulated,* but not justified by this methodology; nor can they be. For methodology, to say it once more, is a normative set of commandments that tell us how to do this and that—in order to accomplish other ends. The "other ends" reside within the world view of which a given methodology is a part. However, we should not be too innocent in thinking that methodology is merely a servant, a deferential tool, glad to be of use. It is an implicit conveyor of the world views with which it is connected.

It needs to be repeated that the methodology of complexity as embedded in arcology is not just another methodology for more complicated problems which are predominantly physical in nature. It is a realm of its own. Its main purpose is to elicit and elucidate the heritage of life not only on the biological level, but also, and above all, on the cultural, on the social, and on the spiritual level. Human habitats are *sui generis* and we must have a methodology which enables us to understand them in their distinctive terms. There is another point to remember, namely, that according to Soleri (and Teilhard) these habitats are a part of the evolutionary endowment leading to the increasing performance of matter, which on the social/human/cultural level means to the increasing spirituality of man.

4. Ontology; Epistemology; Axiology

Soleri's metaphysics is full of enticing possibilities, but it is not a system constructed according to traditional patterns. It is rather a series of impressionistic philosophical sketches. If we were to resort to traditional branches of philosophy: ontology, epistemology, and ethics (or axiology), then it becomes apparent that none of them is developed in Soleri's system of thought. Implicitly, they are all there, but formulated only indirectly.

The ontology that is latent in the idea of arcology, and in the idea of evolution as upheld by Teilhard and Soleri, is akin to Bergson's conception of creative evolution. The universe is constantly evolving. It is in a state of continuous flux. Its nature cannot be simply characterized, however, by the Heraclitean *Carpe diem;* not by resorting to traditional dialectics, either Hegelian or Marxian. The universe and nature have emergent properties built into them. Once a given emergent property actualizes itself, the process is, in a sense, irreversible. This property, along with that previously acquired, codetermines the future of evolution. Evolution is emergent, irreversible, guided by the vector of transcendence.

The emergent nature of the universe does not entail relativism. The physical nature of atoms is not changed by the emergence of the biological, when life comes about as a new layer of the universe. With the layer of consciousness added, still new properties are introduced. With the acquisition of the biological level, our truths about the universe have changed. And so it is with the acquisition of every new level of being and every new important set of attributes which change the nature of living matter, including ourselves, in a nontrivial way. The continuously evolving world (and particularly its living and conscious layers) continuously changes its nature, and the set of important truths about it. The truths of "lower levels" remain the same, but with each new level new truths are being added. Moreover, the number and complexity of the relationships increase, and some of them affect "lower levels" as well. Psychosomatic illness, for example, entails a large number of relationships and thereby truths; some of these relationships shed an interesting light on the somatic via the understanding of the psychic. This can also be seen on the level of the chemical behavior of certain cells, when their reactions are influenced by higher organs and sometimes consciousness alone.

The point of fundamental importance of the emergent ontology, based on the idea of evolution as perfecting itself, is that we cannot conceivably determine the nature of the new emergent properties before they *do* emerge. And furthermore, we cannot possibly rule out the future existence of new properties and phenomena merely because they do not yet exist. In the emergent universe, the past in ontological terms, cannot be the sole arbiter of the future, for if it is, then we annihilate the very emergent nature of the universe. The universe is thus open, mysterious, unknowable, not in the nominal sense but in the emergent sense. And what new truths and new phenomena we are going to discover in it later may be entirely beyond our imagination now; they are certainly beyond the jurisdiction of physical science as we know it now. The universe is also mysterious in the second sense. We have no business in assuming that we know now all there is to be known—with our present and limited cognitive and perceptual faculties, at the present stage of our development. What unfathomable mysteries we may uncover when our cognitive faculties reach new stages of development and perfection only time will tell.

The *epistemology* congruent with the ontology of emergence is partly formulated in various versions of Soleri's methodology of complexity. This epistemology is holistic, organismic, environmentally sensitive, cosmos-centered, continuous and connected, and normative and hierarchical. The contrasting features of the two epistemologies, the present-empiricist and the future-emergent/evolutionary could be presented as follows:

Epistemology

Emergent/Evolutionary	Present-Empiricist
Holistic	Atomistic
Organismic	Mechanistic
Environmental	Physical
Cosmos-centered	Man-centered
Continuous and connected	Discrete and discontinuous
Normative	Descriptive
Hierarchical	Reductionist

A number of evolutionary epistemologies have been developed in recent times not only by philosophical, naïve ecologists and environmentalists, but also by professional philosophers and psychologists themselves, including Karl Popper and Donald Campbell. Those "professional" epistemologists are reluctant to acknowledge the emergent properties of evolution, and particularly the vector of transcendence "leading" to ever increasing spirituality.

The emergent/evolutionary epistemology suggests a spectrum theory of knowledge, as related to the spectrum conception of reality: from the most readily ascertainable physical facts at the most concrete end of the spectrum via the biological and the ecological, the social, the cultural, and finally to the spiritual end of the spectrum. The further we move from the physical end of the spectrum, the larger is the frame of reference and the more complicated the relationships within which we validate the social, the cultural, the spiritual.

The emergent/evolutionary epistemology is a form of creative fusion of logos and praxis in the process of transformation of matter into spirit. Mind, spirit, environment, and the cosmos are not four distinctive and separable elements, but different manifestations of the evolutionary process which, through the creation of mind enables us to separate environment from cosmos and through a further articulation of itself leads into more condensed spirit. Although the language here employed may jar some empiricists, it is perfectly acceptable from the standpoint of the emergent/evolutionary epistemology. We should observe that *all* evolutionary epistemologies represent a dubious proposition from the standpoint of classical empiricist theory of knowledge, still prevalent. However, if the nature of the world is emergent, new epistemologies must be born to express new emergent properties. The emergent evolutionary epistemology contradicting the static empiricist epistemology is only a demonstration of the emergent nature of the world.

Axiology or ethics is another weak point of Soleri's system of thought. There is a conspicuous gap both in Teilhard and in Soleri as far as values are concerned. A coherent and comprehensive system of metaphysics must possess its ethics. There is no such ethics either in Teilhard or in Soleri. There is a vision of boundless horizons, with omega point crowning it all. But there are no values and specific moral imperatives which would enable us to live a fulfilling life *now,* while striving toward greater and greater perfection. The omega vision is inspiring to our imagination, but it leaves our hearts cold now. The conception of residual anguish and the idea of aestheto-genetic imperative help us to become aware of our existential condition but do not help us to alleviate it. A satisfactory moral code not only gives us a sense of purpose and moral support at the time of struggle, but ultimately gives coherence and meaning to our lives, without which future visions, as imaginative as they are, appear to be only an abstraction. And indeed Soleri's entire opus, and the methodology of complexity through which we work, does at times appear too much an abstraction, chiefly because there is no moral code spelled out, no axiology given which would connect the meaning and fulfillment of the individual now with the process of transcendence directed by our vision of omega. What we therefore need is a set of values which, on the one hand, is the summation of the sensitivity of evolution, and, on the other hand, provides concrete guidelines for human behavior in this world now.

I have attempted to outline such a moral imperative in my monograph *Ecological Humanism*. This new imperative comprises and fuses together the three old imperatives: the Promethean Imperative, or the necessity for transcendence; the Kantian Imperative, or the celebration of evolution's highest achievements; and the Ecological Imperative, or the necessity to preserve and enhance the living habitat around us. Specific guidelines for action, in this human world, following from this new moral imperative, could be expressed as follows:

—behave in such a way as to preserve and enhance the unfolding evolution and all its riches;
—behave in such a way as to preserve and enhance life, which is a necessary condition for carrying on evolution;
—behave in such a way as to preserve and enhance the ecosystem, which is a necessary condition for further enhancement of life and consciousness;
—behave in such a way as to preserve and enhance the capacities, which are the highest developed form of the evolved cosmos: consciousness, creativeness, compassion;
—behave in such a way as to preserve and enhance human life, which is the vessel in which the most precious achievements of evolution are stored.

In my opinion, the fully developed arcological philosophy, indeed any philosophy that can be viable in the compassionate universe in the long run, must incorporate a moral imperative similar to the one here expounded as a part of its conceptual and moral modus operandi.

When one dwells on the shortcomings of the idea of arcology, and its metaphysical implications, one must entertain one more fundamental question: Is the city-genesis, the city itself, such a vital instrument in the perpetuation and refinement of spirit as Soleri makes it? Is civilization such an indispensable instrument for further growth and development of spirituality in us and in evolution? Perhaps both the city and civilization are instruments of the past and from now on evolution, and its human branch, will have to find new instruments for perpetuation and enhancement of the latent potential residing in it.

One may grant all the assumptions regarding the spiritual nature of evolution, and even grant the assumption that omega point is the terminal one, and yet one may question that the city is this supreme and inevitable instrument to which we must succumb. Now even if it is the city, why must it be in the shape of arcology? The arcological philosophy must find compelling answers to all these questions. But there is also the other side of the coin: what other alternatives do you have? Individual retreats of the yogi at the foot of the Himalayas, traditional monasteries, the Maoist model in China, the kibbutz in Israel. Arcology actually does incorporate some of the features of each, and integrates them in a most imaginative manner. We can say without much exaggeration that arcology supersedes them all.

CONCLUSION

Philosophy is but an acknowledgment, on an abstract level, of the shapes of things that have already emerged in human society. Philosophy which is "well justified" is always a posteriori; it reflects the shapes of things that "have been"; in a sense, it is a mirror of the status quo.

New philosophies, such as Soleri's, have to struggle for recognition, having at their disposal insufficient conceptual means to express these philosophies to the satisfaction of our discursive reason. We are also plagued by the insufficiency of our experience: we have not had enough experience to test these philosophies in our actual lives. Thus new philosophies which attempt to outline the shapes of things to come can be validated only a posteriori. Soleri's metaphysics, and many similar ones inspired by a post-empiricist and post-industrial vision of the world, will prove viable and will be articulated in detail, if and when people incorporate them into their actual life styles, and only then.

To look at Soleri's metaphysics, and all the philosophical structures and proposals which the idea of arcology implicitly conveys (including its predecessors such as Teilhard de Chardin) through the categories of empiricist philosophy, and dissecting them with the scalpel of present analytical techniques, disinterested, clinical, deliberately narrow, and actually serving the cause of empiricism and covertly the cause of the consumerism-ridden technological society, would be forcing this new philosophy back into the molds it wanted to liberate itself from in the first place.

The first prerequisite of understanding new phenomena is open-mindedness toward their novelty. Open-mindedness is a form of compassion. Is it too much to expect of *philosophers* to be open-minded about new conceptual worlds? I do not ask you to live in an arcology, but I supplicate that you treat its philosophical propositions, and structures they make up, with an open mind.

NOTES

1. Henryk Skolimowski, "Space, Quality of Life and Architecture," in *Architecture Opportunities, Achievements*, RIBA Publications, 1977.

2. Teilhard de Chardin, *The Phenomenon of Man*, New York: Harper and Row, 1965, p. 227.

3. H. D. F. Kitto, *The Greeks*, Baltimore: Penguin, 1951, pp. 123–124.

4. Rousseau, *The Social Contract,* Baltimore: Penguin, p. 149.
5. Walt Whitman, "Democratic Vistas," *The Portable Whitman* (New York: Viking, 1945), p. 427.
6. Warren Weaver, "Science and Complexity," *American Scientist* 36, (1948): 536–544.
7. Ludwig von Bertalanffy, "The Theory of Open Systems in Physics and Biology," *Science* 111 (Jan. 13, 1950): 23–29.
8. Herbert Simon, *The Sciences of the Artificial,* Cambridge, Mass.: MIT Press, 1969, pp. 86, 87, 108.

SKOLIMOWSKI ON SOLERI

Willis H. Truitt

I

Skolimowski has extracted a general philosophical system from the writings of Soleri. Whether or not this is a system that Soleri would embrace, I do not know. I have read his *Matter Becoming Spirit,* which seems to be a metaphysical solution to the problem of a self-propelling and destructive technology—a problem that interests and alarms us all. But the problem with metaphysical solutions is that they are metaphysical. This is not to say that I don't recognize the importance of a totalistic account of technology in the world and in human life. What I am suggesting is that metaphysical solutions cannot easily be linked to human knowledge and practice, and that when they are, the outcome for practice is sure to be sinister. Skolimowski, it seems to me, has built a bridge between Soleri's "spirit," i.e., his metaphysics, and a practical outcome; i.e., he has constructed a *loosely formulated ethical and political doctrine.* And although I doubt that this is a doctrine to which Soleri would eagerly subscribe, it is my business to discuss it. I'm not even certain that Skolimowski is fully conscious of this doctrine, but maybe we will find out

Research in Philosophy & Technology, Volume 3, 1980, pages 153–158.
Copyright © 1980 by JAI Press Inc.
All rights of reproduction in any form reserved.
ISBN: 0-89232-102-4

later. But before getting to it, let me comment on the metaphysics from which it was derived even though that will not be my main concern.

The metaphysics Skolimowski has elaborated is not new. It is the proposition that nature (matter) over time tends toward spiritualization. It is Hegel's theory, only it lacks Hegel's inevitability. But the bridge between matter and spirit even in Soleri's formulation is not, as Skolimowski claims, peculiar to him. It is the evolutionary doctrine of Hegel's philosophy of nature, or more accurately and recently, the doctrine of emergent evolution of Samuel Alexander. Even Alexander's examples of the prior unpredictability of qualities emerging from quantitative complexities is used to support the theory. The Hegelian dialectical antithesis between physical aspects of evolution and the transphysical or spiritual aspects is present also. When I said the system lacks Hegel's inevitability, I simply meant that the assertion that this physical-spiritual dialectic *must* be resolved in time in favor of spirit is not present in Skolimowski (and Soleri). Or to put it another way, our crisis is the real and acknowledged possibility that, and I quote, ". . . when the aesthetic (a spiritual aspect) will be suffocated by the extravagant with which the functional tries to humanize, embellish itself, then it stands to reason that a mathematical world, rational and heartless, will take over."[1]

How might a materialist respond to this business? It would require giving a materialist account of the three transphysical aspects of existence. They are, according to Skolimowski, human consciousness, the aesthetic, and anguish. Obviously, it is impossible to give a detailed account of these phenomena at present, especially since it is not my job to do so. Therefore, let me offer a glib, but I believe realistic, interpretation. First, there is a great deal of evidence of recent accumulation that "the aesthetic" is a basic biological, i.e., physical, human need, or set of needs. I have written about this elsewhere.[2] Secondly, taking Skolimowski's description of miniaturization as a "physical" process, i.e., "making things internally so exquisitely related that little space is required by an organ or an organism to perform exquisitely and sometimes almost miraculously," isn't this an excellent description of the physical-biological evolution of the human brain and human consciousness? As for anguish, it is a condition from which existentialists suffer more, perhaps, than the rest of us and as you might have guessed, I take existentialism to be the last defensive effort to save philosophical idealism—this impossible task may be the cause of the existential anguish.

II

My main job here, though, is to look at the social-political implications of this metaphysics. And so, let me turn to that. I will simply restate these implications as Skolimowski has already done, but not in a redundant way; rather, from another perspective. First of all, let it be noted that the political tendencies of Skolimowski are not necessarily Soleri's. Skolimowski has extrapolated them. I shall proceed by indicating what the politics are and then show what they are not.

What the Politics Are

1. Civic participation: This is, of course, a desirable end. But it is an end and must be worked for. Skolimowski's suggestion is that it can be achieved through the simulation of the Greek polis. I might agree on this if we could avert the institution of slavery. However, slave production might have been important for the polis. We simply don't know.

2. Sharing: By this Skolimowski means the abolition of ". . . rampant, egotistic, narcissistic individualism . . ." and exploitation in which the freedom to plunder and get rich fast are outlawed. Sharing, too, is highly commendable and is a value toward which communism aims. But, will it be engendered by arcological design, by whether or not we live in a polis? Probably not. Were the vices just mentioned absent from the Athenian polis? The problem of engendering a new morality is not purely one of reorganizing living space. It was such an error (emphasis on the techno-environmental aspects of society) that undermined the Soviet socialist revolution. The creation of new values is primarily an ideological task, not an arcological one. If the primary value of sharing is to be achieved, then how will it be nurtured and what values or what ideologies will bring it to acceptance? Skolimowski's answers are:

3. Frugality: This is said to be a metaphysical principle. But basically it is the idea that consumption should be limited. But whose consumption is to be limited? In our society, in this city, that which is frugal for one small class of people would be unbelievable abundance for the great majority.

4. Utopianism: This is the claim that the most direct path out of our present crisis involves a positive spiritual transcendence. This transcendence will allow us to ". . . entertain social orders far beyond anything that has ever been created in human history."[3]

5. Social Darwinism: This is the Nietzschean distortion of Darwinian "natural selection" which teaches that ". . . unfolding evolution, the pursuit of life, towards more and more evolved and accomplished forms, is the pursuit of excellence."[4] Compare this interpretation with that of the great oil baron, John D. Rockefeller, "The American Beauty rose can be produced in the splendor and excellence which bring cheer to its beholder only by sacrificing the early buds which grow up around it. This is not an evil tendency in business (society). It is merely the working out of the law of nature and the law of God."[5] This nineteenth-century conception of

evolution has nothing to do with the biological theory of natural selection by specific adaptation as I have shown elsewhere.[6]

6. Elitism: This is the view that ordinary working people are corrupted by affluence and egotism, steeped in backwardness; that they are orgiastic and don't give a damn about the rest of the world; that they are incapable of making rational decisions and that they utilize the ideal of equality to brutalize and oppress their intellectual betters. This claim is so preposterous that I can only imagine it having been written during the working-class terrorism of the recent coal strike. Fortunately, I live in Florida and was thus able to escape making such a rash expression of outrage.

7. Perfectionism: I have taken this last claim from Skolimowski's ontology. But since I will not discuss the ontology and because it bears on the social issue, let me comment on it here. It is the assertion that evolution is the process of self-perfection (what I identified earlier as Social Darwinism) conjoined to the emergence hypothesis ". . . that we cannot conceivably determine the nature of new emergent properties before they *do* emerge." Quite apart from this being a patent tautology and an archaic form of teleologism, I must say that it has nothing to do with the theory of evolution by means of adaptation. Indeed, what we do know now is that we can determine, within functional limits, what kinds of genetic mutations are likely to occur in a given environment, and what kinds are statistically very unlikely. In any case, the idea that "perfection" is in some sense the "final cause" of evolution is not of any use at all to the scientist. It is far too vague, empirically spurious and functionally empty. "Perfection" seems only to serve as a kind of post facto rationalization of the theory of emergent evolution and politically as an apologetic for Social Darwinism.

What the Politics Are Not

1. They are not an answer to the question: How does a society reconstruct itself (physically and consciously) toward the end of maximum human participation?

2. They are not practical guidelines for the creation of a society or a community in which sharing is to become a shared value.

3. They are not a set of prescriptions which indicate how the vast discrepancy between human suffering and deprivation, on the one hand, and undeserved opulence and affluence, on the other, can be eliminated.

4. They are not a scientific understanding of the existing conditions and tendencies of capitalist commodity production such that we might employ that understanding in revolutionizing society in a progressive and nondestructive direction.

5. They are not instructions on how to rehabilitate those truly desperate

human beings who have been sacrificed to the engines of production for the procurement of profits.

6. They do not provide knowledge by means of which the oppressed and exploited can liberate themselves from the predatory yoke of monopoly capitalism and imperialism.

7. And finally, they are not a science for the guidance of human evolution toward some undefined state of perfection. And until that perfection is clearly, and not metaphysically, defined such guidance is not desirable in my view.

III

In my concluding remarks, I would like to say two things. One about empiricism, the other, surprisingly, about architecture. First, Skolimowski's identification of empiricism with ". . . the consumerism ridden technological society" is as opaque and unsupported as any philosophical claim I've ever encountered. The more obvious choices for a philosophical bogey man, it would seem, are pragmatism and phenomenalism with their inherent subjectivist tendencies. It is these philosophies which by virtue of their methodologies foreclose the possibility of a rational and reflective assessment of technology within the social-environmental totality as I have argued before.[7]

Now for a comment on architecture that I think is in basic agreement with Soleri's theories of design, and here I have very little knowledge and am only guessing. Certainly, architecture cannot be separated from human relations. To a large extent, buildings and their surroundings determine the way we associate one with another. And, also, the visual presence of architecture will affect our sensibilities either for better or worse. This does not mean that entrenched values and value systems can be uprooted and abolished by constructing a new setting. For transvaluation is a result of political, social, and cultural revolution. But part of that cultural revolution is the reconstruction of our environments and I would not for a moment denigrate the importance of architectural reconstruction in this overall task. What I believe is that architectural revolution should reflect the political, social, and moral goals of socialist revolution. We know that the great edifices of the past reflected, or were metaphors for, the values of ruling classes. And we are familiar with the hideousness of much Soviet architecture built during their great revolution and with the repressive characteristics of Mussolini modern and other versions of fascist buildings. Of course, Soviet architecture was a mistake in that it attempted to imitate some of the worst examples of capitalist skyscraper technology. Fascist architecture speaks for itself and is what it is. But in a people's revolution, one would expect to find kinds of construction and distribution that reflect some very elemental human needs. First, this would mean cities of not unmanageable size, and definitely not sprawl. It would mean an abundance of spaces for recreation, and planned minimization of air and noise pollution. And it would also reflect the human sensory and intellectual needs for high variation and structural diversification as opposed to the mechanical uniformity characteristic of development construction and block architecture. This is what I believe Soleri and Skolimowski mean by complexity rather than bureaucratic complication. These last comments are probably not enlightening and are surely not precise, but I think they do have a direct connection with the philosophy of technologies. I cannot say, however, that they have any relation at all to metaphysics.

NOTES

1. Skolimowski, p. 138.
2. "Art for the People," *The Arts in Democratic Society*, D. Mann, ed., Bowling Green, 1977, pp. 58–69 and "Toward a Theory of Aesthetic Culture," *Left Curve* 2,3,4, 1974–75.
3. Skolimowski, p. 141.
4. *Ibid.*
5. From Richard Hofstadter, *Social Darwinism in American Thought*, Boston: Beacon Press, 1955, p. 45.
6. See my articles "Human Nature and the Cooperative Impulse: Biological Foundations," and "Human Nature and the Cooperative Impulse: Philosophical and Anthropological Approaches to Cooperation," in *The Journal of Human Relations* 14, nos. 3 and 4 respectively, 1966.
7. My article "Values in Science," *Research in Philosophy and Technology*, Vol. 1, Greenwich, Conn.: JAI Press, 1978.

PART III: SYMPOSIUM ON JACQUES ELLUL

FROM THE TECHNICAL PHENOMENON TO THE TECHNICAL SYSTEM

Daniel Cérézuelle

In *La Technique ou l'enjeu du siècle,* first published in 1954 and translated into English as *The Technological Society* (1964), Jacques Ellul argues that in the modern world technique has become a unique autonomous force, the principal factor determining contemporary social life. Whereas in traditional societies techniques (in the plural) constitute a fragmentary ensemble that is dominated by and integrated into a civilization within which they take their form, Ellul maintains that in the modern world technique has become a force which more and more rigorously structures both collective and individual social life. This is why he describes our society as technical and technicized—that is to say, ordered in accordance with technical demands. Ellul's point of departure is to base his analysis on the evolution of technique, in order to understand the evolution of modern society. In his first work, therefore, he tries to define the

Research in Philosophy & Technology, Volume 3, 1980, pages 161–170.
Copyright © 1980 by JAI Press Inc.
All rights of reproduction in any form reserved.
ISBN: 0-89232-102-4

characteristics of modern technique seen as a social reality, and then show how these characteristics orient social, economic and political life, and even man in his inner depths, in a specific direction. The result is to demonstrate how technological society tends to form a total, if not totalitarian, order—a society progressively integrated as a function of a technical logic which can only drain away human liberty in the search for power, efficiency, and organization in all areas.

In 1977, Ellul published *Le Système technicien* which returns to, deepens and even revises the perspectives opened up in his first book on technique. Why a new work? What new insights does it offer? Ellul answers that the stage of the technological society which he described twenty-five years earlier has been surpassed. The new fact is that today technique is organized as a system and no longer simply as a coherent ensemble or loosely united group of techniques. From now on, it is as a system that technique structures modern societies, producing a greater and greater integration of each element (including man) as an object. To jump to his conclusions, this transition from the technical phenomenon to the technical system which Ellul describes in the second book only reinforces the alienating character of the impersonal logic of technical organization.

> In its realization without design, the technical system, in every area where it operates, successively produces an objectification which no longer has anything to do with that of Hegel, which no longer belongs to the subject, which does not enter into the subject-object dialectic. Now the active system treats everything that has been seized or incorporated as an object, for the system can neither develop nor realize itself without working on an ensemble of elements that have already been reduced to neutrality and passivity (p. 19).

The purpose of what follows is to provide an overview of the essential points developed by Ellul to support this conclusion. Even if, throughout the 360 pages of the book, one gets the impression that Ellul is anxious about the extent to which technique undermines human freedom, nevertheless, we are dealing with a scientific analysis in which value judgments are kept in the background.

* * *

Ellul's central idea is that it is no longer sufficient to speak of technical or technicized society, for during the past twenty-five years technique has taken on a new fullness and organization. In *La Technique* Ellul had argued that the technical phenomenon was constituted by the means, all the means, considered at a given moment as the most efficient, together with the search for and application of new means in terms of the criterion of efficiency in all areas. But, according to Ellul, this conception has to be revised, because of the observation that

> techniques even in different areas reacted upon each other and it was impossible to study them separately. On the other hand, techniques became more and more numerous and progressively they covered each field of human activity, and from the very fact of their number and their density they seemed to take on a new consistency. Finally, the computer is added to these two factors—the computer which is an element that joins and coordinates numbers of techniques at the same time that its invention is the product of the coming together of various techniques. Thus, we arrived at a new conception of technique as milieu and as system. That is to say that techniques, combined among themselves and involving the totality of modes and actions of human life, took on a qualitatively different importance. Technique ceased to be an addition of techniques to become, through combination and universalization, a sort of autonomy and uniqueness (*Le Système technicien,* pp. 34–35).

The first important point to note is that technique has become the environment in which modern man lives. Ellul emphasizes that, as modern people, we are no longer called upon to utilize techniques, but to live with techniques and in their midst.

> The technical milieu is no longer an ensemble of means which we sometimes use (in order to work or to distract ourselves), but a coherent whole which "corsets" us on all sides, which has become part of us, from which we can no longer remove ourselves: now it is exactly our unique environment of life (pp. 53–54).

In this sense our society is characterized not simply by the proliferation of objects, machines, and technical gadgets which are all passing, but especially by the infinite multiplication of means which mediate every area of life.

The second essential point is that technique not only becomes universal, but also it acquires an organic coherence. It forms a *system* in that it is an ensemble of elements related to each other in such a way that evolution in one part provokes evolution in the whole, and evolution of the whole has repercussions on each part. Modern technique is a network of technical ensembles in interrelationship: the industrial system, the railway system, the education system, the urban system, telecommunications, etc. All these technical ensembles are organized in a coherent way which was not predetermined by any long-term plans of those who created them in the first place. Yet each ensemble interacts with the others and, on the whole, in such a way that each ensemble is now conditioned by the others.

> The technical system is neither abstract nor theoretical; it is only the result of the interrelationship among the multiple subsystems; and its functioning is a measure of

> the combined functioning of each of the subsystems, and of their correct relationship. When there is a short circuit between them or when there is a breakdown in one of the subsystems, the whole is blocked (p. 121).

The elements of this technical system naturally tend to combine more among themselves than with external, nontechnical factors. This involves both a tendency toward changes in the internal causes of a technical order and also a resistance to external influences (of the intellectual order, for example). Finally, this technical system is not fixed: the interaction of the different subsystems produces an evolution of the whole as well as of the parts, an evolution which obeys a strict internal logic. Ellul strongly insists on this point, asserting that if we limit ourselves to a "flat" analysis of the whole arrangement of techniques at a given moment, we miss the essence of the phenomenon, for the technical system, in itself, evolves. "Technique produces its own change . . . . Technique requires, as a specific given, that it implies its own transformation" (p. 92). Thus, it is incorrect to say that it is technique which progresses. It is more a question of a new and independent reality, the *technical system* produced by a union of the technical phenomenon and technical progress.

It is important to recognize that, for Ellul, technique can transform itself into a system only with the emergence of information processing techniques and communications technology whose use and efficiency have grown rapidly during the past twenty-five years. The heart of Ellul's argument rests on the analysis of the intertechnological and unifying role of information processing which he summarized brilliantly in fifteen pages (pp. 103–118), making clear that we should not conceive of the technical system as a mega-machine: the different technical ensembles structuring our society do not involve mechanical interrelationships, but a whole with more and dense interrelationships of information. As Ellul says in a crucial passage:

> Twenty-five years ago, it was *impossible* to speak of the technical system, because all one could establish was a growth of technique in all areas of human action, but an anarchic growth, the areas still being specified by the traditional division of the operations conducted by man, and there was no interrelationship among them. There was the search for the technical means to put them into relationship, but it was impossible to think of anything other than an organization of the institutional kind, since only the institutional way was known to invent procedures or connections among diverse services or separate sectors of activity. It was thus a question of a process of external agreement and of rigid "bolting together," this being precisely what prevented the technical subsystems from developing interrelationships among themselves. Information processing resolved the problem: thanks to the computer, a sort of internal systematization of the technical ensemble appeared, expressing itself through and acting on the level of information; by total and integrated information, the technical subsystems could both form themselves and be coordinated. No human

> being, no human group, no constitution could achieve it. The more technicization advanced, the more the technical sections tended to become independent, autonomous and incoherent. Only the computer could respond (p. 114).

Historically, technique was first applied to nature and then to society for unique, differentiated operations and procedures. But these operations, given the fact of the very coherence of their objects (nature or society), entered into interrelationships to form technical ensembles. Finally

> with the acquisition of techniques of the second and third degree, forming themselves into a truly seamless web, then as a milieu, technique became progressively a thing in itself and independent of its object, a system all its own. At this moment, techniques began to cohere together and to be organized in terms of one another (pp. 95–96).

This change was made possible by the revolution in information. In effect,

> From the moment when each object or technical method did not function only to fill exactly the task for which it has been invented, but also to transmit information; from the moment when each object or technical method started to function not only for itself, but also to register information put out by the whole technical environment (beyond that coming from the natural environment); and finally when each took account of this information; from that moment there was a system . . . . Material production and the movements of physical objects became less important than these nonmaterial activities (p. 104).

Thus in the technical system, there is not a simple interrelation or coordination among the parts: each is a transmitter and receiver of information, and the system flourishes thanks to this network of information which is endlessly renewed—a network from which its flexibility comes and which makes it difficult to grasp at a given moment.

Here we touch a point upon which Ellul often insists: the technical system develops through self-augmentation, independent of all finality (see *Le Système technicien,* third part). Moreover, this process is not carried out over against man in order to possess or dominate him. The system is without intention or objective; nor does man intervene in a decisive manner in its formation to attempt to make a technical system or to bring forth an autonomous technique. All that sociology can ascertain is that there is the emergence of a kind of new spontaneity which shapes the modern world, because it is the technical system that defines the framework for unity in our society. For Ellul, technique is no longer one social factor among others, but "on the contrary it has become not only the determining factor, but even the uniting element" within which our society develops, as he says. As a result, it is no longer possible to hope to resolve one problem of our society in isolation from all others. Our society appears as a whole organized from a homogeneous technical structure, so that it is impossible to question one technique without questioning the whole system, without wishing to destructure the whole. "The responses should be as global as the society itself" (p. 182). This is essential, because the technical system is animated by a dynamic directed to totalization (cf. p. 219) which is expressed through the elimination of everything that is nontechnical in the modern world (cf. p. 274), a dynamic that tends toward total technical organization.

Nevertheless, this process of totalization which defines the horizon of the technical system clashes with the limits which Ellul wants to underline in showing that it is not possible to conceive of technical progress as a pure, logical process. To the extent that technique has to be applied in a concrete world, its possibilities for development, in theory infinite, meet exterior and interior obstacles that cannot be disregarded when considering the future of the technical system.

Having reviewed most of the major writings dealing with the prospects of technical evolution, Ellul emphasizes that it is very difficult to predict the acceleration or the stagnation of the whole of the technical process, and especially that it is not right to make predictions simply by projecting from what exists now. By taking account of social disturbances as well as the direct and indirect costs of progress, Ellul acknowledges that there could be a slowing down in the area of applied physics and chemistry. But this does not necessarily imply a diminution of the rhythm of the growth of the global technical system. The slowing down of techniques producing goods to be consumed does not prevent the technical acceleration from maintaining itself, but only transfers it to *other areas* (p. 342).

On the other hand, if technique is to be considered as a system, Ellul insists on the fact that its rationality is limited in that it is a system lacking "feedback." Although information plays an essential role within the technical system, it is effective only within the technical givens. In contrast, when this system enters into relationship with nontechnical areas—social or natural—it is not sucessful in adapting its course in terms of its effects in a different milieu, and this failure involves serious disturbances in the exterior environments (nature and society) and internal dysfunctions within the system itself. Thus Ellul (p. 313) glimpses the possibility of an augmentation in disequilibrium and dysfunctions which would involve great disorder because of the absence of feedback, and it could result in the deceleration of the whole system. Nevertheless, he does not see the possibility of such obstructions as an encouraging prospect. If it reached the level of theory and consciousness, then it would be possible to foresee a regular and controlled braking of technical growth—if not, there is the risk of an incoherent and dangerous deceleration that would produce serious troubles. This is a point which he insists upon in an important note: only the conscious, reflective determination concerning the limits of technical progress *by man* would allow him to master it.

> The only act of authentic, verifiable and concrete mastery with regard to technique would be to fix the limits of its development, but this is in exact contradiction to the system. Contrary to popular opinion, it is the determination of limits that brings about freedom (p. 328).

Ellul, however, does not enter very far into such considerations. In this book he is mainly interested in analyzing the coherence and the mode of growth of the technical system. But, on the last page, he announces the publication of another volume, yet to appear, on the dysfunctions of the technical system, the absence of feedback, the ambivalence of technical progress, its irrationalities, etc.

* * *

From this summary it can be seen that *Le Système technicien* is an ambitious work, one putting forth a comprehensive description which is strongly structured on a theoretical axis (the autonomy of technique as system considered as the determining social force) and is relevant to the history, sociology, and philosophy of technique. The documentation synthesized is enormous, and the author illustrates his remarks with considerable material. The demonstration proceeds in a methodological manner, although it could be objected that the book becomes dull with too many references to minor authors whom the average reader would not know. Likewise, the concern to convince and to produce a complete demonstration leads Ellul to reply in advance to detailed objections. On the other hand, sometimes the reader is bothered by very general and important affirmations that remain abstract and that would be much easier to grasp with examples (see, for instance, p. 50, symbolization; p. 104, information).

It is necessary to recognize that Ellul has tried to define his conceptual tools with precision, and that he has not hesitated to multiply methodological discussions in order to support the theoretical and conceptual framework on which his analysis depends and to define the theoretical and epistemological extent of his approach. In particular, he is interested in justifying his choice of a global approach. In effect, he says that we cannot understand the different aspects of technical progress and its social impact unless we construct a model of the global operation of modern technique.

Marx wanted to show that it was impossible to understand the socio-economic facts of his age unless they were placed within the perspective of the workings of the capitalist economy, the model and the operational rules of which he sought to define. Only an understanding of the global and abstract schema of the capitalist economy and its social role would allow for an understanding of any particular socioeconomic fact by centering it within the evolving schema of the capitalist mode of production. Without such a model, it would be impossible to understand a mass of essential sociopolitical phenomena. Thus Marx wanted to prove that there is not one universal economy, but socioeconomic types of operation, each having its own laws (Asiatic, feudal, capitalist, bourgeois, socialist, etc.). In the same way, Ellul is interested in showing that technique, in its social role, passes through specific stages—until the eighteenth century, for instance, humanity remained in the stage of "the technical operation." With the industrial revolution, the movement to "the technical phenomenon" began, and for this stage *La Technique* defined the general characteristics and drew up the balance sheet. In *Le Système technicien,* Ellul wants to show that technique is now in a new stage with its particular operation and characteristics. He gives us, then, a sociological theory, one that is synthetic and global, from which the different problems posed by technique can be approached and correctly situated. This being the case, he is able to clarify a mass of seemingly unrelated social phenomena and situate them within a coherent perspective. On the methodological plane, Ellul stresses the indispensable character of such an approach.

> In order to understand the technical phenomenon, to undertake sociology, the first condition is to consider it in its entirety, in its unity. . . . It is a decisive problem of method; it is necessary to study the technical system in itself, and only from there is the study of different techniques possible (p. 102).

It is no longer enough to say that we live in a technical and technicized society. From now on, it is a question of a technical system. What is the extension of this concept? In particular, do we have to identify contemporary society with the technical system? Is the perfection, the universalization, the autonomy of technical progress going to transform society into the cybernetic mega-machine that Mumford has described? In fact, it is not necessary to believe that the problem posed by technique consists in the reduction of society to the state of the machine. It is evident that no society has ever operated in this way. As Ellul writes:

> *We should not confuse the technical system with technical society. The system exists with its own strictness, but it exists within society, living within her, from her and grafted onto her. There is a duality exactly as there is between Nature and the Machine. The machine functions thanks to the products of nature, but does not transform nature into a machine. Society is also a "natural product." On a certain level, nature and culture interpenetrate, forming society into a whole which becomes a nature for man. Within this complex, like a strange, invading and irreplaceable body, comes the technical system* (p. 25, italicized in original).

This system needs society for support, just as it models society in terms of technical necessities; but it continues to exist on the basis of the social, which has its own laws operating on different levels. Between society and the technical system there is not identity so much as tension or even conflict, for just as the technical system disturbs the ecosystems where it is lodged, it also disturbs or questions the sociological milieu in which it develops. In a clear enough way, Ellul repeatedly uses the metaphor of the relationship between cancer and the living organism in which the cancer is born and develops.

The result is that Ellul rejects the monistic explanation which is the great weakness of Marxist sociology. If he declares that technique today is the determining social factor, he also specifically acknowledges that it is impossible to reduce society to technique, because the social is always the product of the convergence of a plurality of qualitatively distinct factors.

If it is impossible to identify the technical system with society, it is even less a question of being simply a formal model of operation without any links to concrete reality. In order to define technique in itself, Ellul first proceeds to a work of abstraction to isolate the concept of technique with regard to many phenomena which, for him, are concretely linked. But the purpose is to understand the concrete better. That is why he writes:

> In order to understand the concept of technique and the technical system, we should not start from the effects of technique on man or on society. It is not starting from sociological or psychological considerations that we can move toward the concept of technique. We must examine the technical object in itself and in its interrelationships. . . . The innumerable studies on alienation, the consequences of television, the organization of work, the effects of mass media on voting, urbanization, etc., can be useful afterwards to understand one aspect or another of the technical system. But these should not be the starting point to elaborate the concept of technique. We must start at the highest level of abstraction, so that then we can *go back to* the reality made up of the relationship between technique and man or society (p. 37, note 3).

Thus, in pinpointing the concept of the technical system, Ellul claims to be constructing a model which effectively takes account of reality, considered from a certain level of abstraction. He wants to define a sort of *ideal type* which allows us to take into account the infinity of concrete manifestations of technique, just as Marx did with bourgeois capitalism or Max Weber did with bureaucracy.

* * *

To my mind, the perspectives set out by Ellul are profitable and convincing. While technique is being recognized more and more as the number one problem of our civilization, we see partial analyses from all sides dealing with the effects of technique on man, on nature, its intellectual origins, etc. Most of these analyses lack a preliminary grasp of the genetic, structural, and sociological dimensions of technique which could serve as a methodological underpinning for all serious reflection. Ellul's work tries to fill this gap, and he gives us a theoretical analysis on a broad synthetic scale, grasping technique globally as a sociological object. On the methodological plane, this work stands in a fruitful perspective, trying to define a basic conceptual tool which most researchers lack. As for the preparation of the project and the pertinence of the proposed model, the result equals the ambitions; Ellul has largely succeeded in that he effectively takes into account the essential aspects of contemporary social reality. It is a masterly and penetrating synthesis.

Translated from the French by
Katherine Temple; edited by Carl Mitcham.

THE ABSOLUTE DIALECTICS OF JACQUES ELLUL

John Boli-Bennett

Ours is a thoroughly dialectical world. Consider the paradoxical reception that Jacques Ellul's work, itself fundamentally structured as dialectics, has received in academic and intellectual circles. In France, where dialectical doctrines of all sorts are dominant, his work is virtually ignored. In the United States, where dialectical thought is largely unknown or rejected a priori, his work has attracted considerable attention, although it is not well understood. Contradictions of this sort are part of the very fabric of life, in Ellul's view. Dialectical relationships in which opposing forces struggle against each other constitute the very core of reality. The example of the reception given Ellul's work is a rather superficial contradiction, but it reflects deeper contradictions that structure much of the world we live in. I will describe Ellul's dialectical stance as "absolute," and at the outset I want to emphasize one aspect of this absoluteness: dialectical relationships are by far the most important phenomena to study in our efforts to analyze and understand reality.

Research in Philosophy & Technology, Volume 3, 1980, pages 171–201.
Copyright © 1980 by JAI Press Inc.
All rights of reproduction in any form reserved.
ISBN: 0-89232-102-4

In this essay I explore the structure of Ellul's dialectics in three stages. First I discuss the dialectics of his sociology, contrasting his conception of the dialectical nature of social and spiritual reality and his dialectical method with that of his predecessors, in particular Hegel and Marx. Then I turn to Ellul's theological dialectics, placing them in the Protestant tradition preceding him (Kierkegaard and Barth), and showing how his theological understanding underlies his sociological analysis.[1] Third, I briefly sketch his analysis of the eclipse of dialectical tensions that the state-directed technical system threatens to achieve in the contemporary world. I close with a consideration of the implications of Ellul's dialectics for his own life as a Christian "in the world but not of it."

Ellul is a persistent dialectician; as Menninger quotes him, he is "a dialectician above all." As will become clear in this essay, all aspects of his work rest on dialectics; his sociological analyses, his theological understanding, the interplay between his sociology and theology, and the method he uses to approach each of his major areas of inquiry are all dialectically structured. Yet at no time does he give a comprehensive exposé of his use of dialectics (in this he follows Marx); in line with the part/whole contradiction to be discussed below, his readers must piece together the structure of his dialectics from many scattered passages on its particular aspects and, more importantly, from a consideration of his work as a whole.[2] Only from the latter vantage point can we understand the task Ellul has set himself. Hence I will not quote Ellul in support of the summary I present. My presentation may be unsatisfying from a rigorous scholarly perspective but it seems to me to be the only way to do justice to Ellul's work and method which themselves suggest the approach I have taken.

The Absolute Dialectic of Social Reality

At the outset we face a serious problem: What is dialectics? The term has been used to cover at least eight different meanings since the time of Zeno in the fifth century B.C. (Hall, 1967). Until the modern period dialectics referred only to thought and philosophy, that is, to the process of reasoning and the methods to be used to try to answer questions about ultimate reality. Dialectics had to do with the way reality was to be apprehended, not with reality itself. For Socrates the dialectic was a method of distinguishing truth from falsehood; through dialogue, or a series of questions and answers, a statement was refuted by extending it to its logical consequences and uncovering a contradiction in those consequences. Plato and Aristotle continued to employ the term to describe processes of reasoning, the former using it more to refer to the construction of logical systems that reason from the universal to the particular or vice versa, and the latter stressing the negating, critical aspect of dialectical refutation that has been revived in the contemporary work of Adorno. In medieval times dialectics came to be synonymous with formal logic (Foulquié, 1969) and treatises on logic frequently bore the title *Dialectica* (Hall, 1967). Kant's philosophy revived the refutational aspect of dialectics as well as the negative connotations associated with the term ever since it was abused by the Sophists in their adoption of dialectics for purely rhetorical purposes.

It was not until Fichte that the focus of dialectics shifted from philosophical systems attempting to understand reality to the object of such systems, reality itself. For Fichte the dialectic was found in society and in history; it was not just a method of reasoning (Gurvitch, 1962). With Hegel we see the first serious development of this idea, and it was Hegel's version of dialectics that initiated the tradition leading (by way of Marx) to Ellul's sociological dialectics. For my purposes the key elements of Hegel's dialectics are these. (1) It is reality itself, especially history (in Hegel's restricted sense of the term; see O'Brien, 1975), that is dialectical in character. This is to say that reality is composed of contradictions, for example, the ontological contradiction of Being and Nothing, or the historical contradiction of Idea and Nature. These contradictions are inherent in reality and are absolutely central to its structure and development. (2) The development of history proceeds by the emergence of contradictions and subsequent "sublation" (*aufgehoben*) of the contradictions, a schema to which Hegel applied the terms affirmation, negation, and negation of the negation. Affirmation and negation represent a contradiction, and it is crucial to note that, while contradictory, they are also necessary for each other's existence; Being has no meaning without Nothing, just as Idea depends on Nature for its meaning and realization (Hartman, 1953). Negation of the negation is the transdendence of the contradiction into a new plane where the contradiction is "suppressed but conserved"; Being and Nothing are transcended through Becoming, but Becoming contains Being and Nothing at its root. (See McTaggart's [1964] classic discussion; my treatment here must be too brief.) (3) The dialectical character of philosophical thought is a reflection of the dialectical character of reality; as thought approaches a valid comprehension of reality it necessarily conforms to reality's contradictory nature. Hence dialectics is not merely a method but both reality and, in the development of philosophy or historical thought, a reflection of reality.

Hegel's dialectic is thus not purely philosophical because he locates the source of contradictions in reality, not in thought. However, the particular content of his dialectic is largely philosophical in the sense that he places his philosophical Idea at the center of all of history and depicts Spirit as the motive force that, regardless of the intentions of historical actors, leads society toward an ever more complete embodiment of the Idea (Wilkins, 1964). His philosophy of history culminates finally in the total and perfected state as the ultimate representation of Reason.

Contrary to Hegel, but accepting Hegel's insistence on the contradictory character of reality, Ellul's dialectics are not philosophical, but sociological. Here he follows Marx, who in certain key respects (but not all) "inverted" Hegel's dialectic. Rather than idealist forces determining history teleologically, for Marx it is materialist forces in the economic structure of society that shape events and the direction of historical development on a more mechanistic basis. As Foulquié puts it (1969, p. 61), in Hegel's view the Idea creates reality; for Marx, reality creates ideas (for example: ideas arise as ideologies projected by classes to further their economic interests).[3]

Ellul accepts Marx's assault on Hegelian metaphysics, but Marx's followers after about 1900 have faltered in attempting to ground dialectics too strictly in materialism, turning materialism itself into an Idea and applying it blindly to all societies at all times. In Ellul's view this reaction against Hegel does not push dialectics far enough: we must consider "idealism" and "materialism" as additional objects for analysis involved in their own dialectics, and we must recognize that one or the other may structure different societies or even that, in a given society, idealism may structure some social movements while materialism structures others. Marxism has not remained sufficiently sociological in that it has lapsed into a priori assumptions about reality rather than allow for the dialectical movement of reality in all respects. As Foulquié (1969, pp. 110–114) says, in dialectical analysis all knowledge is provisional, and no fixed assumptions or principles are permissible. Ellul takes this view very seriously.

Ellul parts company with both Hegel and Marx on another issue: the "negation of the negation" or the synthesis of contradictions. Hegel and Marx projected visions of history that are, in Gurvitch's words (1962, pp. 155–156), "ascending" dialectics. By a series of resolutions of contradictions, history arrives finally (for Hegel) at perfect Freedom enthroned in the State or (for Marx) at the classless communist society where the basic contradiction between thesis (the bourgeoisie) and antithesis (the proletariat) yields to the transcending triumph of the latter.[4] Ellul refuses to allow the logical elegance of such abstract resolutions to shape his sociology, insisting on the irresolvable nature of some contradictions and allowing for the disappearance of contradictions either by the triumph of one of the opposing forces (à la Marx) or by the further development of society that reduces older contradictions to a secondary or even irrelevant role as new, more fundamental contradictions arise (what Althusser calls the "uneven development of contradictions," [1969 pp. 200 ff]).[5] What is important is consistent attention to the actual movement of society—what Ellul describes as "realism," a concept explored further below.

Ellul also rejects Proudhon's emphasis on equilibrium that might arise from contradictions between essentially equal forces in opposition; Proudhon's view has current prominence in functionalism of the Parsonian bent, to the extent that the latter recognizes oppositions at all. Again, it is only in the most abstract sense that we can speak of equilibrium in a world that is increasingly devoted to the generation of discontinuity, rupture, dislocation, novelty, and revolution.[6] The experience of everyday man is one of continual disequilibrium. Indeed, chaos and disorder are major themes of his symbolic universe,[7] as Ellul demonstrates in *Propaganda* (1965) with respect to the mass media. The conflict inherent in the fundamental contradictions of social reality gives rise to innumerable surface conflicts that attack the human being from all sides; in such a situation any sense of equilibrium or balance can derive only from illusory ideologies that veil the deeper disequilibrium.

Let us examine the fundamental sociological contradictions of our day to give substance to this discussion. Here it is essential to stress the radical character of what Ellul considers contradictions, a key aspect of the absoluteness of his dialectics. Contradictions are not the diluted dialectics so frequently analyzed in sociological and historical literature. The forces driving social reality are not the tired workhorses of complementarities or reciprocal relations (as in Berger and Luckman's sociology of knowledge or much of functionalist sociology); nor are they the more formidable locomotives of opposing class interests (Marx, 1965). Rather, they are the geotectonic strains in the earth that account for the shifting of continents and the rise and fall of mountain ranges: fundamental sociological forces locked in acute conflict and acting as the base of all of the other, more transitory and mutable oppositions in society.

What are the geotectonic strains of the contemporary world? Ellul's sociology identifies two primary objects for study: technique and the state. Correspondingly, the basic contradictions of the contemporary world (but not for all time!) are the oppositions between, on the one hand, technique and society; and, on the other hand, the state and revolution. With regard to the first, and using more precise terms from *Le Système technicien* (1977), we find continual conflict between the technical system and the social structure that englobes it. Consisting of the relatively closed structure of techniques that are highly interdependent, monistic, and self-generating, the technical system continually assails society by introducing rapid change and new sources of political power that disintegrate the rather fragile institutional order. In the construction of society man attempts to impose integration, order, and meaning on technique, but the unrelenting *autoacroissement* of technique fractures the social order from within. Given the universalizing and totalizing nature of the technical system, society finds itself in the position of a peasant village atop a volcano erupting from an increasing number of fissures that spew forth unfamiliar and ever more unmanageable forms of lava. Society tries to channel and dam the flow of technique by imposing political, ethical, and organizational constraints on the technical system, constraints that from the point of view of the system are arbitrary, irrational, regressive, and "soft." Technique strains against the walls society raises to channel its progress; the path of autonomous development is clearly visible, but society's obstacles impede its progress. In the past two centuries technique has had society on the run (for the first time in history; see *The Technological Society,* 1964, pp. 64–77, 300–307), due partly to its success in transforming social obstacles to technical development into technical forms that are thereafter integrated into the technical system and act as spurs to further expansion.

The second essential sociological contradiction is the tension between the state and revolution. In the political sphere the state is analogous to society and revolution to technique.[8] The state is the master integrating device of the technicized society. It coordinates the technical system and assumes the task of restructuring society around the demands of the technical system, in the process becoming increasingly technicized itself. The state attempts to elide the older antagonisms of the class structure by recasting all problems in technical, rather than political, terms, not least because this will remove many of the obstacles to technical development. But new political forces continually emerge, and they present conflicting images of both the direction in which technical development should occur and the social relations in accordance with which that development should take place (e.g., "capitalism" versus "socialism"). These opposing forces constitute the source of revolution, which is the attempt to destroy the state in its existing form and substitute new power arrangements contained in a new form of the state. Until its successful completion, revolution is disintegrative, critical, and disruptive of the state's operations, and the technicized state invokes ever more refined techniques to protect itself from revolutionary siege (*Autopsy of Revolution,* 1971). At a less dramatic level this contradiction gives rise to innumerable revolts and rebellions against the state which grow out of the spontaneous anguish of individuals and groups that recognize, often only implicitly, the role of the state in promoting the technical system's drive toward a totally "programmed" society (Touraine's term; Ellul discusses revolt in *De la Révolution aux révoltes,* 1972).

These central contradictions conform, in Ellul's analysis, to Hegel's view of dialectic: the forces in opposition are inseparably locked together and have existence and meaning only in relation to each other. We can separate them analytically: "society" and "technique" are distinct concepts, to be sure. But at the sociological level, that is, in the world we actually encounter, one pole of the contradiction necessarily implies the opposing pole: the two are always found together. Society and technique not only struggle against each other; each makes the other possible. Contemporary society cannot exist without the self-generating technical system, as we are all only too aware; the obverse is equally obvious. Just so, the state is itself one of the main sources of revolution, creating by its very success the conditions that generate political opposition (cf. Marx's phrase that capitalism "contains within itself the seeds of its own destruction"). And, in a particularly significant dialectic, revolution is also a major source of the expanding power and reach of the state, for the outcome of every revolution in the technical age (*viz.* Russia in 1917; China in 1949; Cuba in 1959) is the accelerated growth of the state under the direction of the very forces that previously rallied opposition against state control (*Autopsy of Revolution,* 1971). Revolution and the state go hand in hand, and any attempt to consider them separately can only lapse into abstraction.

Clearly these fundamental contradictions of the technical society permit of no ascending dialectic (recall Hegel and Marx) and obviate application of the notion of equilibrium (Proudhon). The society-technique antagonism has given rise to no synthesis at the sociological level, however many abstract theoretical syntheses may have been proposed in fanciful literature (from Skinner's *Walden Two* [1948] to Huxley's *Island* [1962] and Teilhard de Chardin's Omega Point in *The Phenomenon of Man* [1959]). The state/revolution antagonism similarly stands fast against any synthesizing tide of the sort proposed by Mao, Nasser, or Castro in proclaiming the state as the source of revolution itself. As for equilibrium. . . .

Hence society is absolutely dialectical in two senses: contradictions are at the basis of social structure and movement, and the dialectical character of society is permanent. Particular contradictions may assume greater or lesser importance over time, but whenever one contradiction (e.g., the class struggle) loses prominence it is replaced by another (e.g., the technique/society contradiction). Neither Hegel nor Marx was willing to accept this uncomfortable view. Ellul arrives at it only as a result of his sociological analyses; it is not an a priori assumption but a description of the world we live in.

Methodology of the Absolute Dialectic

In grappling with the dialectical nature of sociological reality Ellul necessarily relies on a dialectical method.[9] Here he freely acknowledges his debt to Marx, Lenin, and other major dialectical practitioners.[10] In this section I present what I see as the major components of his dialectical sociology. Five components demand our attention:

—the contradiction between ideology and reality,
—the contradiction between action, or the intentions behind action, and the consequences of action,
—the contradiction between the whole and its parts,
—the radical ambivalence of action,
—the dialectic between social reality and spiritual reality.

The first three of these derive directly from Marx and the Marxian tradition, though they are extended and more highly developed in Ellul's application of them than elsewhere. The last two come from the neo-Orthodox movement in Protestant theology (known, less commonly, as dialectical theology: Schepers, 1967). This movement began with Kierkegaard and was given its definitive form by Barth in his radical refutation of Schleiermacher's progressive humanist theology that, in Koenker's words, "sanctified the comfortable bourgeois culture of the nineteenth and twentieth centuries" (1971, p. 101). Both of these latter components, and particularly the fifth, have much greater significance in Ellul's work than in that of any other intellectual, and in his development of the fifth component Ellul makes a most striking contribution to contemporary thought.

1. Ideology and Reality

Marx unmasked religion as the obscuring pretext given by the bourgeoisie for its domination of capitalist society. Religion is the ideology par excellence, offering a description of reality having great legitimacy by virtue of its comprehensive answers to life's ultimate questions and uncertainties while it diverts attention from both existential dread and the realities of this world's injustice and violence. One of Ellul's primary tasks is to unmask all ideologies; this is most explicit in *A Critique of the New Commonplaces* (1968). Ellul shows that what is believed about reality, and especially what is believed generally or is promoted by organized political groups, is often only weakly related to reality as it can be analyzed dialectically. As an initial approach to analysis we might do best by assuming that reality is exactly contrary to the ideologies we encounter; from that point we can work backward through a sifting process to separate the kernels of truth that reside in all ideologies from the husks of political interest, self-deception, confusion, and justification that deflect ideologies from reality. As a thorough dialectician Ellul applies this approach to all beliefs, including Marxist ideology itself. Indeed, Marxist ideology is a primary object for his analysis because he sees it as having replaced the very religions it unmasked and it is one of the dominant, and most misleading, ideologies of the technicized world.

The contradiction between ideology and reality arises not solely through the conscious effort of social forces (i.e., political propaganda). It is perhaps even more the result of a growing split between thought and action in general, a split engendered by the triumph of means over ends that Ellul identifies with the rise of the technical system. Marx called attention to this split in his lamentations about the diversion of theory from praxis; Ellul analyzes it in detail in *Propaganda* (1965). It is a general and profound phenomenon. Modern thought (in the form of philosophy, linguistics, hermeneutics, art and literature, and even anthropology) has all but abandoned the world of everyday life and work, losing itself in logic, structuralism, conceptualism, formalism; and in direct reactions, still completely dissociated from concrete reality, against logic, structuralism, formalism. Action, on the other hand, attempts to kill thought, whether by excessive positivism (most noxiously in the social sciences), conditioned-response propaganda, irrationalism, or the more brutal means of the concentration camp, the purge, or genocide.

At the everyday level modern man is hardly capable of reflection on and comprehension of the significance of his actions. In part this is due to the extraordinary complexity of the technicized society and the far-reaching implications of any action in this highly interdependent world (e.g., what, really, are the consequences of my decision to buy an automobile rather than use public transportation?). In part it is the more calculated result of technical means that overwhelm the individual with too much information and propaganda, operating under the assumption that the technical system will function best if its human agents can be controlled mechanically rather than organically. In part, too, it is the result of the bourgeois tendency toward justification: act first, and construct a legitimating rationalization for action later. An inventive mind can always find a convincing excuse.

Yet every ideology is itself subject to a dialectic: it contains both truth and falsehood, and these are inseparably interlocked in that they are both necessary to the success of the ideology. Ideology cannot be totally devoid of truth; Goebbels' "big lie" is a highly ineffective method, too easily unmasked and thereafter useless (*Propaganda*, 1965). But ideology requires illusion in even greater measure. By attaching itself to symbols that deeply stir the individual psyche and by offering integrated, apparently logical and complete analyses of reality such that reality loses all of its dialectical and ambiguous character, ideology gains great power to attract adherents who seek a symbolic universe of order, stability, certainty (see Lefebvre's presentation of Marx's analysis of ideology, 1968, pp. 58–88). As Ellul demonstrates, from the study of ideology we can learn a great deal about reality, but we must maintain an acute dialectical awareness in our study or risk falling under the spell of ideology ourselves.

2. Action and Its Consequences

Things turn into their opposites. I have already given one example of this component in mentioning the religious role assumed by Marxism after it had begun from a radically antireligious standpoint. It is not simply a matter of our inability to predict the consequences of our actions (though this is a factor); nor is it only a matter of consequences contradicting the avowed intentions of action, intentions that are ideologically expressed and in fact hide the true intentions behind action. Rather, it is the nature of social reality that with great regularity our projects turn upon themselves with complete negation (Adorno develops this theme at length). Ellul therefore tests reality with the tentative assumption that the assumed or desired consequences of a sociological endeavor will finally be contradicted by its full development. Christianity provides a telling second example: originally an unrelenting destroyer of the very basis for man's religions, it entered into doctrinal and institutional forms and became the most religious endeavor of the Western world until recent times (*The New Demons,* 1975). As a third example: the efficient operation of the technical system depends heavily on the principles of control, knowledge, and simplification. An efficient technique extends our ability to control some aspect of reality through (positivist) knowledge derived from study of that aspect in a highly simplified and manageable environment. But the technical system itself vitiates the possibility of control in a larger sense: we construct the system and are indispensable to its further growth and elaboration, but we cannot control it at all. Similarly, the technical system prohibits knowledge of its structure and operations, not least because the ubiquitous simplification of reality becomes hopelessly complex when individual techniques are joined together in multilayered, highly interdependent networks. Control turns into impotence; knowledge turns into ignorance; simplification turns into unapproachable complexity.

Can we account for the production of such antinomies, such self-negations, as these? Ellul's accounting relies mainly on his understanding of the theological realities of existence, but at the sociological level we can say this much: the struggle for power, being completely autonomous from all constraints of an ethical or moral nature (*The Political Illusion,* 1967), places thought and action in a precarious context. Everything is subject to politicization and, once politicized, thought and action obey the law of politics: power engenders power in opposition. Commonly enough, opposing power triumphs and drastically subverts the efforts of the old order. This process occurs at every level, from the individual to the state: power can only result in its own negation, a lesson we learn from *The Politics of God and the Politics of Man* (1972).

The contradiction between action and its consequences takes classic form in the tension between means and ends. The usual formulation of this tension is the issue of whether the ends can justify the means, which is tantamount to asking whether the unintended ends actually obtained by means are worth suffering for the sake of the intended ends that also result. For Ellul this formulation is too simple. Means do not inevitably achieve the ends intended, for the character of means often determines the ends obtained. We may find that it is only the unintended ends that are obtained, and the intended ends disappear from sight. As one example, he shows in *Violence* that unjust means cannot serve just ends. Violence cannot establish peace, freedom, or equality; rather, it "begets violence—*nothing else*" (1969, p. 100). In the technicized society the tension between means and ends tends to become lopsided because our means (technique) have become an end in themselves and the question of ends draws less and less consideration. This is one aspect of the eclipse of dialectics to be discussed more fully later.

3. The Whole and Its Parts

Ellul returns time and again to an insistence on the contradiction between what we can uncover by looking at the multifarious parts of a social phenomenon and what we can see by considering the phenomenon as a whole (here his view is in the tradition of Stalin and Lukacs: Foulquié, 1969, pp. 62–63). It is on this dialectic that he bases most of his criticisms of positivism, and it is worth calling attention to his lengthy critique of positivist approaches to measuring the effectiveness of propaganda in the Appendix to *Propaganda* (1965); there he gives very concrete expression to ideas I can discuss here only abstractly. In essence Ellul begins with Durkheim's sociological stance: there is an emergent social reality that cannot be deduced from the lower-level elements that make it up (*The Technological Society,* 1964, pp. xxviii–xxx). This emergent reality has its own existence. It develops according to its own properties and according to the relationships among its various parts. *La Technique* is not merely a mélange of innumerable individual techniques; the propagandee is not simply a person who encounters a large number of propaganda messages each day; the state is not just a collection of bureaucrats, soldiers, politicians, and police who direct various aspects of social life. Rather, to take only the example of propaganda, the propagandee is one who exhibits a habitual response pattern in which thought and action are highly divorced from each other and the individual is moved to action (or inaction) by a process best described as operant conditioning. The propagandee is a role we adopt when exposed to propaganda, regardless of whether or not we have been so overloaded with propaganda that we exhibit this trait as a "permanent" personality characteristic. We automatically react to propaganda, either in acceptance of its urgings to specific action or in rejection and out-of-hand dismissal. In either case the link between thought and action is very weak. At a higher level of analysis, Ellul inquires into the effects of propaganda on democracy. He shows that propaganda necessarily makes openness to compromise very unlikely, tending to promote rancor and intolerance instead. This is so whether or not particular propaganda campaigns employ themes of rancor and intolerance.

The whole, then, is different from the sum of its parts. We cannot analyze the technical system by looking only at the many techniques and subsystems that comprise it; we must look at the way these subsystems are interlinked, at how they affect each other, and we must also look for the characteristics of the system as a whole that are not to be found in its parts. Self-generation, monism, automatism of technical choice, universalization of technique: these are properties not of individual techniques but of the system as a whole. And the holistic properties of emergent social reality often precisely contradict the lower-level properties of the parts. Hence, from the point of view of any individual technique, technical development seems to bring us greater freedom, more choice, and more control over our environment. But from the level of the whole system technique instead encloses us with more constraints, offering less choice in anything but a superficial sense and decidedly less control over the new environment that is the creation of technique itself.

Ellul makes use of the dialectic between the parts and the whole in his general approach to identifying the basic contradictions of society. He works backward, we might say, beginning with the multifold contradictions that appear on the surface of everyday life and then proceeding deeper to more general levels in searching for subsurface contradictions that generate those at the surface. By repetitive application of this process he burrows through level after level until at last he can go no further: the fundamental contradictions have been reached. Here he differs sharply from positivists (especially those in the Popper tradition) who tend to take a deductive approach that begins with general "assumptions" and reasons forward to concrete instances of the abstract concepts contained in the assumptions. Yet Ellul works in both directions at once: every excavation to a deeper level is constantly evaluated by referral upward to surface events, to assure that the deeper abstractions maintain sufficient contact with all levels of reality.

This method depends on the parts/whole dialectic in that the deeper contradictions characterize the whole under consideration while the surface contradictions characterize only the parts of the whole. From this perspective we can appreciate the manner in which Ellul has incorporated the Leninist distinction between primary contradictions and secondary contradictions and Mao's development of this dialectic to maturity (*On*

Contradiction, 1953; cf. Althusser's discussion, 1969, pp. 200–218). Secondary contradictions are more partial and superficial; primary contradictions are fundamental and holistic. Similarly, the primary aspects of a given contradiction belong to the whole, while that contradiction's secondary aspects pertain to the parts alone. For example, the primary contradiction between the state and revolution produces such secondary contradictions as the conflict between traditional American distrust of government ("rugged individualism") and the insistent demand for action that every organized interest group makes on the state. Or, the contradiction between technique and society leads to such apparent absurdities as continued U. S. production of large, inefficient cars when political considerations (dependence on foreign oil sources) point to an imperative need for more efficient vehicles. Hence the primary contradictions subsume and help us make sense of secondary ones. It is in operating analytically at many levels simultaneously, considering the whole, the parts, and relations among the parts, that we can explain the dynamics of complex, structured contradictions (cf. Althusser's discourse on the uneven development of contradictions and "overdetermination," 1969, pp. 87–116, 200–218).

The contradiction between the whole and its parts leads Ellul to be extremely critical of positivist sociological endeavors that insist on measurement as part of the methodology of social analysis. What is measurable is not in general what is most important. At best we can measure some of the parts, but the emergent whole is often completely unmeasurable, sometimes for only practical reasons but often by the very nature of things: some things are inherently unmeasurable. Unmeasurability is most frequently encountered at the level of spiritual reality, but even on the material plane we cannot measure such things as the quality of life, spontaneity, the autonomy of technique, or man's capacity for moral judgment (*Le Système technicien,* 1977, pp. 131 ff).[11]

Ellul also criticizes the positivist tendency to distort the natural use of language, a tendency that reflects the positivist emphasis on parts at the expense of the whole. When words are treated solely as isolable, separate parts, they only too easily can be redefined or stripped of their natural-language meaning and narrowed into an artificial concept that is inadequate to cope with the richness of reality. Words are more than parts: they are structured in complex ways at the emergent level of the whole language, and breaking language into parts is breaking those structures so radically that the meaning of the linguistic whole is lost (see *Métamorphose du bourgeois,* 1967, pp. 292–293; *Autopsy of Revolution,* 1971, p. ix). This reliance on natural language leads Ellul to avoid explicit definitions of terms. He chooses instead to indicate the general meaning of terms and let the reader induce the fuller and more natural meaning from the context of his use of the term. Such is the treatment he applies to such key terms as "technique," "the bourgeoisie," "propaganda," "politics," and "revolution."

4. The Radical Ambivalence of Action

Kuhns (1971) has appropriately called attention to this component in his summary of the works of Ellul and several other "post-industrial prophets" (see Gurvitch's discussion as well, 1962, pp. 203–207). Because of the inherently dialectical nature of reality, virtually all action has ambivalent effects. We can identify both good and bad effects, positive and negative consequences, in everything we do, regardless of the moral outlook we apply in making such assessments. Ellul pushes this component to the limit, insisting that this radical ambivalence is not merely theoretical but sociological: the positive and the negative are not only coincident; they are inseparably joined together and the one does not arise without the other. In the political realm this idea seems almost too obvious: by de Jouvenel's (1963) analysis, all political problems by definition involve contradictions between competing interests that cannot all be satisfied at the same time. Any political compromise is sure to contain both positive and negative consequences from the perspectives of each party to a dispute.

With respect to technique this component is less easily accepted, but in Ellul's view it crucially describes the impact of technique on man. Every technique has both good and bad effects; it both liberates and enslaves us, it extends human capacities while it closes man up, it brings certainty and assurance while it contributes to the systemic insecurity of the technical whole. This is true primarily because every technique is used for both good and bad ends; in fact, it is used as completely as possible. It is terribly naïve to take the common position found among idolaters of technique: "It isn't technology, it's the use we make of technology—it's man who chooses to do evil with technology, technique is neutral" (Mesthene, 1970; Marcuse, 1964; McHale, 1969). In a purely abstract sense this may be true. We can imagine a society in which man "chooses" to use technique only for good ends, but unfortunately that is not the world in which we live. Control of atomic fission leads to both the bomb and nuclear power plants; advanced computer technology leads to both better diagnosis of diseases and more comprehensive state surveillance of populations; general availability of automobiles leads to both ease of movement and frequent fatal collisions on the highways, and we never have the one without the other.

We can take this contradiction still further. Every use of technique, even if for apparently good ends, contains undesirable effects as well. The dangers of nuclear power plants are well known, as are the pollution problems caused by automobiles. Computer diagnosis of diseases contains risks as well. It may bring too-complete bureaucratization of medical care such that physicians no longer apply their experience and skill in face-to-face attention to patients. Lack of personal attention opens the way to improper and injurious treatment for unusual and complex maladies because the computer system, working from abstract knowledge only, prescribes incorrectly. Every technique has ambivalent effects.

What goes for technique goes for other action as well. In Ellul's view, if we are honest and thorough in our assessment of our actions we must accept the fact that there are very few clear-cut, unambiguous choices in our lives. Life is complex, and dialectical analysis tries to take the full complexity of life into account.

5. Social Reality and Spiritual Reality

Finally, Ellul's work rests very fundamentally on his acute awareness of a contradiction between social and spiritual reality. In the social structures that man constructs we find a persistent and universal urge to create a world of control, certainty, permanence, and inevitability. Society is, in a sense, a tissue of deceptions, masks, and illusions generated by man to excoriate the demons of uncertainty, impotence, temporality, and contingency that confront him at the spiritual level. Here Ellul is essentially in agreement with the existentialist perspective, especially as developed by Kierkegaard (1939): "fear and trembling" seize man's soul, and when man looks inward at his basic spiritual condition he finds that nothing on the human plane can calm his torment. Despite our elaborate technical system, our mastery of nature, our improving ability to analyze society as a machine and employ the state to direct it, man realizes that in fact there is no spiritual hope for him (*Hope in Time of Abandonment,* 1972). The symptoms are all around us, if we only admit their significance: all of the technical and political power we have amassed, and all of the power we may yet amass in coming eras, cannot bring man any closer to security and peace within himself.

Hence we find that social reality constantly attempts to deny the relevance of spiritual reality. This phenomenon is perhaps more marked in the modern era than ever before (though there are, dialectically enough, plenty of "spiritual" movements that go to the opposite extreme in denying the relevance of social reality to spiritual reality). Science can answer all of the really important questions (Comte's positivism); Marxism has dethroned religion permanently, leaving man free to struggle for the perfect society ("modern man has come of age"—Bonhoeffer, 1953; Cox, 1965); what counts is how much food the Brazilian peasant has on the table, not whether he "finds meaning in life." In *A Critique of the New Commonplaces* (1968) Ellul surgically dissects numerous everyday aspects of this social onslaught on the spiritual. Yet spiritual reality is remarkably intransigent. Its uncertainty and impermanence lead man to transform science and Marxism into religions, to make "history" a mythological god that he willingly falls on his knees to worship, to make materialist and empiricist ideologies into ritual. All of these attempts to escape spiritual reality can only be made in vain, for spiritual reality has just as much existence and impact on man as does the social reality he constructs to escape it.

There is, then, a dialectical movement between social and spiritual reality. It is in the interplay between these two levels of reality that the truly crucial issues facing man are to be found, and it is for this reason that all of Ellul's work is oriented around this dialectical movement. We can even say that, for Ellul, the most fundamental contradiction of all, at the human plane, lies here: man's constructive, forward-looking, disciplined social enterprises derive from his need to settle his troubled soul, while his destructive, annihilating, rebellious tendencies derive from the violence inherent in his religious impulses and the agony of never being able to succeed in his social endeavors.[12] Both the contradiction between society and technique and the contradiction between revolution and the state have their roots here.

The five dialectical components presented here lead Ellul to a position he describes as "realism" (*The Ethics of Freedom,* 1976, pp. 332–355). By this term he firmly separates himself from the Hegelian idealist tradition, which in this context refers especially to the construction of an elaborate philosophical *system* that applies a priori principles to the study of reality (see Malantschuk's discussion of Kierkegaard's attack on system-building, 1971, pp. 169 ff). Realism implies rejecting all systematic thought and grounding all analysis in the world as it is. Ellul also separates himself from the Marxist tradition, which to Ellul is equally as idealist as Hegel's work (cf. Gurvitch's discussion, 1962, pp. 155–156). Marxist idealism takes form, first, in its projection of a utopian future that Ellul believes cannot develop from the sociological context in which we live; and, second, in its dogmatic refusal to remain firmly rooted in changing reality, betraying Marx's own principle of constantly referring theory to praxis so that analysis does not become abstract and divorced from reality (Sartre is a favorite target of Ellul in this regard; consider his *Critique of Dialectical Reason,* 1976, in which he attempts to deny the importance of any level of analysis but that of the existentially situated individual). Realism means that one must face the facts: contradictions are inherent in social reality, and if there are any sociological regularities that operate with lawlike force we are bound to admit that utopian futures are to be found exactly where the word itself implies: nowhere (*Betrayal of the*

West, 1978, pp. 148–169). Nothing on the human plane can alter the absolute dialectic; such is the world we inhabit.

The Underlying Dialectic of Spiritual and Theological Reality

So far I have presented only the sociological aspects of the absolute dialectic. In this section I turn to the dialectics of theological reality that underlie the sociological dialectic, beginning with a final sociological aspect—the dialectic of desacralization/sacralization—and then focusing on Ellul's Christian understanding of the ultimate source of dialectics in God's radical No and Yes spoken to man.

1. *The Realm of the Sacred*

Man cannot live without a sense of the sacred (*The New Demons,* 1975). He faces an inherently anxious and uncertain spiritual situation that offers no objective basis for generating answers to the ultimate questions of existence. His response to this precarious condition is his largely unconscious attribution of sacredness to aspects of everyday life. In whatever environment we find man throughout history, whether it be a mostly natural environment or one that is largely the product of his own artifice, he fixes upon elements of his environment and designates them as repositories of ultimate truth and irreproachability. The sacred then serves as a global orientation device, a lamppost or lighthouse that shines its beam on the pathway toward proper action and valid meaning. The sacred gives rise to more superficial but highly significant phenomena, such as myth and ritual, which are the concrete representation of the sacred and reaffirm the ultimate authority of the sacred on a routine basis.

Maintenance of a given sacred order is not automatic. Like all other aspects of sociological reality, the socially constructed sacred is subject to transformation or displacement. Such displacement Ellul calls desacralization, the process by which the sacred order gives way to iconoclastic forces that shatter it by exposing it as one form or another of ideology. Christianity was one such desacralizing force, shattering the pantheistic hedonism of Greco-Roman culture with the Word of a Wholly Other God who loathes man's self-centered religions. Marxism was another iconoclast, shattering the bourgeois Christian ideology that masked severe exploitation and alienation of the lower classes in society. But, as these very examples illustrate and as I have intimated before, a powerful dialectic operates in the sacred realm as well: every desacralizing force becomes, in its turn, the core of a new definition of the sacred (*The New Demons,* 1975). Thus Chrisitianity was sacralized to become a dogmatic, closed, hypocritical religion that eventually contained almost nothing but ideological justifications for the bourgeoisie. Thus also Marxism was sacralized to replace Christianity and blind the world of the twentieth century to new forms of exploitation and alienation. Thus, too, the great progressive iconoclast of technique is sacralized even as it debunks innumerable sacred orders of traditional societies and religions. For, despite the common belief to the contrary, man's need for the sacred is stronger today than ever, precisely because so many desacralizing forces are at work.

The sacralization of iconoclastic forces undergirds many sociological contradictions, for the sacred is invariably employed for all manner of ends. Appeals to the sacred justify murder and war as easily as they justify generosity and peace; think of what has been done in the name of God, or the proletariat, or progress (technique), or revolution. Once again, these ambivalent uses of the sacred are welded together and cannot be separated. And, once again, we find that each movement that shatters the sacred realm contains within itself radical contradictions. When we tap the sacred and unleash the powerful forces structured around it, allowing them to stir sociological reality (whether at the level of the individual, the group, the class, or the nation), all hell breaks loose: the ramifications are unpredictable and complex, spreading out in all possible directions at once.

In probing even further into the nature of the sacred Ellul uncovers another major dialectic. The sacred, at least in the modern world, is itself composed of bipolar axes that contain contradictory forces locked in acute tension. In *The New Demons* (1975) Ellul identifies the axes of the modern sacred as (1) a tension between technique and sex; and (2) a tension between the nation-state and revolution. Both technique and sex are sacred. They are unquestionably desirable; we accept them as part of the fundamental character of human endeavor; they are permanent and unchallengeable. So too with the nation-state and revolution. Yet sex is at the same time the transgressing counterpart to the constructive, anti-entropic character of technique, just as revolution is the transgressing counterpart to the nation-state; these pairs are not so much opposites as they are two sides of the same coin, reinforcing each other and making each other possible (Temple, 1976). Technique binds man in steel grips that he cannot sever. He turns to sex as a liberating force, finding release and freedom in the private world that technique yet permits him. The state closes man up in a thickening network of regulations, enforcement agencies, and systemic technical necessities, but revolution offers man a vehicle with which to crash the roadblocks of the state and burst forth into true humanity (cf. the theology of revolution: Shaull, 1967; Gollwitzer, 1970; Blanquart, 1970).

Man's sacred, which is a sociological phenomenon operating at both the social and spiritual levels, is thus doubly dialectical. From the bipolar axes of the sacred derive the incoherence and self-contradiction of so much of what we observe in daily life. Some of the best examples appear in propaganda, which continually draws on the axes of the sacred to maximize its effects on us. In advertising we see the simultaneous referral to both poles of the sacred axes: high-technology products proffered by stereotypically seductive women; sexual stimulants promoted for their proven technical superiority. In political propaganda we find, along this same line, that the state proclaims itself revolutionary, that obedience to the state is the road to freedom, and that revolution implies seizing control of the state and expanding it to create the better, freer society. All of these contradictions-in-terms appear senseless (is man really that stupid?) until we realize that the sacred is composed of these opposing but interlocked tensions.

2. The No and the Yes of the Theological Dialectic

We come, finally, to the deepest level of dialectical reality. Here Ellul leaves the sociological realm entirely, and it is here that he opens himself up to categorical rejection of his sociological analyses because of the general suspicion we modern intellectuals have of anything that we recognize as based on faith (forgetting that we all, necessarily and without exception, base our ultimate beliefs on nothing more—nor less—than faith). What is the ultimate cause of man's dialectical situation? To put it most succinctly: man's rejection of God. The Fall represents a catastrophic rupture in the order of creation. God created man to love and obey him. When man refuses to accept God's order (Adam eating of the tree of knowledge—*To Will and To Do,* 1969; Cain building the city—*The Meaning of the City,* 1970; Jonah fleeing God's word—*The Judgment of Jonah,* 1971), man enters into what Ellul calls the order of Necessity (*To Will and To Do,* 1969). This is precisely the realm of sociological determinations and dialectical contradictions.

The order of Necessity is what remains to man in the aftermath of God's response to his efforts at autonomy: a resounding, radical, and absolute No! Man is utterly condemned for rejecting God. Without qualification God rejects man's self-reliance because man thereby chooses not to live in the love and harmony created by God for him. God freely gives man that choice. But, given the choice, man must also be prepared to accept its consequences: the will to power and the will to death (the "secret host," *l'idéologie du néant;* see *Métamorphose du bourgeois,* 1967), and the dialectical implications of these demonic forces so well discussed by Nietzsche. The order of Necessity is, as Temple suggests, a sort of minimally tolerable mode of being which God makes available to man so that he can at least live, for man would be doomed to total self-destruction if left completely to his own devices after his rejection of

God. Hence, as long as man rejects God he is subject to Necessity. Sociological forces determine his actions, emergent reality develops according to its own laws completely outside of man's control, and contradictions are routine and ubiquitous. In the order of Necessity man's projects invariably turn upon themselves, and the theological foundation for this profound tendency toward dialectical development is the contradiction within man himself: made to love and obey God, he nevertheless seeks to love and obey only himself and the social structures he constructs. Man's soul is torn by the contradictory situation that he himself chooses. Self-negation is only too inevitable.

But God not only condemns man totally; he redeems him with perfect and complete mercy at the same time. Jesus Christ is God's Yes! spoken to man without qualification. By taking human form and suffering the ultimate humiliation of persecution and death (and, in the Reformed tradition, the death of Jesus really is the death of God) God goes so far as to sacrifice himself completely for man. God thereby assumes the burden of man's rejection and all of the sins that man's rejection implies. Thus man's life in the order of Necessity is made tolerable: salvation is permanently available; God has given man his promise and there is no possibility of God withdrawing it. Here, for Ellul, is the source of man's inquietude. The way lies open before him; God has both spoken and acted in the most dramatic way conceivable to show his boundless love for man, yet man consistently refuses God's love. And we must stress this point: over and over, in the present day through technical development and state expansion as surely as in biblical times, man strives for autonomy apart from God. Yet he is bound to fail.

Man thus lives within the tension of God's No and Yes, between man's rejection of God and man's awareness of his rightful place in obedience to God (cf. Boehme's early development of this idea: Koenker, 1971, pp. 56–57). At the deepest level, then, man lives the absolute dialectic, and all of the other contradictions at higher levels are founded on this dialectic. The tension between the No and the Yes is permanent, unless man returns to God or unless God decides that the moment has come to rid man of his will to power and bring him back to his proper relationship with God.

* * *

For Ellul the manifold contradictions of sociological reality derive from the theological dialectic in which man lives. It would be a mistake, however, to assume that Ellul's sociological analyses are therefore biased or distorted by his faith (or what in current parlance we prefer to call his "values"). Ellul accepts the distinction between fact and value, between analysis and understanding, quite readily, and he maintains that any sociological analysis is worthwhile only if it self-consciously attempts to transcend the values and prejudices of the analyst. One way he attempts to realize this position in his own work is through a dialectical analysis of his own sociology: considering it as a possible ideology, does it mask reality, exclude relevant aspects of reality, serve as self-justification? How well he succeeds in maintaining contact with reality and avoiding the construction of yet another ideology is a judgment I must leave to readers of his work.[13]

The Impending Eclipse of Dialectics

Technique and the state represent, in an ultimate sense, nothing new in that they are but the contemporary manifestations of man's will to power and drive to autonomy. As such they fall in line with virtually all of human history. We might therefore be tempted to say that the triumph of technique and the state is only quantitatively different from previous attempts to close God out from the human plane; but, like other dialecticians, Ellul finds that quantitative change often translates into qualitative change. Such is the case here. The threat posed by technique and the state is not so much that they limit man's freedom severely, for freedom is only quite rarely achieved in any society. Rather, the danger is that they will eclipse the dialectical process entirely.

Technique is characterized by universality and totalization. The technical mode penetrates everywhere, spreading to all aspects of life and all parts of the globe so that nothing is left outside the system of control, measurement, and demands for efficiency. Technique threatens to destroy all alternative approaches to action and thought: morality, art, politics, religion are increasingly reduced to technical terms. In short, no real opposition to technique exists any longer. It is only a matter of time before everything is absorbed into the technical system (*The Technological Society,* 1964).

The same can be said of the state. The state penetrates society ever more thoroughly, working hand in hand with technique to incorporate all groups, activities, and forms of expression within the state-directed technical complex. Resistance against the state is limited and transitory, and most resistance has the unwitting consequence of further strengthening the state in the long run. Again, it is only a matter of time, given the present structure of sociological forces at work in our world, until the state absorbs all. And the growing unity between the state and technique (a unity that is new, for historically one of the important dialectics was that of the opposition between technique and the state; we now find this opposition quite intolerable) points toward the time when, everything human having entered the technique/state complex, all opposition of any fundamental sort will be totally impossible. The issue is not whether, in this situation, people are comfortable, happy, satisfied; the triumph of the bourgeois *idéologie du bonheur* (*Métamorphose du bourgeois*, 1967) suggests that these are precisely the adjectives that will apply, quite validly, to people in this future end-state. The issue is, rather, the development of the end-state itself: a state of being in which life comes to an end.

For Ellul, vitality lies in dynamism. The dialectical process of human existence is, above all, one of movement and change. Dialectical movement is not, in itself, good; the good lies only in loving and serving God. But dialectical movement at least maintains an openness to finding another way. It allows social actors, whether individuals, groups, or nations, to deal separately with the questions of existence. In other words, dialectical movement permits man to struggle toward genuine choice and to find breaks in the fences of sociological determinations through which he can force his way to freedom. The closing up of the human world in the completely monolithic technique/state complex is the end of the possibility of freedom. It is thus the end of any possibility that man can live as God intends him to live: in the struggle for freedom. Certainly the technique/state complex gives man choices; indeed, it kills man with too much choice. But all of the choices that it leaves man are completely superficial and artificial. He can choose among a variety of automobiles, toothpastes, lifestyles, or political ideologies, but they all rest on the same fundamental assumptions about human existence.

The technique/state complex is homogenizing above all. It eliminates differences, oppositions, contradictions at every level. But, because of the severe split between ideology and reality in the contemporary world, we live under the illusion that the technicized society is in fact more heterogeneous than any previous society. We see the mingling of races, cultures, value systems, religions, and political philosophies, and we believe that we have greater choice, more options, and, hence, greater freedom. But there is a dialectic at work here. The underlying homogeneity, arising from technique's ability to transform all historical options into the technical mode and from the state's ability to politicize everything and bring it into the arena of state jurisdiction, is practically indiscernible because of the superficial diversity of our daily lives. We have developed highly institutionalized mechanisms for maintaining this indiscernibility in the mass media. It is only the unusual, the different, the bizarre that constitute "news" or "matters of interest"; but almost nothing of the content of the mass media has anything to do with the fundamental lines of development of the technicized society unless we learn to interpret this surface chaos with an analytical framework. As Ellul puts it, it is in the face of this deeper erosion of differences that technical man, in desperation, most emphatically accentuates the superficial differences that remain (see the discussion of choice in *Le Système technicien,* 1977, pp. 354–360).[14]

But homogeneity is death. Where there is no opposition, there is no struggle; where there is no struggle, there can be no transcendence (on the indispensability of "tensions": *The Political Illusion,* 1967, pp. 206–223). Hence the triumph of the technique/state complex means the closing up of the vitality of human history, and the better perfected the system, the more complete the closure. Utopian thinking, by seeking the "perfect" (but stagnant) society, threatens the dialectical process and would purge freedom permanently from the world. Here is the final mistake of Marxist analysis: it negates its own fundamental reliance on dialectics by seeking to kill the dialectical process, an ultimate contradiction itself. Even the Church tends in this direction by jumping on the bandwagons of secular political movements and statism, and Ellul severely criticizes the Church for thus unwittingly aiding the threat to dialectics (*False Presence of the Kingdom,* 1972). The Church, and all of us, must be warned that in the order of Necessity man is not truly man except in a situation of tension; all else is an illusion or a tragedy.

Personal Implications for Ellul: the Dialectics of the Christian Life

The world is inherently dialectical, but at the human plane technique and the state threaten to eclipse dialectical tensions completely. What is the Christian's response to this imminently disastrous situation? It is precisely in dealing with this question (but by no means in "answering" it) that Ellul makes his most significant contribution to contemporary theology.[15] Barth produced a *Church Dogmatics* (1936–1939) that is quite complete and satisfactory, by Ellul's understanding; what Barth and all of his followers to the present time have failed entirely to do is speak to the daily life of man in the technicized society. Their failure rests on their lack of analysis of the sociological structure of this society, and Ellul harshly dismisses most Christian ethical writings for the weakness of their sociological foundation (Cox comes under frequent attack from this perspective). Let me briefly present Ellul's position.

The role of the Christian in the human world is best captured by the scriptural commandment to be "in the world but not of it." Ellul founds his entire outlook on the tension of this commandment, the importance of which Kierkegaard emphasized in his critique of liberal theology. Ellul reformulates it in terms of the "necessary but impossible" tasks to be undertaken by Christians. Notice the radical contradiction involved here. The Christian must of necessity live in the world; there is no choice, and he betrays God if he seeks only an other-worldly escape into mysticism or the monastery. Yet the Christian cannot possibly live in the world, for it is the site of man's unrelenting sin in rejecting God and seeking autonomy. Ellul explores two main aspects of this necessary but impossible Christian task: living a Christian ethic (*To Will and To Do,* 1969) and playing a revolutionary role as a Christian (*Presence of the Kingdom,* 1967).

A Christian ethic is necessary because God has spoken to man. God has told man in no uncertain terms what his rightful place in existence is. He calls on the Christian to come to that place and actualize it in human life. Man is to love and serve God, and the Christian is to help other men love and serve God just as the Christian tries to do. It is only in loving and obeying God that man can find true freedom, that is, freedom from the rigid sociological determinations of the order of Necessity; and a life of freedom is precisely what Christ shows to man as the Way. Hence the Christian must consider the relationship between all forms of human action and Christ's way. He must learn to distinguish between those actions that love and serve God and those that do not, and he must incessantly struggle to strengthen the former and oppose the latter. So the Christian must enter the realm of ethics.

Yet a Christian ethic is impossible at the same time, both in principle and as a practical matter. It is impossible in principle because, in the words of Paul, "All things are lawful" (*The Ethics of Freedom,* 1976). Christ brings man freedom from the law, considered both as the Mosaic law of the Old Testament as well as all of the other laws that operate at the human level: legal statutes, sociological determinations, the self-negation of politics, the self-destruction of violence. Ethical laws lie at the root of religion. They close man up in blind, dead systems just as the technique/state complex closes man up, and they attempt to eclipse the dialectical situation that man must necessarily live in. Christ brings liberation from religion. He brings it, not in an abstract sense, but in the sense that he shows man the true way of life as necessarily antireligious, dynamic, vital, transcending. Hence any attempt to formulate a "Christian code" is, for Ellul, directly counter to God's word; it is yet another form of man's drive toward autonomy. In a word, it is nothing but sin itself.

A Christian ethic is impossible as a practical matter because man cannot know the good; or, rather, man knows only that the good is obedience to God. But God has given to man no set of moral prescriptions to guide his efforts in discerning the good and seeking to actualize it in the world. All things are lawful, and the difficult task is to understand which things are, again in Paul's words, "expedient and edifying." Given the radical ambivalence of action, the traps of ideology, and the contradiction between action and its consequences, man is always in a state of fear and trembling when he risks judging capitalism, communism, the labor movement, bisexuality, or even technique and the state. These human endeavors are not good or bad in themselves. They are to be judged only in light of how well they may lead man away from the will to power and the will to death into the circle of God's love. Man is hardly equal to the task of making such judgments.

A Christian ethic is impossible in yet another sense that overrides these two. Only God can really change man's condition. All of man's efforts to act autonomously are doomed to failure because man lives in the order of Necessity, whether Christian or not. Hence man can only struggle with all of his abilities, attempting to see the relationship between human action and God's word and giving of himself as fully as possible while yet realizing that, in the end, he can only say, "Not my will, but thine, be done." There is no easy way out of this profound dilemma. The Christian cannot live in the world, but he must.

Let me add the final dimension to this problem. The Christian lives not only in this world at this time; he also lives, every moment with all of his being, in the time of the return of Christ (*Apocalypse,* 1977). The Barthian tradition is fundamentally eschatological. At any moment God could decide, as he has promised, to enter history in a commanding way and vanquish sin entirely from human life. The Christian attempting to live an ethical life must realize that all of his actions, and all of his words, are inevitably small and impotent. He must take life seriously, but there is really nothing serious except God's final entry into the world and the action God will take to restore man to his rightful place. Thus the Christian's life is radically relativized, despite its seriousness. Relative to God, man can do nothing but offer of himself and pray that his efforts are not in vain (this is Barth's position and is generally counter to the tradition of Boehme, Hegel, and such modern theologians as Tillich and Teilhard de Chardin; see Ellul, *Prayer and Modern Man,* 1970).

Ellul views the Christian revolution from this same perspective (*The Presence of the Kingdom,* 1967). Despite the death of revolution in the twentieth century there is still a necessary but impossible revolution to be made.[16] Man must be brought to love and obey God; God has spoken. Achieving this goal is clearly impossible. It has always been impossible, but in the present world it is, if I may so abuse language, more impossible than ever because of the unprecedented power of technique and the state to further man's efforts at autonomy. It is not simply that the ideologies of technique and the state give us the lie that the solutions to all human problems are within our grasp. It is also, and more so, that the technique/state complex has brought to virtual perfection the bourgeois legacy of the Protean society (Lifton, 1968), able to absorb all forms of opposition or contradiction and integrate them into the technical mold (*Métamorphose du bourgeois,* 1967). The effort required to break the will to power and the will to death is stupendous. Nothing less than a total transformation of human activity, at every level, is called for, and this is a revolution of such depth and comprehensiveness that even the contemplation of it must seem like madness (see the discussion in *De la Révolution aux révoltes,* 1972, in which Ellul shows that only rebellion—anguished, desperate, futile, suicidal thrusts against the technique/state complex—is possible). The Christian revolution cannot be made. But God has spoken; it is his implacable will that Christians throw themselves into this project, and of necessity the Christian life must be revolutionary in this radical sense.[17]

How, then, does the Christian proceed, in Ellul's outlook? Contrary to all that common sense tells us, the first step lies in refusing to offer solutions. Ellul's readers are baffled and frustrated by his persistent "failure" to point out the "right" way to act and think in relation to the world. Some critics accuse him of being nothing but a nay-sayer, offering no constructive programs of his own. But for Ellul the work of a Christian is inherently a personal work. Each person is called by God; each person must choose whom he will serve, God or the order of Necessity; each person must work out his own ethics and his own view of the way to achieve the necessary revolution. For God speaks to men and women, not to Man. God's love is directed at people, not at Humanity or History or Progress or any of the other abstract myths that rule human society. Is man to surpass God's will and assume that he can know good and evil and impose his personal views, however well founded, on his fellow men (even unwittingly)? Is Ellul to offer up a program that can then be dogmatized and transformed into a religion, thereby suffering the same self-negation that all systematic programs (not least Marxism, but also liberalism, Catholicism, Maoism, Hegelianism, and all the rest) have suffered?

Because Ellul is so acutely aware of the dangers of programs, he only very rarely (in, say, *The Ethics of Freedom,* 1976, and in *L'Homme et l'argent,* 1954) states explicitly the ethical positions his own personal reflection suggests to him. What he advocates is an intensely personal involvement with the main struggles of our time, an involvement that constantly seeks to strike a balance in the dialectic between the necessity of revolution and the imperative of personal freedom. Any program that does not seek this balance will suffer self-negation, building revolution at the expense of the individual *qua* individual and the individual before God. In his own work he therefore tries mainly to provoke individuals to serious thought and the will to join action to thought so that the split between ideology and reality can be lessened. He seeks to unmask the illusions of our time so that, more cognizant of the real issues and fundamental problems, people can begin to work toward their separate versions of the path (or, better, paths) to follow. In this sense his work is only prolegomenon, and it is highly nondirective prolegomenon that assumes that there is an infinite number of paths that man can follow if only he will let God open them up to him. Thus he clearly distinguishes what he as a Christian intellectual is trying to do from what other Christians, in very different life circumstances, might try to do. For intellectuals, yes: he feels that he has set a general course of action that needs to be pursued further than he can take it. But even here his program is at most very schematic. Continue the work of unmasking illusions, continue the work of recognizing and strengthening the dialectical tensions in society, continue the work of trying to maintain some dynamic openness in a society that threatens to close all avenues of movement and freedom. To prescribe anything more would be to negate his own contribution.

Conclusion

I will close with a few words about the dialectics of Ellul's own work. His adherence to what I have called the absolute dialectic takes concrete form in the dialectical structure of his own work. The inability of most of his readers to appreciate this dimension of his work accounts for many of the misplaced criticisms to which he has been subject. For one thing, a dialectic operates between the two main areas of his work, sociology and theology. He works in tandem, treating a topic from both points of view in separate volumes. It is not quite correct to say that two works on a given topic are complementary; rather, they confront each other. As he describes the process in *Apocalypse* (1977, p. 13), he uses the theological study as an *example* to show the true significance of sociological reality. The theological study explores the foundation of a sociological phenomenon; the example acts as one instance of what God reveals about the phenomenon under study, providing a discourse that man can ponder for an indication of God's will with respect to that phenomenon. At the same time the sociological study confronts theology with an analysis of the crucial issues, on the human plane, with which theology must come to grips. At no time, however, does the interplay between the two works provide a program of action to be followed; their total effect is only to situate man at the heart of God's No and Yes. The resulting tension is the razor's edge on which man must live.

Second, Ellul incorporates the contradiction between the whole and its parts quite fully into his work. As demanding a task as it may be, his work really must be read in its entirety or not at all. For example, he is known as an outright opponent of technique, a "technophobe" of the first order. This characterization has no basis in fact, absolutely none, though it is not an unreasonable inference from reading only *The Technological Society*, 1964. As he himself has said, what he opposes is the religion of technique, not technique per se, just as he opposes the religion of the state or the religion of Christianity or the religion of revolution. The danger of technique is that it becomes a god we worship in place of the true God; but all things are lawful, and in its proper place technique has as much to offer man, even in working toward salvation, as any other human endeavor. As this example shows, we simply cannot understand the complexity of Ellul's thought unless we grapple with his entire work. If we do, there is much we can learn.

The place of the dialectician in the technicized society is a difficult one. With the triumph of Marxism, dialectics has become a fashionable pursuit, but most of what passes for dialectics is little more than a mask for one or another movement that seeks the end of dialectics and a smoothing out of all of the rough spots in human existence. There are some rare exceptions; the work of Gurvitch (1962) comes to mind, as well as that of Adorno (1973) and some of his followers in the Frankfurt school. In general, though, we prefer to live in our comfortable, unidimensional, illusory ideological frameworks, especially in this thoroughly progress-oriented technical society that we call the United States. This may be a blessing in disguise. It may be that the only way dialectical thought can continue to serve any purpose at all is by remaining outside the mainstream as a constant irritant and nagging reminder of the deeper realities of life. Were it to enter the technical system fully, one of the last remaining sources of genuine tension would be gone, and the technique/state complex would only have come that much closer to the resolution of all contradictions that it so resolutely seeks.

NOTES

1. Here I follow Temple's distinction between sociological analysis and theological understanding, a distinction stressed by Ellul in conversation but never fully discussed in his work. Throughout I will use the term "sociological" to include both social and spiritual reality, which for Ellul are aspects of the human plane of existence; they are to be sharply distinguished from theological reality, which has to do with man's relationship to God.

2. Thus even the terms "dialectics" and "contradiction" are very rare in his work, though in correspondence and interviews he more frequently employs them.

3. It is important not to push this notion of inversion too far. Marx was by no means a strict materialist, despite the many Marxists who have taken this path; indeed, his critique of Feuerbach was devoted mostly to a refutation of strict materialism.

4. Gurvitch rightly stresses the contradiction in Marx's work in criticizing Hegel's ascending dialectic while producing one of his own. Proudhon, even more acerbic toward Hegel's ascending dialectic, made the same mistake.

5. Here is where contemporary Marxism reveals its fatal weakness: the older contradiction between bourgeoisie and proletariat has been superseded by new antagonisms, described below, but Marxists continue to believe that nothing essential has changed since Marx's day.

6. These terms should not be taken at face value. Underlying the bewildering disequilibrium of rapid change and abrupt departure, the fundamental contradictions have great continuity. "The more things change . . ."

7. I borrow the term from Berger and Luckman.

8. We should not push this analogy too far.

9. It is with some trepidation that I use the term "method" here. Ellul's focus is strictly on the dialectics of reality, not on method. The components I discuss below are general properties of reality; they are sociological realities, not characteristics of a method. His "method," then, consists in approaching his object of study with these components in mind, but with the awareness that the object may not in fact display all of these components; it may even contradict them, in some cases.

10. Ellul has high praise for Althusser's discussion of dialectics (*For Marx,* 1967), and I find many parallels between Ellul's use of dialectics and the "hyper-empiricist" approach of Gurvitch (*Dialectique et sociologie,* 1962), despite the latter's unfortunate choice of terms to describe his approach.

11. Put briefly, unmeasurability obtains because the links between emergent reality and specific empirical indicators are, in many cases, extremely indirect and complex. We can see the links theoretically but sensory observation cannot penetrate to the core of situations; it is locked outside.

12. Stated in this simple form, this description appears to open Ellul to the charge of reductionism. But this contradiction is sociological, not merely psychological, as we can see by viewing it in the context of Ellul's entire work.

13. As should be clear, my view is that he succeeds remarkably well. I am preparing a fuller treatment of this issue in a longer study of Ellul's dialectics.

14. Any traveler can readily see the homogenizing effect of technique. Even in the poorest societies most removed from the mainstream Western culture one encounters technicized sectors that are "just like home"—the Hilton, the banks, heavily trafficked streets, French restaurants, business suits, and so on. One also sees quite readily how determined the Third World is to enlarge its technicized sector as rapidly as possible, until it absorbs all of society.

15. And, I would add as a non-Christian, to contemporary ethics as well. Ellul succeeds well enough in separating his theological understanding from his sociological analysis of the human plane of existence that he speaks forcefully to the atheist and the agnostic as well as the Christian believer. We can all find much food for thought in his ethical reflections.

16. At the sociological level Ellul discusses the problem of the necessary (but impossible) revolution in *Autopsy of Revolution* (1971). The necessity reflects two conditions: "Man must sense to some degree that he cannot endure life as it is"; and "The basic social structures must be . . . incapable of acting to satisfy express needs" (pp. 238–239). The character of the necessary revolution discussed here has no specifically Christian context; it is to be made wholly at the human level.

17. This is not to say that the Christian is to abandon himself to what passes for "revolutionary" movements in the present world; these movements only hasten the growth of technique and the state. The Christian must search for another way.

REFERENCES

Adorno, Theodor. *Negative Dialectics.* Trans. E. B. Ashton. New York: Seabury, 1973. Pp. xxi, 416.

Althusser, Louis. *For Marx.* Trans. Ben Brewster. New York: Pantheon, 1969. Pp. 271.

Barth, Karl. *Church Dogmatics.* 4 vols. in 13. Trans. G. T. Thomson. Edinburgh: T. T. Clark, 1936–1939.

Berger, Peter, and Thomas Luckman. *The Social Construction of Reality.* Garden City, N.Y.: Doubleday, 1966. Pp. vii, 208.

Blanquart, Paul. "The Christian Faith and Marxism in Revolution." In IDO-C, ed., *When All Else Fails.* Philadelphia: United Church Press, 1970.

Bonhoeffer, Dietrich. *Prisoner for God: Letters and Papers from Prison*. Ed. Eberhard Bethge. Trans. Reginald H. Fuller. New York: Macmillan, 1953. Pp. 190.

Cox, Harvey. *The Secular City*. New York: Macmillan, 1965. Pp. 276.

Durkheim, Emile. *The Rules of Sociological Method*. 8th ed. Ed. George Catlin. Trans. Sarah Solovay and John Mueller. Chicago: University of Chicago Press, 1939. Pp. ix, 146.

Ellul, Jacques. *The Presence of the Kingdom*. Trans. Olive Wyon. New York: Seabury, 1967. Original 1948. Pp. 153.

———. *L'Homme et l'argent*. Neuchâtel, Switzerland: Delachaux & Niestlé, 1954. Pp. 218.

———. *The Technological Society*. Trans. John Wilkinson. New York: Knopf, 1964. Original 1954. Pp. xxxvi, 449.

———. *Propaganda*. Trans. Konrad Kellen and Jean Lerner. New York: Knopf, 1965. Original 1962. Pp. xviii, 320.

———. *Métamorphose du bourgeois*. Paris: Calmann-Lévy, 1967, Pp. 302.

———. *The Political Illusion*. Trans. Konrad Kellen. New York: Knopf, 1967. Original 1965. Pp. xxi, 258.

———. *A Critique of the New Commonplaces*. Trans. Helen Weaver. New York: Knopf, 1968. Original 1966. Pp. vii, 303.

———. *Violence: Reflections from a Christian Perspective*. Trans. Cecilia Gaul Kings. New York: Seabury, 1969. Pp. 179.

———. *To Will and To Do*. Trans. C. Edward Hopkin. Boston: Pilgrim, 1969. Original 1964. Pp. viii, 310.

———. *The Meaning of the City*. Trans. Dennis Pardee. Grand Rapids, Mich.: Eerdmans, 1970. Pp. xix, 209.

———. *Prayer and Modern Man*. Trans. C. Edward Hopkin. New York: Seabury, 1970.

———. *Autopsy of Revolution*. Trans. Patricia Wolf. New York: Knopf, 1971. Original 1969. Pp. x, 300.

———. *The Judgment of Jonah*. Trans. Geoffrey Bromiley. Grand Rapids, Mich.: Eerdmans, 1971. Original 1951. Pp. 103.

———. *De la Révolution aux révoltes*. Paris: Calmann-Lévy, 1972. Pp. 379.

———. *False Presence of the Kingdom*. Trans. C. Edward Hopkin. New York: Seabury, 1972. Original 1964. Pp. 211.

———. *The Politics of God and the Politics of Man*. Trans. Geoffrey Bromiley. Grand Rapids, Mich.: Eerdmans, 1972. Original 1966. Pp. 199.

———. *Hope in Time of Abandonment*. Trans. C. Edward Hopkin. New York: Seabury, 1972. Original 1964. Pp. 211.

———. *The New Demons*. Trans. C. Edward Hopkin. New York: Seabury, 1975. Original 1973. Pp. 228.

———. *The Ethics of Freedom*. Trans. and ed. Geoffrey Bromiley. Grand Rapids, Mich.: Eerdmans, 1976. Pp. 510.

———. *Apocalypse*. Trans. George Schreiner. New York: Seabury, 1977. Pp. 283.

———. *Le Système technicien*. Paris: Calmann-Lévy, 1977. Pp. 361.

———. *The Betrayal of the West*. Trans. Matthew O'Connell. New York: Seabury, 1978. Original 1975. Pp. 200.

Foulquié, Paul. *La Dialectique*. 7th ed. Paris: Presses Universitaires de France, 1969. Pp. 127.

Gollwitzer, Helmut. *The Christian Faith and the Marxist Criticism of Religion*. Edinburgh: St. Andrews, 1970.

Gurvitch, Georges. *Dialectique et sociologie*. Paris: Flammarion, 1962. Pp. 242.

Hall, Roland. "Dialectic." In Paul Edwards, ed., *The Encyclopedia of Philosophy*. New York: Macmillan, 1967, pp. 385–389.

Hartman, Robert. "Introduction." In G. W. F. Hegel, *Reason in History*. Trans. Robert Hartman. New York: Liberal Arts Press, 1963, pp. ix–xi.

Hegel, Georg W. F. *Reason in History*. Trans. Robert Hartman. New York: Liberal Arts Press, 1953. Pp. xlii, 95.

Huxley, Aldous. *Island*. New York: Harper, 1962.

Jouvenel, Bertrand de. *The Pure Theory of Politics*. Cambridge, England: Cambridge University Press, 1963. Pp. xii, 220.

Kierkegaard, Søren A. *Fear and Trembling*. Trans. Robert Payne. New York: Oxford University Press, 1939. Pp. xv, 192.

Koenker, Ernest B. *Great Dialecticians in Modern Christian Thought*. Minneapolis: Augsburg, 1971. Pp. 158.

Kuhns, William. *The Post-Industrial Prophets*. New York: Weybright and Talley, 1971. Pp. 280.

Lifton, Robert Jay. "Protean Man." *Partisan Review*, Winter, 1968.

Lefebvre, Henri. *The Sociology of Marx*. Trans. Norbert Guterman. New York: Random House, 1968. Pp. 208.

Malantschuk, Gregor. *Kierkegaard's Thought*. Ed. and trans. Howard Hong and Edna Hong. Princeton, N. J.: Princeton University Press, 1971. Pp. x, 388.

Mao Tse-tung. *On Contradiction*. New York: International Publishers, 1953. Pp. 61.

Marcuse, Herbert. *One-Dimensional Man*. Boston: Beacon, 1964. Pp. xvii, 260.

Marx, Karl, and Frederick Engels. *Manifesto of the Communist Party*. Trans. Samuel Moore. Peking: Foreign Languages Press, 1965. Pp. 83. Original 1848.

MacTaggart, John E. *Studies in the Hegelian Dialectic*. New York: Russell and Russell, 1964. Pp. 255. Original 1896.

McHale, John. *The Future of the Future*. New York: Braziller, 1969. Pp. ix, 322.

Menninger, David C. "Jacques Ellul: A Tempered Profile." *The Review of Politics* 37 (1975): 235–246.

Mesthene, Emmanuel G. *Technological Change: Its Impact on Man and Society*. Cambridge, Mass.: Harvard University Press, 1970. Pp. ix, 127.

O'Brien, George Dennis. *Hegel on Reason and History*. Chicago: University of Chicago Press, 1975. Pp. 184.

Sartre, Jean-Paul. *Critique of Dialectical Reason*. Trans. Alan Sheridan-Smith. London: NLB, 1976.

Schepers, M. B. "Dialectical Theology." In *New Catholic Encyclopedia*. vol. 4, pp. 842–843. New York: McGraw-Hill, 1967.

Shaull, Richard. "Revolutionary Change in Theological Perspective." In Harvey Cox, ed., *The Church and Revolution*. New York: Association Press, 1967.

Skinner, B. F. *Walden Two*. New York: Macmillan, 1948. Pp. xvi, 301.

Teilhard de Chardin, Pierre. *The Phenomenon of Man*. Trans. Bernard Wall. New York: Harper, 1959. Pp. 318.

Temple, Katherine C. "The Task of Jacques Ellul." Unpublished doctoral dissertation. Hamilton, Ontario: Dept. of Religion, McMaster University, 1976. Pp. xi, 466.

Wilkins, Burleigh Taylor. *Hegel's Philosophy of History*. Ithaca, N. Y.: Cornell University Press, 1974.

TECHNOLOGY AS THE SACRED ORDER

David Lovekin

Through the recent works of C. G. Jung, Ernst Cassirer, Claude Lévi-Strauss, Mircea Eliade, and Jacques Ellul, it is evident that primitive or mythical thought has become a major concern for an important segment of the modern world not merely confined to the discipline of anthropology. All of these thinkers are convinced that primitive thought plays an important role in the cultural world of modern man, a startling conjecture if we are to regard the modern world as essentially the result of scientific, mathematical, and technological activity. Common sense dictates that the modern world is rational and that the primitive world is irrational, but this view presupposes, among other things, that the primitive world is structured around myths, taken as false accounts, and that the modern world is structured by science, taken as truth. The above thinkers, however, have done much to nullify this view.

I will here briefly discuss the relationship between primitive thought and

Research in Philosophy & Technology, Volume 3, 1980, pages 203–222.
Copyright © 1980 by JAI Press Inc.
All rights of reproduction in any form reserved.
ISBN: 0-89232-102-4

the modern world from perspectives suggested by these men. They all agree that primitive thought as a specific form of consciousness is ordered around the sacred object and that it must be distinguished from scientific thought which is ordered around the desacralized object. In this sense, scientific thought and primitive thought are in opposition. It is important, however, to determine the extent of their opposition. Lévi-Strauss, for example, sees mythical consciousness as providing the ground for classifying activity which ultimately leads to scientific classification, much as the two differ; myth begins with the whole and derives the part, while science begins with the part and derives the whole.[1] Eliade,[2] like C. G. Jung,[3] understands primitive consciousness as providing modern man with certain all-pervading structures of consciousness which may differentiate modern man from primitive man, while at the same time unifying them. For Eliade and Jung, however, these structures must be raised to the level of self-consciousness, lying as they do in the unconscious. These structures, then, provide a basis for all theoretical activity. Cassirer, following the lead of Rudolf Otto, argues that the root of primitive experience is centered in the sacred, around which the world is ordered in an immediate totality, uniting the subject of experience with the object of experience as the sacred object becomes all-pervasive.[4] Primitive myths and rituals point back toward this center of unity which often appears as a pole of the profane. For Cassirer, this original unity of consciousness was spontaneous and vital, but as he notes in "Judaism and the Modern Political Myths,"

> Myth [in modern politics] was no longer a free and spontaneous play of imagination. It was regulated and organized; it was adjusted to political needs and used for concrete political ends. What formerly appeared to be an ungovernable unconscious process was subjected to a severe discipline. It was brought under control and trained to obedience and order. Myths were brought into being by the word of command of the political leaders. They could be made at will, becoming an artificial compound manufactured in the great laboratory of politics. The twentieth century is a technical century. It invented a new technique of myth and this invention proved to be decisive in the final victory of the National Socialist party in Germany.[5]

From Cassirer's statements, we might suppose that myth is now in the hands of the modern political state, used for conscious ends which benefit the heads of state. Jacques Ellul, while seemingly in agreement with all of the above views of the sacred and the myths which vitalize it, would disagree that this return of consciousness to the immediacy of a sacral world is the result of an order dictated by an Orwellian Big Brother. Rather, Ellul would argue that the reappearance of the logic of the sacred follows a natural progression dictated by the logic of technique itself, as technique begins in rationality and ends with autonomy.[6]

For the rest of this paper, I will consider the relationship between primitive thought and the modern world from a perspective suggested by Jacques Ellul who has convincingly argued that modern society should be considered a technological society instead of a merely scientific and rational society.[7] While all of the above thinkers have provided thorough examinations of primitive consciousness, only Ellul has developed a full-blown notion of technological consciousness. My purpose in this paper, then, will be to examine Ellul's claim that technological and sacral consciousness overlap. Stated simply, Ellul's position seems to be that for an object to be desacralized, consciousness must separate from it. An object is sacralized, on the other hand, when consciousness identifies with it. Objects in the natural world are de-objectified with technical advance, but objects produced by technological consciousness become more real than natural objects: they become sacralized as reality is identified with the technician's vision. Next, I will raise a concern that is outside of Ellul's analysis. Ellul's illuminating account of the technical phenomenon has led me to the conclusion that the body is co-opted in technological activity, and that this co-option contributes to the return of sacral consciousness.[8] More specifically, I will suggest that the role the body plays in the conscious act may help to explain how it is possible that consciousness reunites with its object, in a return to sacral consciousness. In short, it will be held that sacralizing consciousness returns once the body is no longer a possible object or horizon for one's experience. Finally, I will briefly consider the possibility of two sacred orders in the attempt to distinguish more clearly the sacred as it appears in the humanities from the sacred which appears in the sciences. This distinction is warranted by Cassirer's observation that modern myths are strangely inhuman, manufactured, as it were, in the laboratory of technique.

Jacques Ellul's perspective on technology and myth may be found in *The Technological Society, Propaganda,*[9] *The Political Illusion,*[10] and *The New Demons*. I will summarize that perspective in the following manner: the technological society has become a monolithic society admitting only one point of view, the point of view of technological consciousness. Because technology is usually associated with science and with other abstract "rational" activities, technology's nonrational, nonscientific side has escaped us. But, in *Propaganda* and in *The Political Illusion,* Ellul says that the "image"—the manufactured, mediated fact—has replaced the world of immediate facts, restructuring the citizen's view of the state.[11] Modern states rarely rely upon alienating, inflammatory propaganda such as was used during World War II to stir up anti-Japanese sentiment, but now employ unifying propaganda which placates the citizen's world rather than disrupts it. In short, modern propaganda functions to remove the distance between the citizen and his cultural world. Ellul notes that technology's effect on language and communication has been to remove the tension between the signifier and the signified, the word and the object, the speaker and the spoken to, as the speaker's world becomes a plenum of technical phenomena, of mediated artifacts and symbols.[12] Words now merely mean other words, symbols mean other symbols, and Coca-Cola *does become* the *real* thing. The subject and predicate become identities. In short, the image has become reality, and the transcendental aspect of symbolizing has been removed when what we say is what we mean, nothing more, nothing less. Traditionally, symbols carried us beyond their incarnation in words or speech: Coca-Cola, we might say in a more traditional milieu, was *like* the real thing, leaving us with a true decision of whether or not we will purchase the "reality." In summary, it seems to be Ellul's contention in these works that the natural world, the world of nature, has been replaced by the cultural world, the world of technique, and that the distance between the modern citizen and his technological world has collapsed. But there is a necessity in all worlds, either natural or technological, for a sense of being and origin, for a sense of direction in time which points back to what the culture has been as well as forward to what it might become, and for a sense of absolute fulfillment. In the pretechnological world, the natural object often provided the sense of the sacred and its attendent myths, but in the technological world, the natural object, the object not made by man, is replaced by the technical phenomenon which, with an acute dialectical twist, attains the status of the sacred.

II

Before considering Jacques Ellul's account of technology and its appearance as the sacred of the contemporary world, it is necessary to examine the development of *technique* as it evolves toward *technology* and to a completely technological civilization, a process which, as stated above, involves the separation of consciousness from the objects of the natural world. In general we may say that technique refers to any specific means or method used to accomplish a specific end, while technology refers to the general aggregate of such methods rationally arrived at and having absolute efficiency as their goal (*TS*, pp. 19–21). In primitive, nontechnological cultures, techniques usually involve technical operations that are simple extensions of the body. For example, the means or technique employed for turning the soil in a garden would be a hoelike object or tool which effectively extended the arm toward the ground. But, while a study of technique usually centers on such material technique or tool-using, such a study is incomplete without the awareness of its grounding in the spiritual techniques of magic (*TS*, pp. 26–27).

The historic decision of man to affect and alter his destiny by objectifying his desires in the world entailed the recognition that there was no necessity to the natural world order. But it is absurd to think that man came to this decision instantly. Ellul notes that for the primitive magician the elements in the world were all of a piece, revealing, upon inspection, a world order (*TS,* p. 26). The magician's job was to coerce that order to work for him, even though he had to presuppose the necessity of the order before he could use it. It is because of this that magical techniques tended to die within the cultures in which they were born. Any alteration of the technique or ritual could have disastrous effects if sacred necessity were violated. If a magical technique failed to produce a desired result (i.e., if a rain dance did not produce rain), it meant either that the magician performed the ceremony improperly or that the sacred forces were unwilling to acknowledge the magician's request. It would be assumed, however, that the technique was correct, mirroring, as it did, *necessity*. In order for techniques or technical operations to develop into technical phenomena and into the techniques of efficient ordering of a technological civilization, technical objectivity had to prevail. As Ellul notes in *The New Demons,* "The modern world . . . is a world transformed into an object from which man thinks to withdraw himself so as to act upon it. He expels it in order to control it. He isolates himself from it in order to calculate its techniques. There really is no more sacred" (*ND,* p. 63). In other words, consciousness must separate itself from its object, a project tantamount to desacralizing the natural world and ridding it of its necessity.

In *The Technological Society* Ellul shows technological consciousness as exhibiting seven specific stages of separation from or unification with its object. These stages are: rationality, artificiality, automatism, self-augmentation, monism, universalism, and autonomy.[13] Rationality and artificiality refer to the qualities of the technical mind which demand a mathematics-like exactness and a formal methodology for all hitherto spontaneous activities. This type of ordering, however, having absolute efficiency as its goal, removes the technical man from the natural world and its primitive technical operations and places him within an artificial world of his own making. Characteristic of this stage of technical development is the tendency of natural objects to become technical processes and of technical consciousness to become objectified, taking on a reality of its own. I shall consider the decision to turn the soil, to penetrate the Earth Mother, from the primitive's perspective. First, I must separate the earth from me, disarm her of her powers of harming me, or perhaps consider that she wants me to penetrate her; in either case, I have, in part, asserted myself by objectifying and giving qualities to the earth.[14] I further objectify the earth by taking my conception of it to be more important than the earth itself. This is accomplished when I take a time-honored technique and try to improve upon it, seeking the one best way, thereby implying that the former way is not "*the* way." The whole cloth of the primitive universe develops a cosmic tear. "The way" is now "my way" as long as I seek it rationally and abstract the good qualities of the technique from the bad (that is, the efficient from the inefficient). Employing technical reason, suppose I develop a bulldozer which allows me to move as much soil as I wish. Now I have turned myself, the how, and the ground into abstractions as my attention is focused on the machine which now does the work. As technical consciousness develops and as the body and other natural objects retreat to the perceptual and conceptual horizon, the object of technical consciousness becomes technical consciousness itself. This is the case even though technical consciousness does not know this; technical consciousness must attain self-consciousness for this to happen.

With the development of technical rationality, decisions to deal with a specific problem in a technical way are made automatically and unconsciously, indicating that what may have been, at one time, self-consciousness has now become merely unreflected consciousness. For example, the decision to measure the distance from a worker's arm to his drill press and to construct a technique (perhaps to build a machine or to develop specific arm exercises enabling him to reach that machine more efficiently) leads to an unconscious and automatic decision for efficiency in all areas, even in the world of leisure, as indicated by the recreational possibilities of CB radios, which were originally designed to accommodate emergency situations. This leads to the self-augmentation of techniques wherein useful objects must become use itself: each useful object, say a shovel, has an indefinite number of uses, but as each use an object might have is abstracted and technically refined, a technical phenomenon will result, which will, in turn, engender an indefinite number of other technical phenomena. Thus, a mere shovel yields a bulldozer which may lead to a back-hoe or to gadgets to make those devices more efficient, etc. In this sense, technical development is unlimited as well as unpredictable as it now assumes an unconscious, anonymous, almost magical life of its own. This leads to the characteristic of monism, wherein use is inseparable from being and valuative judgment is inherent in the technique. It is in this way that what can be done will be done, what can be accomplished will be attempted, and attempts to block technical progress will be regarded as obstructionist. Using a technique immorally comes to mean not making any technical use of it at all, all of which reveals technical consciousness becoming one with its object. Technical universalism is indicative of technique's ability to become a culture wherever it is introduced. In pretechnological cultures, culture was determined by geographical location and the amount of natural resources, and by the political, religious, aesthetic, and philosophical attitudes of its people. Because of this, cultures lacked uniformity, but in the modern world, all cultures are developing technological cultures as the natural landscape becomes the technological landscape.

Technology assumes its most monolithic proportions, Ellul states, when it achieves autonomy, i.e., when it assumes the character of the sacred and the necessary, a long progression culminating in the unconscious objectification of technological reason. In this stage, consciousness, now separated from the objects of the natural world, unites with its own conceptions, taking them as ultimate realities. All other transcendental mysteries are banished.

III

Going beyond his discussion of technology as an autonomous sacred in *The Technological Society,* Ellul, in *The New Demons,* provides a logic of the sacred, which further clarifies the autonomy of technique. Ellul maintains that it is the function of the sacred to provide a culture with its direction and meaning. Everything is seen in relation to it. Further, it affords a locus in time and a sense of beginning as well as of transgression—that is, all temporal events flow from it. And, finally, as the sacred structures the time and place of the individual, it integrates the individual into the group. Now, this must be considered in greater detail.

The sacred is the ultimate mystery which is at once threatening and comforting, a *mysterium tremendum,* as Rudolf Otto wrote in *The Idea of the Holy,* but it is an unknown which provides a sense of orientation; all proceeds from it. As Ellul puts it, "The sacred is an organization of action in space and at the same time it is the establishment of a geography of that space in which the action can be undertaken. It is a veritable general topography of the world, involving all aspects of the latter, material and spiritual, transcendent and close at hand" (*ND,* pp. 51–52). For the primitive the sacred was the essence of the natural world, but as Ellul is later to show, the sacred may reside anywhere. Not only does the sacred function as a point of focus and order, but it also functions as the extraordinary, as that which is beyond temporal ordering; sacred time is the time of ecstasy, of the festival of transgression and, hence, of disorder. Ellul states, "It is a time between the times, a silence between words, a plunge into the absolute origins, which one must *come out of* in order to begin. It is a plunge into chaos which one must come out of if the order is to have force, virtue and validity. It is a delimitation of the time during which the dark powers can act, an opening into that which man distrusts but cannot eliminate" (*ND,* p. 52). The sacred serves an integrating function as well. It is the ultimately transcendent which is, at the same time, the wholly immanent, the tie which runs throughout the social collectivity, making its unity possible.

According to Ellul, the above functions of the sacred appear within specific forms: the sacred appears as unpredictable, dark and destructive, a concentration of all that threatens man, but it is that which saves man as well, providing a sense of absolute meaning and justification; the sacred and the absolute values and rites of commitment attendant to the sacred are always incarnate in a person or in some object, and in this sense provide a transcendent that is immanent; and finally the sacred is the unity between antitheses, between such opposites as the ". . . pure/impure, holy/blemished, cohesion/dissolution, profane/sacred, respect/violation, life/death" (*ND*, p. 57). Ellul goes on to suggest, as I have anticipated above, "In the world of the sacred, man is related to the world directly. There is a lack of distinction of subject and object, an immediacy of relation, an experience of totality" (*ND*, p. 65). This unity of subjective and objective aspects of consciousness is reflected in the symbolism of the sacred which intends as identification between subject and predicate. This unity, like the conscious unity mentioned above, is not a self-conscious unity; the symbol is not known to be a "symbol," taking "symbol" in its common sense as a word, idea, or object *made to refer* to another object—in this case the symbol *is* the object.[15] Once the symbol is known to be a "symbol" the symbol loses its power and is no longer able to vivify the sacred.

It is important for Ellul's case that we distinguish between the sacred and the myths which vivify this form of consciousness. The logic of the sacred is a logic of identity between subject and object, between the world and a particular account of it. Myths may, on this view, be seen as spin-offs of this mentality. Ellul adds, however, that, "Henceforth the myth is seen both as the subconscious stand taken by the human collectivity toward the structures, and as the meaning which it attributes to them" (*ND*, p. 94). Thus, myth needs to be seen as an expression of the sacred and not as the structuralist Roland Barthes would view it, i.e., as a form of rationalization. Barthes is aware of the vitalizing force of myth as a power of joining, say, subject and predicate in the metaphor, but his analysis of language presupposes a language where such a split has already occurred (*ND*, p. 93). That is, Barthes is able to speak of modern myths but not of myth as a direct expression of the sacred or as a form of mythical or sacralizing consciousness. For Ellul myths appear as unauthored sacred symbols, losing their power only as they become detached from mythical consciousness (*ND*, p. 97). Myths as vivifications of the sacred are neither revolutionary nor conservative, inasmuch as they are tied to the sacred (*ND*, pp. 96–97). And from these facts we may infer that "The organization of the sacral world is an organization of the actual world in which man lives. Myth, on the other hand, is a fictive statement about a reality in connection with a given portion of that world. Hence, the sacred keeps man constantly at the level of the real, whereas myth leads him into a fictive universe" (*ND*, p. 121). In other words, Ellul is saying that the sacred is composed of immediate experience and that myths are mediated or, at least, that they point man to the mediated. It is in this sense that they lead man away from the "real."

Seeing that technology has become the sacred is possible with a consideration of the following: technique has been allowed to remain one of the major mysteries of our time, a mystery that we have decided to respect (*TS*, pp. 141–142). Nature, on the other hand, is an object for exhaustive study as a potential for conquest. That which was once taken as real must now become rational in a technical sense. Technical development, however, takes on a life of its own as all must conform to its demands. It is well known that industrialization was taken on with great forboding, that, initially, the worker's attitude toward the machine could have been described as admiration mingled with terror. It may, in fact, still be so, for it is not uncommon to hear modern man deprecate his technological life and its evils of pollution, overcrowding, depersonalization, global extinction, etc., while at the same time shaking his head and replying: "What can one do? It's the way things must be." This attitude, Ellul notes, is similar to the primitive as he faces the absolute, the *mysterium tremendum* (*ND*, pp. 50–52 and pp. 70–74). The gods are fearful, the gods are harsh, and their demands are exacting, but without them the universe would have no meaning. According to Marxist interpretation, technology was viewed as the proletariat's salvation, indicating an almost Messianic respect for this force. Today, however, all political forces see their economic and political futures in terms of technical innovation, a true ground of being.

The mass media reinforce technique as the sacred. With television the miracles of technology are rendered visible, while retaining the essential mystery; the world of immediate reality pales before the image from Pakistan or New York, the faraway is produced here and now—a denial of space and time itself—but in this denial, the mystery is preserved as the transcendent is made immanent. Any time, any space may be produced at the technician's will, resulting in a true eternal moment. But, in this sense, the mysterious also becomes the familiar:

> This somewhat mysterious, yet completely scientific power, which covers the earth with its radio waves, wires and paper, is to the technician an abstract idol which gives him a reason for living, and even joy. One indication, among others, of man's sense of the sacred in technology is the care he takes to treat it with familiarity. It is well known that laughter and humor are frequently a person's reaction in the presence of the sacred. That is true of primitive peoples, but it is also the reason why the first A-bomb was called "Gilda," that the giant cyclotron at Los Alamos was named "Clementine," and that radioactive contamination is called a "burn." The technicians at Los Alamos rigorously banned the word "atom" from their vocabulary. All that is significant (*ND*, pp. 73–74).

The familiar, however, is no longer the sensory immediate of an individual with his own past, present, and future but is the manufactured, eternal present of transformed mass-man. It is in this sense that the media as pure technical phenomena serve an integrating function, connecting each individual and forming the fabled global village of modern, retribalized man, a community grounded in the ephemeral process of technique, a creation of realities out of mere symbols, a true creation *ex nihilo*.

Through the impact of technique, modern man comes to see himself as his own creation, as *Homo faber* (*ND*, p. 74). In his role of *Homo faber* man comes of age; man of pretechnological cultures is, thereby, not man at all. Civilization becomes equated with technology. This leads, as Ellul notes, to some bizarre views of the past, for example, to those of Von Daniken, author of the best-seller, *Chariots of the Gods?*, who, with a true technician's mentality, explains that the so-called mysteries of the past, such as the Sakkara pyramid in Egypt, were really built by technologically superior spacemen from another planet.[16] Obviously it is impossible to build such grand edifices without the aid of advanced technology—or so the faith of technology would have it.

So far I have briefly suggested how the sacred, technique, may be regarded as the mysterious, the awe-inspiring, the center of being, the beginning and the future of time, and the principle of integration of the individual into the group. It now remains to show how the sacred incorporates or unifies order with transgression, i.e., to mention how the sacred is incorporated into societal rituals as well as how it becomes incarnate in an individual. The ritual, of course, makes this incarnation possible.

Sex, according to Ellul, is one transgression pole of technique.[17] Sex is desacralized when it is no longer considered a valuable activity, meaningful in and of itself. More importantly, however, "Sex is no longer a natural, free sphere of activity. It is an instrument of strife, a struggle for freedom" (*ND*, p. 77). More specifically Ellul notes that sex, at least during the late 1960s and early 1970s, functioned as political technique. Referring to the student take-over of an administration building, Ellul notes: "In May of 1968 I saw in a faculty council room a very significant inscription: 'This place has been desacralized. These chairs have been fucked on' " (*ND*, p. 77). As may be noticed, here we have a transgression that is not a transgression but the unification of seemingly opposing poles, a traditional function of the sacred. One may have transgression but not transcendence, for the sacred must remain immanent. In the above, the seemingly oppressive and rigid order of the university is attacked with sexual technique and not with free sexual expression. Pursuing this matter further and extending Ellul's account, as sexuality leaves the natural world and is dominated by the cultural world, Eros loses its grasp on the whole, which in the primitive world was grounded in the natural world, and falls back to the world of the particular and to the world of technique. Now, as sex is no longer meaningful as a natural act, it must be made meaningful by correct performance in accordance with the one best way. This view is corroborated by the proliferation of sex manuals, indicative of a culture that no longer knows if sex is pleasing.[18] In addition, we see that spontaneous and bodily sex must be transformed into a rational activity if it is to be one of the poles of the sacred, and that the nature of the opposition is really technique versus body, an opposition which, as I have indicated, concerns technique greatly.

It could be said that technicized sex has become a major ritual leading to the sacralization of the individual and his acts, but Ellul sees the activity of the consumer as providing richer territory. Following a clue left by Lévi-Strauss in *The Raw and the Cooked,* Ellul notes that consumption may be connected to man's most primitive religious attitudes. I would extend the notion of the consumer, further, to regard the tourist as the consumer's consumer, whose sacralizing activity consists of viewing a self-contained system of sights and markers and of honoring the space that technological rationality has laid out for it, space which has become a technical phenomenon. Sight-seeing, in addition to its honorific pose, helps to relieve the alienation caused by the former industrial society by the "museumization" of work.[19] That is, alienating labor (which, in an industrial society, sets men against themselves and each other by removing the laborer from an intimate relationship with the product of his own labor, by reducing him and his fellow laborers to functionaries, and by numbing his senses with dull, repetitive tasks) is resolved in the sight-seeing act. The sacred and the profane, the technical and the bodily, have been united. This is evidenced by such diverse phenomena as the Museum of Science and Industry as well as by a large picture window on a pizza parlor, enabling passers-by to watch work being done. Productivity becomes sight-seeing, as machines do the work and as man is reduced to a mere spectator in the process. Sight-seeing as a full-time, sacralizing activity relieves the possible frustration inherent in this seeming life of inactivity. Sight-seeing and consuming are often nicely combined in enclosed and self-contained shopping centers, the possible worlds of the future. In this way one is able to move about and be near the buying without actually having to buy. Sacred space, it may be hoped, rubs off.

Myths, as vivifications of the sacred, have to be viewed as unique to the culture in which they appear. These myths, either primary, secondary, tertiary,[20] etc., must appear as the culture's collective unconscious which, in our technological culture, is well represented by the mass media which are anonymous yet personal at once: "Anonymity can no longer be assured by ancestral tradition in a society geared to the future and rejecting continuity with the past. The anonymity is now assured by the mass media. The someone who carries the story to all, the someone who is completely known and completely anonymous and is assimilated to the 'no one' speaking in the myth is, par excellence, the television announcer'' (*ND,* p. 97). The mass media do not, in themselves, produce myths but, rather, guarantee their authenticity (*ND,* p. 97). As McLuhan and, now, Ellul have shown, the mass media and their cultural disruptions, resulting in the almost irrational return to the immediate and in the rejection of traditional, linear thinking, and in the loss of memory, serve to show that man has not demythified. Rather the irrational, appearing in the logic of the sacred—a logic of identities without difference as a result of mind unifying with its objects—is very much a current reality (*ND,* p. 97). According to Ellul, the two major myths are history and science (*ND,* pp. 98–109). History becomes mythical as it loses any sense of meaning outside of the mere flow of time itself. History becomes a mere accounting of events. It no longer begins with Christ's incarnation into the world, with man's discovery of the truth—a project which has been denied in the realm of the merely finite—and with man's coming to terms with necessity in his attempts to become free. The meaning and structure of the world can now only come from history itself, and at this point history becomes mythical as the truth simply becomes nothing but man's own account of things. History has a meaning simply by having a structure (*ND,* pp. 99–101).

Science, like history, becomes mythical when it comes to believe in the absolute verity of its own accounts, when the scientist no longer finds his experiments verified by the alleged truth but when his experiments become truth (*ND,* pp. 100–101). As Ellul states: ''To the degree in fact, to which objectivity stems from pure methodology, then becomes a state of consciousness, an attitude, an ethic, it becomes a value judgment, an exclusion of every other mode of apprehending truth. That relation to truth introduces us into the mythical. But, more than that, objectivity presents itself as a value which synthesizes all science. It is just that to which the mythical discourse lays claim . . .'' (*ND,* p. 53). Science, then, becomes mythical when it is reduced to scientific method and, therefore, to a technique, and when truth is reduced to the sensually immediate and cut off from a transcendental realm.

IV

Ellul does not consider how it is possible that consciousness separates or unifies with its object. That is, technical consciousness in the first stage of technical reason must separate itself from its object in order to work on it. In other words, it must see the object as truly independent of it, but it must also be viewed as an object with no intentionality of its own or, at least, with no intentionality of any consequence, so that it is able to take on technical intentionality. When the object becomes identified with technical intentionality, however, sacral consciousness has arrived. Now, we need to suggest how that is possible.

As I have argued elsewhere, following Ellul's analysis, technical consciousness is a form of consciousness which attempts to unite itself with the world in such a way that its ideas of the world *are* the world and not merely both similar to and different from it.[21] A quest for such certainty, perhaps as old as consciousness itself, clearly begins with Descartes' fourfold method, a paradigm of the method of technique.[22] First, it is necessary to deal only with the clear and the evident and then to proceed to break down complexities into simples. These simples are reassembled, and the process is perpetually repeated. If this method were followed, certainty could be ours, a certainty modeled after what Descartes took to be the scientific method. As a result of this abstractive process applied in the *Meditations,* Descartes concluded that the human condition contained two irreparably bifurcated elements, mind and body, elements which Descartes tried to unite with recourse to a benevolent God. Further, Descartes concluded that nothing could be known with as much certainty as thought itself, establishing introspection as the principle of certainty. Descartes' primary problem, and this is also the technician's problem, is to explain the relationship between mind and body, idea and object. If the principle of order, i.e., the criterion of similarity, between, say, the mind and the body, is wholly independent of their relationship, as was Descartes' God of *Meditation III,* then the principle is too abstract. How, for instance, would such a God be knowable? The condition for such knowledge was Descartes' *Cogito,* an immediately self-consistent and wholly immanent principle which now suffers from being too concrete. "I think, therefore, I am," is true now that I am thinking it, but is it still true weeks from now? God was the guarantee that it was, but knowledge of God was made possible by this wholly immanent criterion. Hence, the famous Cartesian circle is, perhaps, the result of trying to order a system in such a way that the principle of order was neither hopelessly abstract nor swept away by contingency.

That this is the technician's problem may now be obvious. Once he has separated himself from the world, and once he has separated all of the elements of the world from each other, how will they be put back together? It seems that either the ties are too abstract (art, religion, or philosophy) or too concrete (science, and technology itself). When the ties are too concrete, sacral consciousness is the result.

One way out of both of these dilemmas is to begin with the notion of an embodied subjectivity, of a subject that is objectified, with both aspects being necessary conditions of our knowledge of the world. It is because of my existence as an embodied subjectivity that I am part of the world while being apart from it, a being that both is and is not; my awareness, because of my body, is situated and perspectival. I am not everywhere at once, but I am here and now, and it is partially for this reason that my knowledge is only of the world; it is not identical with that world. However, I wish to suggest that it is my awareness of this condition which may account for my notion of a transcendent principle at all. Because my "I" is objectified, I am both similar to and different from other objects (embodied or not) and from each moment of my own existence. In short, I am both immanent and transcendent, both a being-in-itself and a being-for-itself. Existence, in this regard, is considered a project of being and process in which the subject becomes embodied and the body becomes transcended. The object becomes a resistance for the subject, and their interaction becomes part of a necessary "labor of the negative," to borrow a phrase from Hegel. Thus, mind and body, whatever we take them to be, become conditions for knowing and for awareness. I wish to stress, however, that the notion of a transcendent at all is a result of this laboring negativity and to stress that I am not interested in arguing whether there is a truly transcendent object, or Absolute, or not. Simply, my embodied condition is a condition possible only for a transcendentally immanent object. The experience of "Otherness" is a result of a being which is surrounded by nonbeing. This is, of course, not to say that I, as an aware object, invent Otherness but, rather, to say that Otherness is, in this way, a condition of my awareness.

V

Otherness, whatever its ontological status, appears in experience in a variety of ways, a problem well considered by William James in *The Varieties of Religious Experience*. I am concerned, however, with that form in which a sacred order appears and in which a transcendent is taken as wholly immanent. James sees this as mystical experience and cites St. Theresa's revelation as paradigmatic. Considering her account of the "prayer of union," James quotes: "In short she is utterly dead to the things of the world and lives solely in God . . . I do not know whether in this state she has enough life left to breathe. It seems to me she has not; or at least that if she does breathe, she is unaware of it. Her intellect would fain understand something of what is going on within her, but it has so little force now that it can act in no way whatsoever. . . . Thus does God, when he raises a soul to union with himself, suspend the natural action of all her faculties. She neither sees, hears, nor understands, so long as she is united with God."[23] Avoiding a discussion of the problem of the true nature of mysticism, and, rather, bringing this fascinating account in line with my concerns above, from St. Theresa's revelation we witness nonperspectival experience. More precisely, her body no longer provides an object or horizon—a background—for her experience. Her "I" is no longer similar to and different from other objects in experience. In a sense, objectivity vanishes. Otherness collapses into experience as she becomes it. It is interesting that for St. Theresa, and a perusal of mystical literature reveals similar experiences, the body or some natural object becomes a springboard for such experience. For this reason, however, I am led to conclude that it is the body as a possible or actual object for consciousness which keeps sacral consciousness in abeyance; that is, it keeps Otherness as a phenomenal experience at a distance.

However, it is not clear, in reconsidering the above, that the mystical experiences of St. Theresa would be accurately described, in Ellul's terms, as a return to sacral consciousness, although there is a definite unity between the subject and the object of experience. Her experiences were formed in a world from which the primitive gods had been driven, a world nearly impossible for us to imagine. That is, those who saw the air, the water, the earth, and fire as gods and who lived, in this sense, in a sacral world in which the world was merely a projection of their own embodied conditions, a projection of which they were not conscious, is far removed from our experiences of the world, and possibly from St. Theresa's. But, Ellul wishes to remind us that there are new gods, gods again formed by the projections of man's embodied condition, resulting in technical phenomena with the difference that these projections are, initially, the result of reflective and abstractive thought. How, then, are we to compare the ecstasy of St. Theresa with the awe and fear of primitive man, and what, further, are we to say of the technician's fascination with his own ad hoc versions of the sacred? Ellul provides no clear epistemological key.

Ellul has convincingly shown that the logic of a technological society in no way excludes the logic of a sacral society. He has, in fact, indicated that the destroyer of the sacred is destined to become the sacred, but he has not adequately explained how this dialectical twist occurs. In this regard, I have suggested that the relation of the body for consciousness is a factor in a return to a sacral order which results from the mind abstracting features from the natural world—with technical phenomena as the result—and from the sacral order prior to that activity. The technical phenomenon and the shaman's pole are both examples of attempted bodily transcendence and both, if Ellul is correct, are the results of a sacred ordering in which transcendental Otherness collapses into immediate experience. The technical phenomenon, however, would seem to be the result of a scientific and abstractive mentality, whereas the shaman's pole seems to exhibit another form of intentionality rooted in the cultural, religious, and philosophical perspectives of that society.

I believe it is necessary, in light of the above, to consider the possibility of a logic of the humanities which runs counter to, or at least parallel with, the natural sciences and with a dialectic of its own. Ernst Cassirer in the *Logic of the Humanities* concludes that knowledge is divisible in terms of nature concepts and culture concepts and that the former are the result of the perception of objects as things while the latter are derived from expressive perception. Cassirer states: "The world which the self encounters is in the one case a thing-world and in the other a person-world. In the one case, we observe it as a completely spatial object and as the sum total of temporal transformations which complete themselves in this object; whereas, in the other case, we observe it as if it were something 'like ourselves.' In both cases the otherness persists; but this very fact reveals a characteristic difference. The 'it' is an absolute other, an *aliud;* the 'you' is an *alter ego*. There can be no mistaking the fact that, always, as we move in the one direction or the other, perception takes on a different significance for us and a distinctive coloring and tone . . ."[24] In the cultural world these two world views are always in competition. When one threatens to overtake the other, a "tragedy of culture" results.[25] When "thing perception" predominates, the object loses its "aura," its ability to return the gaze of the observer, and its ability to "look back."[26] When expressive perception predominantes, the subjective world overtakes the objective. However, if these two directions of spirit are irreducible, and if a certain dialectic between the two is essential to the life of culture, is it not reasonable to assume a dialectic within each spiritual direction? The advantage lies in being able to understand that culture concepts, central to the humanities, obtain, of necessity, a different intention and direction from the nature concepts of the natural sciences.

We must recognize that the subject in each world may lose itself in its own objects (Otherness would collapse into the immediate); the poet may lose himself in his images, and the scientist may be taken up by his concepts. That is, each may forget that ". . . the truly great works of culture (whether in the humanities or the natural sciences) *never* confront us as things absolutely fixed and unchanging, shackling and stifling the free motion of the spirit in fixity. Their content has being for us only by virtue of the fact that they must continually be possessed anew and hence continually recreated."[27] The sacred order arises, then, when the object moves back to the subject, when the Absolute appears absolutely. I wish to urge, in conclusion, that the Absolute may appear in two fundamentally different forms in relation to two separate directions of spirit and that the appearance of the Absolute as merely particular is contingent upon the role of the body in relation to consciousness.

ACKNOWLEDGMENT

I wish to thank Professor Carl Mitcham for a number of valuable stylistic revisions, and to thank Dr. Gerald Stormer for many insights and remarks which helped to clarify my thoughts. Finally, I must thank Professor Donald Phillip Verene for his notions of cultural philosophy which have provided a base for much of my work.

NOTES

1. See Claude Lévi-Strauss, *The Savage Mind,* trans. George Weidenfeld (Chicago: University of Chicago Press, 1966). Especially note the following: "We have already distinguished the scientist and the 'bricoleur' [mythmaker] by the inverse functions which they assign to events and structures as ends and means, the scientist creating events (changing the world) by means of structures and the 'bricoleur' creating structures by means of events" (p. 22). In short, for the mythmaker, events are the means for creating structures which are the ends, but for the scientist, structures are the means for creating events which are the ends. Putting this in another way and connecting this insight to a distinction between rituals and games, Lévi-Strauss notes the following: games have a disjunctive effect, while rituals have a conjunctive effect. More specifically, "The game produces events by means of a structure; and we can therefore understand why competitive games should flourish in our industrial society. Rites, and myths, on the other hand, like 'bricolage' (which these same societies only tolerate as a hobby or pastime), take to pieces and reconstruct sets of events (on a psychical, socio-historical or technical plane) and use them as so many indestructible pieces for structural patterns in which they serve alternatively as ends or means" (*ibid.*, p. 32). The unity and reality of the event is presupposed by mythical consciousness which further attempts to unify, say, divergent sense data to vivify that original unity. For example, when football was introduced to the Gahuky-Gama of New Guinea, the game was played for several days running so that each side could reach the same score; an original equilibrium was presupposed to exist between the sides (*ibid.*, p. 32).

2. For example, in Mircea Eliade, *Myths, Dreams and Mysteries,* trans. Philip Mairit (New York: Harper Torchbooks, 1967), note Eliade's realization that man has always seemed to hold sacred a primordial time, a time before time when mortality and sexuality were no problem. Further, the seemingly perennial belief in the power of memory to achieve reunion with that time helps to account for Plato's theory of anamnesis, Freud's unconscious, and Jung's archetypes (pp. 52–54). These views which account for the unity of all men connect each man with men of his present world as well as with men of his past. Consider the concern for the past and for memory of it from another perspective. That is, "In many religions, and even in the folklore of European peoples, we have found a belief that, at the moment of death, man remembers all his past life down to the minutest details, and that he cannot die before having remembered and relived the whole of his personal history. Upon the screen of memory, the dying man once more reviews his past. Considered from this point of view, the passion for historiography in modern culture would be a sign portending his imminent death" (*ibid.*, p. 234).

3. For Jung's theory of the collective unconscious and its manifestation in modern culture see C. G. Jung, *Modern Man in Search of a Soul* (New York: Harcourt, Brace and World, 1933). Also see *Collected Works of C. G. Jung: Civilization in Transition,* Vol. 10, trans. R. F. C. Hull (Princeton, N. J.: Princeton University Press, 1964). Of particular interest in this volume is Jung's "Flying Saucers: A Modern Myth." Jung reasons that flying saucers, which have suddenly reappeared in the skies of modern man, may be interpreted as archetypal mandala figures which, in primitive cultures, served as expressions of unity and harmony. Men at all stages of cultural development need such symbolic expression.

However, "It is characteristic of our time that the archetype, in contrast to its previous manifestations, should now take the form of an object, a technological construction, in order to avoid the odiousness of mythological personification. Anything that looks technological goes down without difficulty with modern man" (p. 328). Men in primitive times, according to Jung, saw lights in the sky during troubled times, and saw them as signs from the gods, but in modern times, apparently, modern man's collective unconscious finds solace in flying saucers, the technological chariots of the gods.

4. Ernst Cassirer, *The Philosophy of Symbolic Forms,* Vol. II, trans. Ralph Manheim (New Haven and London: Yale University Press, 1955). Particularly note Part III, "Myth as a Life Form: Discovery and Determination of the Subjective in the Mythical Consciousness."

5. *Contemporary Jewish Record,* VII, April 1944, p. 116.

6. For the logic of technique see Jacques Ellul, *Technique ou l'enjeu du siècle* (Librairie Armand Colin, 1954); Jacques Ellul, *The Technological Society,* trans. John Wilkinson (New York: Knopf, 1964). All references will be to the English text, hereinafter cited as *TS,* and will appear in the text followed by a page reference.

For the logic of the sacred, see Jacques Ellul, *Les Nouveaux Possédés* (Libraire Artheme Fayard, 1973); Jacques Ellul, *The New Demons,* trans. C. Edward Hopkins (New York: The Seabury Press, 1975). All references will be to the English text, hereinafter cited as *ND* and will appear in the body of the text, followed by a page reference.

7. Ellul never clearly distinguishes the relationship between scientific thought and technological thought, i.e., he has not offered a thorough epistemological account of both, but he does suggest that science, functioning as science and not as technique, respects truth, taken as a transcendental concept. When the truth becomes merely immanent, science becomes technique. See *TS,* pp. 7–10, 142, and *ND,* pp. 98–102.

8. This present analysis, however, is to be taken as an extension of my article, "Jacques Ellul and the Logic of Technology," *Man and World,* 10 (1978), pp. 251–272.

9. *Propagandes* (Paris: Armand Colin, 1962); Jacques Ellul, *Propaganda: The Foundation of Man's Attitudes,* trans. Konrad Kellen and J. Lerner (New York: Knopf, 1967). All references are to the English edition.

10. *L'Illusion politique* (Paris: Robert Laffont, 1965); Jacques Ellul, *The Political Illusion,* trans. Konrad Kellen (New York: Knopf, 1967). All references are to the English edition.

11. *Ibid.,* pp. 49–66 and pp. 96–102. Also see *Propaganda,* pp. 3–87. Although Chapter I of this work deals specifically with the characteristics of propaganda, the integrating function of propaganda to which I have alluded is developed throughout the entire work.

12. *The Political Illusion,* pp. 209, 215.

13. See *TS,* pp. 79–147. Also see my article "Jacques Ellul and the Logic of Technology" for a more complete discussion of these characteristics.

14. Consider in this regard Mircea Eliade's notion that technical activity is, at first, inseparable from mythical perspectives: "It is not our technical discoveries in themselves, it is the magico-religious acceptance of them which has altered the perspective and the content of the religious life of the traditional societies. Nor must it be supposed that agriculture itself, as a technical innovation, could have had such repercussions upon the spiritual ambience of archaic humanity. Within their horizons there was no separation between the tool—the real, concrete object—and the symbol which gave it value; between a technique and the magico-religious operation which it implied. Let us not forget that the primitive spade symbolized the phallus and the soil the telluric matrix; the agricultural action was assimilated to the generative act; in many Austro-Asiatic languages the spade still has the same name as the phallus" (*Myths, Dreams and Mysteries,* p. 138). It is interesting to note that primitive technical activity was, essentially, bodily activity and that it often unashamedly carried with it sexual connotations. However, for the primitive, the world, like

him, was essentially sexual, and man's sexual activity was seen, therefore, to be in harmony with the natural order. But, as Rollo May has indicated in *Love and Will* (New York: W. W. Norton, 1969), pp. 43–45, the modern world is rapidly becoming desexualized, even though sexual technique has become a prime concern. Further, Ellul has noticed that, in the modern world, sex becomes the transgression pole of technique. This paradox, I believe, is the result of technique having co-opted the body.

15. *ND*, p. 69. The notion of the mythical symbol as being identical to its object and as being fundamentally different from conventional linguistic symbols which merely indicate objects perhaps owes its origins to Giambattista Vico; cf. *The New Science of Giambattista Vico*. I have in mind Vico's notion of the "imaginative universals" which have been admirably examined in D. P. Verene's "Vico's Science of Imaginative Universals and the Philosophy of Symbolic Forms" in *Giambattista Vico's Science of Humanity* (Baltimore: The Johns Hopkins University Press, 1976), pp. 302–311, and in Verene's "Vico's Philosophy of Humanistic Imagination: Its Significance for Theory of Knowledge" (forthcoming in *Social Research*).

16. Ellul refers briefly to Von Daniken's account as: ". . . side-splitting in its scientific pretense . . ." (*ND*, p. 135). A direct examination of Erich Von Daniken, *The Chariots of the Gods?* (New York: Bantam Books, 1970), however, suggests that Ellul's remark is an understatement. It is, though, a useful book for uncovering the faith of the technical mind. Consider the following:

"Where did the narrators of *The Thousand and One Nights* get their staggering wealth of ideas? How did anyone come to describe a lamp from which a magical genie spoke when the owner wished?"

"What daring imagination invented the 'Open, Sesame!' incident in the tale of Ali Baba and the Forty Thieves?"

"Of course, such ideas no longer astonish us today, for the television set shows us talking pictures at the turn of a switch. And as the doors of most large department stores open by photocells, even the 'Open, Sesame!' incident no longer conceals any special mystery" [*ibid.*, p. 65].

The conclusion is implied that, since these ideas had to have been suggested to the "ancients," therefore, they were visited by spacemen from a technologically advanced world. Otherwise, such ideas could never occur to anyone; or, such seems to be the technical mind's concept of the imagination. It must also be remarked that this book has sold millions of copies.

17. According to Ellul, the modern sacred has two axes with two poles, one of which represents transgression: one axis is technique/sex, while the other is the nation-state/revolution (*ND*, p. 71). In this paper I confine my discussion to the axis of technique and sex. It is unclear why Ellul separates these two axes, seeming to suggest that the nation-state is, somehow, separate from technique, which is a view that would depart significantly from what he has maintained in, e.g., *The Political Illusion*. If it is Ellul's intention to suggest that these two axes are not, indeed, separate but are merely two different sides of the same phenomenon we call technology, then there is no problem. Perhaps this is why one pole is termed "technique" rather than "technology," even though Ellul is not always consistent in making these linguistic distinctions.

18. Consider Rollo May's remarks in *Love and Will*, p. 40. Of particular interest is the section entitled "Salvation Through Technique" (pp. 43–45). Further, May notes that the rise of a new Puritanism responsible for the lack of pleasure in modern sex is due to the following: *"First, a state of alienation from the body*. Second, *the separation of emotion from reason*. And third, *the use of the body as a machine*" (p. 45). Also see "Eros in Conflict with Sex" (pp. 64–98).

19. See Dean MacCannell's *The Tourist: A New Theory of the Leisure Class* (New York: Schocken Books, 1976). "Museumization" is the term used by MacCannell to indicate work

and labor, or the lack of it, in the post-industrial society. That is, as he notes, sight-seeing has become a major occupation that restores an individual's feeling of totality, which became fragmented in the industrialized world (pp. 17–56). Man can no longer feel at home in nature; his home is now in the technological world, and "tourism" makes that possible. Also see my essay "Degenerate Travel: The World of the Tourist in Technological Society" in *Essays in Humanity,* eds., David Lovekin and Donald Phillip Verene (Dixon: Sauk Valley College), pp. 167–189.

20. *ND,* pp. 96 and 109. Although myth and science are primary myths in that they most strongly vivify the sacred, an analysis of tertiary myths would, likewise, be important. An exact criterion for determining these is not provided by Ellul. He simply indicates the following as tertiary: "At this level we can enumerate (and each would require its own explanation) the myth of the machine, those of hygiene and health, the myth of bourgeois, those of justice and peace, that of the actor or star, of the hero, those of oil and productivity. There are many others" (*ND,* p. 109).

21. This is a major point which I develop in my article "Jacques Ellul and the Logic of Technology."

22. I am here considering Descartes' remarks in the *Discourse on Method* in the *Philosophical Works of Descartes,* trans. and ed. Haldane and Ross (New York: Dover, 1955), pp. 87–93.

23. *The Varieties of Religious Experience: A Study in Human Nature* (New York: Collier Books, 1961), p. 321.

24. Ernst Cassirer, *The Logic of the Humanities,* trans. Clarence Smith Howe (New Haven and London: Yale University Press, 1960), p. 93.

25. *Ibid.,* pp. 182–217.

26. I am using a notion suggested by Walter Benjamin. Of particular interest, in this regard, is his essay, "The Work of Art in the Age of Mechanical Reproduction," in *Illuminations,* trans. Harry Zohn and ed. Hannah Arendt (New York: Schocken Books), pp. 217–251.

27. *Logic of the Humanities,* pp. 193–194.

THE SOCIOLOGY OF JACQUES ELLUL

Katherine Temple

> Since 1935 I have been convinced that on the sociological plane, technique was by far the most important phenomenon, and it is necessary to start from there to understand everything else.[1]

For more than forty years, M. Jacques Ellul, professor of the history and sociology of institutions at the University of Bordeaux, has delineated the world in which we live, by describing the phenomenon of technique. A number of his sociological books have attracted those who have felt a gut reaction against the unquestioned optimism attached to the unrestrained growth of science and technology, and yet who also felt they had no toehold from which to start to think about it. These people welcomed Ellul's stark and sweeping exposure of the forces dominating modern society. As early as 1954, he wrote *La Technique ou l'enjeu du siècle,* and its translation as *The Technological Scoeity*[2] remains his most important book for the English-speaking world. Since then, he has pub-

Research in Philosophy & Technology, Volume 3, 1980, pages 223–261.
Copyright © 1980 by JAI Press Inc.
All rights of reproduction in any form reserved.
ISBN: 0-89232-102-4

lished further specific studies with remarkable consistency and with a combination of breadth and precision. To many, therefore, his most recent book, *Le Système technicien*[3] will generate considerable interest, for here he reworks his original position almost twenty-five years later. To date, it marks the culmination of the task he started in 1935.

The purpose of this chapter is to present a bibliographic study of Ellul's sociological corpus. At the outset, I stress that I am limiting myself to his sociological books, for some confusion surrounds his status as a sociologist. On the one hand, as he himself says, ". . . I explicitly take a partisan position in a debate between schools of sociology."[4] Because Ellul's brand of sociology, influenced by Karl Marx and Max Weber, is at odds with major North American trends, often he is dismissed as either a charlatan or an eccentric, or else he is reclassified as a social philosopher. On the other hand, because he also writes theological reflections, his analyses are cited as Christian apologetics. Many people make the mistake of assuming that one has to be Christian to assess Ellul's sociological descriptions. Such is not the case. Whether or not his accounts are always accurate, he himself frequently underscores his occupation as a working sociologist.[5] Even a cursory reading indicates that his writings as a sociologist warrant detailed attention. As a partial introduction to this substantial thinker, here I want (1) to present the main points of his view of technique, with special reference to *The Technological Society*; (2) to locate his other major social books within that framework; (3) to review the highlights of *Le Système technicien*; and (4) to present some considerations for any attempt to assess Ellul's contribution as a social analyst.

1. Technique

> The term *technique*, as I use it, does not mean machines, technology, or this or that procedure for obtaining an end. In our technological society, *technique* is the *totality of methods rationally arrived at and having absolute efficiency* (for a given stage of development) in *every* field of human activity.[6]

In the preface to the American edition of *The Technological Society,* Ellul offers this succinct definition of technique and it is the one commentators refer to most frequently. Unfortunately, many of them do not take seriously the warning in the first sentence which denies that "technology" and "technique" are synonymous. In fact, the distinction is lost even in the translation of the title of the book. Furthermore, the concise definition of technique itself is more subtle and complex than first meets the eye. One short sentence combines three different aspects—all of which are necessary for an appreciation of Ellul's argument.

In the first place, technique involves the primacy of reason as the highest and governing priority of human activity. What anybody means by "reason" is never an easy matter, and Ellul perhaps does not facilitate matters by spelling out his own perspective on it. From his different writings, however, one can glean what he thinks reason is. The most important thing to note is that he sees it as a tool that carries out a function. As a tool, it can work only within a particular and limited sphere. In addition, it always operates in the same way, albeit in the hands of different masters. It is the faculty by which people are able, first, to put the internal world of the passions into order and keep them in check and, secondly, to put the external world into order. The former, very closely resembling Freud's concept of the *ego*, is the prerequisite of the latter. This second ability means that it constitutes the analytical organ of the mind. It can order and organize—make divisions of experience into concrete observations, distinguish among observations; delineate patterns and relationships among discrete facts; relate means to ends; see options; make inferences, deductions and comparisons; assess internal coherences and make projections from them. From looking at this list of possibilities, it is clear that, for Ellul, reason is limited to the practical realm and also it cannot stand on its own or justify its own activity. It must be directed by something outside itself, in whose service the operations of reason work and find their meaning. Reason, in short, is a relative faculty, but not a nonfaculty.

In order to confront reality, people *need* to use reason, for Ellul thinks that, in Nietzsche's terms, the straight Dionysiac is more dangerous than the Apollonian. To give up the function of reason altogether leads to the retreat into the irrational. At the same time, he warns against the tendency to try to force reason to exceed its limits. Reason becomes *de-raison* when it oversteps its bounds to claim that everything can be totally ordered or totally explicated by reason. The rigorous use of reason avoids the extremes of the irrational fallacy and the rationalistic fallacy. The latter has been the tendency in modern society. Nevertheless, an account of the possibilities of reason is not sufficient to encompass technique, for people have been differently rationalistic at times in the past. To repeat, he says that reason must appear under some guise, must have a direction that cannot be found within itself. What, then, does Ellul see as the factor presently molding the faculty of reason?

This question brings us to the second part of Ellul's definition—efficiency. In everyday language, the notion (if not the practice) is not difficult. It is the procedure of the *Homo faber,* the inventor who streamlines all his methods to a single end. Ellul suggests that the purest and most easily discernible form of efficiency is the machine. The most efficient solution is a form of mechanization that will guarantee the simultaneous working together of the parts, to come up with the desired effect, without leaving any area open to chance, human error or foible. Efficiency is concerned with practical utility, applicability without waste, the insurance of results, systems building, simplification and systematization. In itself, there is nothing dangerous about efficiency and Ellul would never challenge its limited usefulness nor advocate inefficiency as an antidote to modern ills.

The first two components of Ellul's definition of technique are fairly straightforward. The technical phenomenon exists in any society and certainly is not distinctively modern. It is one of the structures of society, the structure of the totality of technology, i.e., machines and machine-like operations, that follows its own patterns of growth and development. At this point, one could well discuss the purpose of the efficiency, such as pleasing the gods, money or happiness. Nevertheless, Ellul explicitly states that contemporary technique has a third dimension that takes it beyond the machine stage, which, by itself, is more or less obvious to all.

The third dimension refers to the phrase "in every field of human activity." For Ellul, this final part introduces something entirely new into our society: for the reader, it supplies the most problems. In order to come to grips with what it means, we have to separate, in Ellul's writings, descriptions of two closely related processes. The first is the sheer proliferation of technique, for a variety of reasons, in our era. There is no need to go into his whole argument, except to say that the net result is the spread of the principles of machine technology into virtually every human activity and reflection. Technical reasoning is so prevalent that people are incapable of thinking outside its boundaries in their search for security and meaning. That is to say, reason in its technical form has become *de-raison.* It has come to the point where it is hard to see that efficient reason or rationality need not be the only form of reason or the only way to organize the world.

Even the multiplication of techniques, taken by itself, does not point to what Ellul considers to be the uniqueness of modern society. The difference emerges when the quantitative change brings about a qualitative change. For this second, allied process, it is not a question of the internal workings of technique, so much as the relationship between the technical phenomenon and the other sectors of the society in which it has arisen.

> But let the machine have its head and it topples everything that cannot support its enormous weight. Thus everything had to be reconsidered in terms of the machine. And that is precisely the role that technique plays. . . . The machine could not integrate itself into the nineteenth century; technique integrated it. . . . Technique has enough of the mechanical in its nature to enable it to cope with the machine, but it surpasses and transcends the machine because it remains in close touch with the human order. The metal monster could not go on forever torturing mankind. . . . Technique integrates the machine into society. It constructs the kind of world the machine needs and introduces order where the incoherent banging of machinery heaped up ruins. . . . All-embracing technique is in fact the consciousness of the mechanized world.

> Technique integrates everything. It avoids shock and sensational events. Man is not adapted to the world of steel; technique adapts him to it. It changes the arrangement of this blind world so that man can become part of it without colliding with its rough edges, without the anguish of being delivered up to the inhuman. Technique thus provides a model; it specifies attitudes that are valid once for all. The anxiety aroused in man by the turbulence of the machine is soothed by the consoling hum of a unified society. . . .
>
> When technique enters into every area of life, including the human, it ceases to be external to man and becomes his very substance. It is no longer face to face with man, but is integrated with him, and it progressively absorbs him.[7]

Technique, in integrating the machine into society and vice versa, moves from being an instrument or an intermediary between man and his external environment (an instrument that helps him from being powerlessly engulfed by that environment) to being the main force behind and running throughout reality itself. Put slightly differently, technique has moved from being one aspect of society within a wider framework to being the whole basis of society—the source and standard of all social relationships. In this sense, Ellul says that technique has taken over from nature as the accepted touchstone of all reality. This does not mean that all of life has become technique: it does mean that all relationships vis-à-vis each other and the technical phenomenon have altered. Because of this change in the role of technique, contemporary people are left with almost no opportunity to find another intermediary for assistance in dealing with the new configuration of reality that shapes how they live.

> Obedience to the plow and plane was indeed the only means of dominating earth and wood. But the formula is not true for our techniques. He who observes these techniques enters another realm of necessity. This new necessity is not natural necessity; natural necessity, in fact, no longer exists. It is technique's necessity which becomes the more constraining the more nature's necessity fades and disappears. It cannot be escaped or mastered.[8]

Ellul thinks (and many reviewers have missed his basic point) that most people have failed to grasp the significance of the radicalness of the change that has taken place within the very fiber of society. Because technique has become *the* social fact, overarching and permeating all other facts, relationships and values (as nature, when it was the milieu of society, used to do), one cannot easily separate it out for description. Nevertheless, it is necessary to see this transformation in order to appreciate what is at stake in our society. Hence the French title. All of *The Technological Society* concentrates on the two sides of technique: its intrinsic characteristics and the unique results stemming from its extrinsic relationships with other structures of society. To bear in mind the parallel between the milieu of nature and the milieu of technique is helpful both in understanding Ellul's definitions and in holding together how he coordinates his vast array of material.

* * *

Before simply pointing the reader to the book, I should comment on some misconceptions about what Ellul is saying. For example, he never says that society has (or even will) become completely technicized. He does distinguish between *la technique* and *la société technisée,* even though the two are tightly linked in that the one leads to the other. "In fact, technique is nothing more than *means* and *the ensemble of means.*"[9] That ensemble gains ascendancy when it is accepted and fostered in all areas of life. Only through such a route has society changed. Since technique and the whole of society are so closely related, there is no contradiction in putting them together. In the perspective of basic problems, however, Ellul sees the possibility of a technicized society as the crux of the issue. As a sociologist, using the faculty of reason, it is a legitimate part of his trade to extrapolate from present trends. Therefore, if he implies that the entire society is already machine-like, it is only to demonstrate the probable outcome, should present direction continue unchecked. It is almost as if he is saying that people collectively tend to achieve what they desire. The question is, "Do we realize what is finally involved in what we now desire?" The picture can be painted only by pushing it to its logical extremes. We have to take seriously his denial that the system or the unification is complete. At the same time, we have to take seriously his claims about the characteristics of the ensemble of means. The two must go together, before it is possible for us to assess the accuracy of what he says.

But how is it possible to have means without end? One can easily dismiss the whole concept of technique as a use of words with no meaning, for the central key, "efficiency," includes within itself the need for a desired goal or effect. It is simply impossible, so the argument goes, to speak of a system that is nothing but means. No matter how rationalized, mechanized, systematized a society may become, it must be at the service of some end—if only comfortable self-preservation or the alleviation of man's estate through productivity. As already indicated, efficiency does start with a goal, but Ellul argues that those goals are finally absorbed, so that technique can continue to function in its own way, regardless of the original motive. "Whatever the differences in its development in England and France, however, the technical consciousness that appeared was identical in both countries."[10] Technique does start with what went before, going back to the original mystery of invention; nevertheless, once in open motion, technique follows only its own set patterns. "It evolves in a purely causal way: the combination of preceding elements furnishes the new technical elements. There is no purpose or plan that is being progressively realized.''[11]

As already indicated, efficiency becomes its own goal and source, with everything else falling by the wayside. In *The Technological Society,* Ellul zeroes in on the description of what is happening, stresses the monolithic tendencies with the problems inherent in them. Only in other writings does he express the final impossibility of finding what people are looking for in this particular allegiance to technique. He refers, for example, to ''the gargantuan of means at the service of the absolute of nothing.''[12]

Whereas he obviously sees that the modern project can exist, he also recognizes that clinging to efficient operation in all phases of life is actually a clinging to the abyss and death. The search for security through attachment to nothingness is a contradiction at the heart of modern society. He does not sound this warning philosophically in the manner of the existentialists nor moralistically in the manner of the Children of God. He states it as sociological projection that, because of the impossibility at the heart of technique, people will not be able to achieve the security and well-being they originally desired, through the search for means ad infinitum. In this part of Ellul's description, he moves beyond saying that it is enslaving (as is a dominant tone in *The Technological Society*) to predicting that the exaltation of technique (including the wild outbursts that are in no way its opposite) can lead to a complete negation of meaning and an eventual collapse. The search for unlimited power is always tyrannical and then self-destructive in a terrible way and Ellul wants to warn about the unrestrained excess of power without purpose of which we are all a part.

The warnings that Ellul sounds throughout his writings lead us to one further consideration, and that is the question of what he wants to accomplish through his sociology. Certainly, he does write with a passion usually missing from detached ''academic'' reports. Still, since he refuses to discuss solutions, it is possible to argue that he is so persuasive about his thesis that he inhibits all action in response. Perhaps his description is so chilling that it leads only to resignation and despair, or else an ''eat, drink, and be merry'' attitude. How does he stand up to charges of undue pessimism—predictable in a Calvinist?

> I am neither by nature nor doctrinally a pessimist, nor have I pessimistic prejudices. I am concerned only with knowing whether things are so or not. The reader tempted to brand me a pessimist should begin to examine his own conscience, and ask himself what causes him to make such a judgment. For, behind this judgment, I believe will always be found previous metaphysical value judgments, such as: ''Man is free''; ''Man is Lord of creation''; ''Man has always overcome challenges'' (so why not this one too?); ''Man is good.'' Or again: ''Progress is always positive''; ''Man has an

> eternal soul, and so cannot be put in jeopardy." Those who hold such convictions will say that my description of technological civilization is pessimistic. I ask only that the reader place himself on the fact level and address himself to these questions: "Are the facts analyzed here false?" "Is the analysis inaccurate?" "Are the conclusions unwarranted?" "Are there substantial gaps and omissions?" It will not do for him to challenge factual analysis on the basis of his own ethical or metaphysical presuppositions.[13]

This plea points to Ellul's insistence on the fact-value distinction in the field of the social sciences—a dimension of his work I shall discuss later. For now, this quotation is important to pinpoint his understanding of the limits and the role of sociology.

> I do not presume to put chains around man. But I do insist that a distinction be made between diagnosis and treatment. . . . Technique presents man with multiple problems. As long as the first stage of analysis is incomplete, as long as the problems are not correctly stated, it is useless to proffer solutions. And before we can state the problems correctly, we must have an exact description of the phenomenon involved.[14]

The distinction between diagnosis and treatment is important, for Ellul's overriding concern, the one that moves him to the sociological endeavor, is a healthy society. Unlike many sociologists, he believes that sociology both can and must pose the problems. It cannot, within its own limitations, provide the solutions. Optimism or pessimism, for Ellul, are words simply not applicable to the discipline. They might be applicable to the understanding of the reality described, but that question is of a different order. The human possibility of intervention into the present course of society is a question of allegiance and attention and not of description. He sums up his own purpose, sociologically, as "a call to the sleeper to awake."[15]

The language of "awake and asleep" is a metaphor to stress the primacy Ellul gives to consciousness and clarity. Clarity is not only the *raison d'être* of any science: it is also a virtue, without which action tends to be at random, counterproductive and eventually destructive. He insists that to give up the attempt to describe technique is to submit to its power, without any challenge. The consciousness of the desirability of alternatives derives only from clarity about what is happening and clarity about what is now the ruling consciousness. In fact, one could say that, for Ellul, clarity (however arrived at) may be the only human equipment left to provide even the possibility of a new intermediary between people and their new experience of reality as technique. Without clarity, there is the danger of being overwhelmed by that reality, even as premodern man felt threatened by the power of nature. At least, previously, people were aware of the dangers and collectively took steps to make society livable.

In our new situation, Ellul emphasizes, we have very little awareness of the dangers and, therefore, also of the possibilities. And that is our problem.

2. Other Books Within the Framework of Technique

In Part 1, I have devoted considerble attention to Ellul's concept of technique,[16] because the rest of his writings constitute neither an advancement nor a revision of that thesis. Neither are they mere repetition. Rather, they provide the refinement of it, by honing in on the impact of technique in relation to specific areas of society—the areas that could only be touched upon in *The Technological Society*. For example, while in the original book he limited discussion of propaganda to approximately twelve pages, in 1962 he published a full-blown study of modern propaganda as it becomes integrated with technique. Actually, most of his output, up to *Le Système technicien,* is an elaboration of some of the themes introduced in Chapters III, IV and V of *The Technological Society,* entitled "Technique and Economy," "Technique and the State" and "Human Techniques."[17] Although each piece of writing stands on its own as a massive sociological undertaking, the overall framework or, in Ellul's own words, "la grille de lecture" is introduced in the first book. Each subsequent book helps to flesh out the argument that society is increasingly shaped by the forces of technique.

Apart from substantiating the original concept of technique, throughout all his works runs a second common thread. That is, his express purpose to debunk the clichés abounding, both in the popular mind and in more scholarly circles, about modern society. With a witty and devastating style, he plays the intellectual gadfly most explicitly in *A Critique of the New Commonplaces*. Here, he takes on twenty-six catch phrases, such as "The main thing is to be sincere with yourself" or "Cultivate your personality: be a person!" and dissects them to demonstrate how much they distort reality and lead up blind alleys. Similarly, in his other descriptions, he goes to some pains to query other influential but misleading accounts of what is going on. He does so neither from the desire to score academic points nor from an abrasive personality! Quite the contrary, in his search for clarity in the interests of all of us who must live in the twentieth century, Ellul wants to strip away dangerous illusions that shroud the real world. Therefore, in this brief survey, I shall focus on the specific areas he deals with and the shibboleths he wants to lay to rest.

* * *

Politically, Ellul sees the single most important trend to be the unrestrained growth of the power of the state. The state now extends itself relentlessly through being ever more tightly enmeshed with technique.

> But our society is more than technological. As the various techniques developed in one ever-broadening sector, at first along parallel, then along converging lines, the power of the state also developed. . . . Because the two phenomena tended to pervade the entire society, they were bound to meet and become assimilated. We can no longer say that "politics controls technology," or envisage the state dissolved into class structures. Each has been altered by the other. Technology has become the basis of political policy; the state has become technologized extensively. . . . For the system has two structural levels: the spontaneous unconscious one on which the various branches of technology develop and interact, pursuing their own life cycle with no intent to produce a particular type of society; and the conscious and voluntary one on which the state organizes society for the purpose of coordinating and utilizing techniques to the greatest possible extent.[18]

In the political sphere, his most important study is *The Political Illusion,* where he presents a twofold argument. First of all, *all* issues have become political issues and all problems are submitted for political solutions. This observation, he says, may be a truism; nevertheless, this development distorts the *relative* nature of politics as the *relative* domain of public interests with several solutions. In our age, the state is considered not just the arbiter, but more so, the definer of what is good and proper in all fields. In order to fulfill this role, the state must be increasingly rationalized, with a double effect—the autonomy of politics as the efficient, planned state and the abstraction of politics away from the concrete toward the realm of easily manipulated images. All facts become political facts, viewed through the spectacles of rational, state efficiency. The second focus in *The Political Illusion* is the illusion inherent in the belief that the modern state is primarily defined or determined either through the ballot box or through citizen participation. The net result is the growth of an all-pervasive, centralized bureaucracy to the detriment of any genuine democracy.

Without going into the details of his argument, we should note Ellul's view of democracy. Obviously, he thinks we are moving in the direction of tyranny, but he does not necessarily think it will be painful. Elsewhere, he refers to "the happy slave."[19] Also, there is no one group at whom to point the finger, for the forces bringing about the technological society, however abstract, are not external to man. They result from collective desires for control and security which we all share. The dilemma consists in the fact that the desire for freedom and control through participation have somehow brought us security without freedom.[20] Ellul indicates that he personally prefers democracy and would prefer movements toward decentralization of governmental authority. An examination of the actual situation, however, shows *no* such movement. All the forces of society tend toward a centrism that precludes other options—except in the wishful thinking of the imagination. Small may be beautiful, but it is not likely!

His own predilection for democracy serves to illustrate Ellul's insistence on the fact-value distinction of the modern social sciences. He constantly refuses to evaluate what he sees as a sociologist (a statement many readers find hard to swallow!). The reason for this stance (in which he follows Max Weber) is an understandable fear. He is concerned lest the establishment of a fact automatically makes it "good" or else the importation of "values" into a description turns one away from the debilitating trends, trends that idealistic rhetoric can only camouflage and encourage. For Ellul, it is the illusions about society that contribute to its untrammeled expansion.

The definition of technique includes, to repeat, "every field of human activity." Bureaucracy or even the state does not villainously creep around in a vacuum. Each is supported and nurtured in the public mind through the media. According to Ellul, all the instruments allegedly designed to enlighten and liberate people have turned to abet the monolithic tendencies in the marriage between state and technique. *Propaganda* underscores the transformation of information, vital for a well-functioning democracy, into its inverse—manipulative propaganda. The subtitle, "The Formation of Men's Attitudes," shows that propaganda is the device whereby facts are used to integrate people into the technical milieu.

> In propaganda we find techniques of psychological influence combined with techniques of organization and the envelopment of people with the intention of sparking action. This, then, will be the broad field of our inquiry.[21]

In this book, Ellul goes perhaps even further than McLuhan in asserting that the medium is not neutral. Propaganda, for Ellul, follows its own laws of effectiveness and directions that do not vary according to different groups using them.

> Whatever the doctrine promulgated by propaganda, its psycho-sociological results are the same. To be sure, some doctrines are more coherent subject matter for propaganda than others, and lead to a more efficient and consistent propaganda; other doctrines—republican and democratic—are rather paralyzing and less suitable. But the only result is the progressive weakening of the doctrine by propaganda.
>
> Conversely, what gives propaganda its destructive power is not the singleness of some propagated doctrine; it is the instrument of propaganda itself. Although it acts differently, according to whether it promulgates a closed system or a diversity of opinions, it has profound and destructive effects.[22]

In short, the results of the expanding media in a technical milieu have not been the "open sesame" of liberalism, so much as a way of adapting people who might otherwise have been ambivalent and leading them to embrace technique.

In a little-known monograph entitled "De la Signification des relations publiques dans la société technicienne'' (The Significance of Public Relations in the Technological Society), Ellul ripped into the processes of group dynamics long before they became a cult in North America. Again, despite all claims to the contrary, he maintains that the practices of ''human relations'' techniques are depersonalizing, render intolerable situations tolerable and lead people to accept their fate willingly.

> [T]he leader progressively draws the group to a certain conclusion and in doing this the group brings pressure on any resistant element which finally yields and agrees because it *does not have the impression* of being commanded but of participating freely in a group discussion. To substitute group pressure for that of the leader certainly has the effect of *facilitating* things but not of increasing individual liberty.[23]

This example pinpoints not only the use of human techniques in propaganda, but also the primacy of adaptation in our society. Although he has been accused of losing his case through overstating it, here Ellul does sound a concrete warning—one we would have done well at least to heed!

* * *

It is clear that what we are adapted to conforms more to Weber's prediction of the spread of rationality and bureaucracy than to Karl Marx's liberation through class struggle and revolution.[24] This observation brings us to Ellul's most sustained comment to date on the Marxist interpreation of history in *Autopsy of Revolution*. This book combines a description of the course during which modern revolutionary movements have ground to a halt, with a critique of the Marxism that inspired them in the first place. One should be careful to explain that Ellul does not here go into a detailed analysis of the man himself, but rather he challenges Marx's view of what constitutes real change. Ellul argues that the belief in accelerating the forces of history can lead only to a tightening of those forces and never to radical change. The debate comes on the level both of the nature of freedom and the concrete outcome of modern attempts at revolution.[25] Primarily, Ellul discusses the way in which Marx's followers have appropriated his system and argues that the excesses and failures of Communist revolutions cannot be dissociated totally from Marx's own position.[26] On this tack, Ellul sides largely (though not 100 percent) with Albert Camus's brilliant polemic, *The Rebel*; for example, he draws heavily on Camus's distinction between revolt and revolution and they share observations about the modern state. In terms of political theory, they both propound versions of the classical anarchist critique of Marx. In terms of practicalities, Ellul believes that the Left has been responsible for spreading Western technique globally and for rendering genuine revolution impotent.

A sequel to *Autopsy of Revolution,* called *De la Revolution aux revoltes* (From Revolution to Revolt), details specific movements in Latin America, Africa and China to demonstrate that none has developed into a revolutionary force. Even though each began as a radical reaction against oppression, the final product is the hastening of a premodern society into the domain of modern technique. Predictably, this position has brought forth much protest from liberals as well as from Marxists (that is, those who read him at all). It goes against the grain. A careful reading, however, reveals that Ellul's position is more radical, and no service is done by distorting facts to fit preconceived molds. Ellul never attacks the struggle against oppression, for it is one impetus for his own studies. He does question, for example, the assumption that a radically different society is emerging in China. As for the argument that he must therefore be a rightist imperialist, one could refer to his refutation in "Anyone Who Says He Is Neither a Rightist nor a Leftist Is a Rightist" in *A Critique of the New Commonplaces.*

Yet another study, *Metamorphose du bourgeois* (The Metamorphosis of the Bourgeois) upsets the apple cart of the Marxist class analysis. Accepting the centrality of the bourgeoisie in the nineteenth century, Ellul defies the *sine qua non* of modern revolutionaries that a simple class analysis still holds in the twentieth century. He is not so naïve as to think there are no longer the rich and the poor, the more exploited and the less exploited. ("Class" in classical Marxist terms has never had such ill-defined connotations.) Rather, he maintains that the bourgeoisie gained ascendancy through their use of the new means of production, through their adherence to technique. It was not the motivation of money (as Weber so ably demonstrated) nor the centrality of capital that won the day. It was the will-to-power actualized through technique and veiled by the ideology of happiness (which he also shows evaporates as lacking in content). Because of their ability to assimilate all that stood over against them, the bourgeoisie did gain hegemony and succeeded beyond their wildest expectations. They succeeded so well that they ceased to exist as a separate class. What has finally arrived is the *embourgeoisement* of a whole society. It is no longer possible to distinguish one class that can be pitted against another for the control of the means of production. In the common desire for an evanescent happiness, to be achieved through the domination of human and nonhuman nature, technique itself has emerged triumphant. Its criterion of efficiency has run roughshod over any other consideration, including either happiness or money. Although some people remain uneasy with Ellul's lack of precision in defining "class," this book is a major examination of the difficulty in discussing class in a technical milieu. If the Marxist class analysis falters, then revolution will have to come from a source other than class struggle and virtually all the major schools of the Western intelligentsia must be reassessed.

* * *

A final blow to the way in which modern educated man sees himself comes in *The New Demons* which combines Ellul's arguments about freedom in the technological society and his opposition to Marxist practice. He approaches the problem via a refutation of the prevalent belief that, as a free and advanced species in control of our life, we are no longer shackled by religion. It is true, he argues, we are not held by the same sacreds, myths and religions as people were in the past. The reason, however, is not an emancipation. The change comes solely from the fact that our environment has changed from a natural to a technical milieu. In detailed chapters, Ellul delineates the modern sacred axes of technique/sex, the modern myths of history and science and, finally, the political religions in which people invest their whole trust, from which they derive norms and through which they are impelled to action. Using a phenomenological approach, Ellul seeks to demonstrate that the structure of belief now is exactly parallel to the structure of belief in the past.[27] The main difference, apart from milieu, is that we have no perspective on our situation, for we feel that we are (or someone is) in control of the technique we ourselves have developed. Because we delude ourselves, it will probably be more difficult for us to establish healthy and stable societies than it was previously. A look at what we have come up with does not inspire confidence.

Whether or not Ellul's account of religion (one in which Marxism, for example, turns into a religion) is adequate is a question beyond the scope of this chapter. It should be noted, in passing, that he does not refer even to the major non-Western traditions that have been called "religious"—most notably Buddhism—and his premodern examples all come from what have been called "primitive religions." His argument, at any rate, does take up themes alluded to in *The Technological Society* and does throw out of balance major sociological axioms that our society is pluralist and that it has "come of age" religiously. Also, he reaffirms the importance of beliefs in holding together society and the interrelatedness of belief and material factors in the development of any society. Perhaps the need to believe that this society is superior to anything that went before is the major stumbling block that prevents people from taking Ellul seriously and, more important, prevents them from looking at the actual situation.

So far, I have examined the fundamental works that comprise Ellul's sociological output. Apart from *Le Système technicien,* one remains and that is *The Betrayal of the West,* a relatively short book that helps to put everything else into perspective.

In *The Betrayal of the West*, Ellul challenges the even more recent intellectual fad of glorifying every tradition except one's own and putting the blame for *all* the evils of the world on the West. Although he himself has launched many a ruthless campaign to expose current trends, nevertheless he still believes that there are unique features worth preserving; for example, hope, the individual, the defense of the weak by the strong. Throughout, he declares his own love for the Western tradition. The problem lies not so much in Western "values" as in the failure to preserve and nurture them. Within this context he attacks the left in a chapter entitled "The Truly Poor and the End of the Left." It is not for him a question of turning to the right wing, for they have no resources at all and no viable alternatives. Only the Left has inherited the mantle of the West and has argued and struggled passionately in the name of those values which have attracted the rest of the world. Therefore, the success or failure of the Left is pivotal. What does he see?

The title of the book and the section called "The Shipwreck of the Left" indicate his pain and disappointment.

> The organized left turned into the kind of general for whom the troops are solely a means of winning the battle. The human reality of the soldier who suffers under him is completely unknown to him. That is exactly the attitude of the left to the poor; in this, the left is only imitating Lenin. On this point, and this point alone, I disagree radically with the left. The trouble is that everything else depends on the right or wrong attitude to the poor!
>
> The left has now become as much of a liar and hypocrite as the bourgeoisie, because it continues to proclaim what it regards as its own great virtue: the defense of the poor. It continues to portray itself as representative of the classes that live in wretchedness, but it is lying. The left defends and supports only what can be of service to it, only the people whom it can use for its propaganda or some form of direct action. It uses the poor exactly the same way capitalism does: it exploits them.[28]

Through the ascendancy of tactics and strategy, the truly poor have been betrayed and, in that betrayal, lies the failure of the Occident. The supremacy of technique, which was to bring liberation from oppression, has turned precisely into the opposite.

> Technique is the present cause of man's enslavement. It is not simply that, of course. Hypothetically, technique could be wholly a cause of man's liberation, just as, hypothetically, the state could be a source of security and justice, and the capitalist economy could be the source of happiness and of the satisfaction of needs. But all that is hypothetical. In reality, the state, the economic structure, and technique have been sources of alienation. Man now feels dispossessed of himself.[29]

The domination of technique, he maintains further, has led also to the betrayal of reason, history, the individual, liberty and love—all that the West has stood for. In their place stands a dead mechanism.

* * *

The decline of the West does not mean the end of the world with the coming of the Apocalypse. This book, contrary to the appearance of some passages in it, is *not* a theological one. It is a sociological treatise dealing with the present evolution of Western civilization.[30] Therefore, we should now turn to his view of the functioning of a healthy society. For any society to remain alive and well, he argues the need to maintain dialectical tensions within it. The history of "dialectics" is complex, and I do not mean to demonstrate Ellul's place within political philosophy. In general terms, he refers to "dialectics" as a view of the whole which consists in relationships among different or contradictory elements which remain held together in an inseparable bond of tension. Each aspect can be seen and analyzed or described only in terms of its opposite pole, while the whole includes the interaction or the field of force between the poles. It is important to note that Ellul does not see dialectics as a specific mode of reasoning, but rather as a description of the relationships involved in the way things are. The result for Ellul's work is that he continually looks to challenge, movement and the forces of relationship in any attempt to articulate the whole.

Sociologically, he considers that societies live through the dialectical play of the inner contradictions of the forces at work in society. Although the content of these forces will vary from civilization to civilization, the common danger remains that the tension and interaction among them will be completely resolved. When the major forces become increasingly unified into a monolith, when society ceases to evolve, it dies. Society then stagnates and will either crumble internally or else collapse under its own weight—with no guarantee of the outcome—or else succumb to external attack. But why does he think a dialectic within society's forces is necessary for its continued existence?

In response to this question, he has spoken first from the perspective of micro-sociology, using the example of the relationship between husband and wife. When there are no more fruitful tensions or no more differences within a common relationship, or when the two become increasingly similar, then the members have nothing more to say to each other; there is no more growth and communication. In other words, the social unit has broken down. Without any internal reasons for existence, the marriage continues only as a result of external pressures: it disappears if the external (usually juridical) constraints disappear.

Analogously, Ellul says that within larger societies there are similar effects. For any society to maintain itself, it must find challenges to realize why it exists, and these challenges stem largely from the interplay of the major forces. For example, he says that France appears less and less important to many Frenchmen, so that the relationships among groups are increasingly those of retreat rather than those of dialectical tension. This route, he sees as one way a society can deteriorate. The second route comes on a wider scale when the major forces that have provided drive behind society become so rigidly unified that society exists only as a mechanism. In both cases (and they usually go hand in hand), there is no internal meaning or challenge, so that society goes into decline. For Ellul, society can find a collective meaning only through dialectical relationships among individuals, among groups and among the dominating forces. When these relationships of fruitful tension disappear, a political system can maintain itself for only a limited time against internal disintegration or external threats. Otherwise, it must be so enclosed that it is no longer a living society. For example, he has predicted that conditions are becoming increasingly right either for anarchy (in the sense of social chaos) or dictatorship. Although he believes that a dead society can continue for some time through force, he does not think it could survive indefinitely. Unlike Marx, he thinks that the collapse does not necessarily lead to a higher stage, and it always proves catastrophic for those involved in it.

As I have already indicated, Ellul states that our society is distinguished by the transition from the use of technology as a tool, through technique as *a* structure, to rationality as *the* definition of reality which tries to transform everything into its own image. In the middle stage, it was possible to discern healthy dialectical relationships between the force of the technical phenomenon and political forces in the incipient democracies. By now that polarization and relationship have integrated to the point where technique and politics are virtually indistinguishable. For Ellul, obviously it is not an issue whether or not technocrats hold office (as one could conclude from *The Political Illusion*); the issue is the criteria for politics. In addition, technique now holds an unprecedented position, for it has undergone apotheosis and is now the dominant axis of the sacred. The double effect ensues that the possibility of dialectical relationships and the likelihood of apprehending what is happening both tend to diminish.

> This determination to eliminate contradictions . . . has eliminated whatever would surely lead to a contradiction. The passion for unity simply annihilated whatever continued to resist assimilation or gave rise to new questions.[31]

Our society, then, becomes increasingly closed, abstracted and stagnant, so that alternatives are continuously more limited.

> We live in a world of limitless repetition, which we like to believe is inventiveness, novelty, a constant new beginning. . . . This inability to innovate except by ringing the changes on signs (not symbols!) is for me the proof that the end of the West is upon us.[32]

Because he sees the danger of society becoming a closed mechanism, he continually confronts technique in order to arouse awareness and to stimulate a determination to keep alive all healthy relationships still possible within society. He thinks that at least the acknowledgment of the implications of unchecked technique is a sign of potential health. Precisely because sustained social dialectics are not inevitable, he wants to warn of the dangers before it is too late. The only antidote he suggests, sociologically, is the encouragement of any dialectical tensions within society to prevent the trend toward total unification.

> What is needed is groups capable of denying the state's right—today accepted by everybody—to mobilize all forces and all energies of the nation for a single aim, such as the grandeur or efficiency of that nation; we must not give in to blackmail as such. If all intellectuals were not used by the state, and if all resources were not concentrated in the hands of the state, and if all private interests did not subordinate themselves to the state, we would never have the Pierrelatte establishment. All right—so much the better. What is needed is groups capable of extreme diversification of the entire society's fundamental tendencies, capable of escaping our unitary structure and of presenting themselves not as negations of the state—which would be absurd—but as *something else,* not under the state's tutelage, but equally important as solid and valuable as the state.
>
> They must, that is, be poles of tension confronting the state, forcing the latter to "think again" and limit itself to considering real political problems without being in a position of omnipotence.
>
> It would obviously be dangerous if groups of that kind were to emerge, and would in a certain sense perhaps reduce the power of the nation, the growth of technology, the economic and military competition with other nations. But this is the condition for life itself. . . . Confronted with such terms, people will shrug their shoulders and the more strictly scientific minds will say: "What is all that?" It is, *for example*, not to bandy words about, not to feed on illusions—as many strictly scientific minds do in our day! Creating such poles confronting the state, one must not forget that opposition factors would be part of the system; to put it differently, they would not aim at disregarding the state or tend to destroy it, but, by returning an autonomous vitality to certain parts of society, would make possible a political life that would be something else than mere illusion. This would provide the state with the possibility of true life. Here the objection will be: "All this is entirely utopian, and in fact it is an illusion to believe that the establishment of such organisms would be possible." I have never said that it is possible. I have only indicated what I consider to be the basic condition for social and political life and the *only* way to escape the political illusion. If one does

> not want to follow it, so be it. The future is clear enough under such conditions. More or less quickly, the political illusion, which is transitory in nature, will dissolve into ashes, and what will be left will be an organization of objects run by objects.[33]

3. Le Système Technicien

> Technique is not happy *to be*, and, in our world, to be the determining or *principal factor:* it has become System. This is what we will try to show in this analysis.[34]

The dangers implicit in the unification of society into a nondialectical mechanism lead directly to Ellul's latest book. Now, he calls technique not merely a phenomenon or an ensemble, but expressly "a system." The implications of this shift form the core of *Le Système technicien*. Whereas the other books I have discussed represent a refinement of the second part of *The Technological Society,* this book involves a thorough reexamination of the original thesis about the intrinsic characteristics of technique itself.

Before moving directly to the basic content, some preliminary remarks are in order. The published text constitutes only one half of the manuscript Ellul prepared as his reworking of the thesis. The second part, yet to appear, will focus on the dysfunctions of the technical system. The next book will cover three parts:

> [First] the absence of feedback in this system with particular reference to the ambivalence of technical progress, the irrationality of the system, the harmful effects, the pollutions, etc. Secondly, the inadequacy of proposed remedies (in particular, the uselessness of the thought of Marx in this domain). Finally, the orientation of the system itself toward the institution of a feedback, for the passage from a process of growth to a process of development.[35]

For many readers already familiar with Ellul, these are the very questions for which they have been waiting. They may be somewhat disappointed and feel it is only an exercise in updating, recapitulation and refinement. To a certain extent, this assessment is justified—but only to a certain extent. Given the nature of the topic and the controversies surrounding it, the twenty-three years intervening are hardly insignificant!

For example, he incorporates and comments on a number of other theories of society that have emerged, with varying impact, in the interim. The introductory chapter stands as a model for a review of literature and scattered throughout the book is fruitful dialogue with thinkers such as Richta, Habermas, McLuhan, Illich, Daniel Bell, Kuhn, Wiener, Kahn. This continued discourse with recent scholars greatly enhances the quality of the original debate. The increase in the number of footnotes can slow down the reading. Yet, greater care will prove an asset, for many readers tend to underestimate the subtlety of Ellul's arguments. One does get the impression, if only in style, that Ellul has perhaps grown more mellow toward his opponents. This is not to say that he has in any way lost his cutting edge. I do mean, however, that his flashing and formidable style (which, incidentally, belies the man) has modified somewhat, so that his discursive method has more direct impact on the reader. Similarly, he deals with certain topics that had not even emerged for discussion in 1954. Most notable is the development of the computer and information retrieval and storage systems. With the exception of Joseph Weizenbaum's *Computer Power and Human Reason: From Judgment to Calculation,* this book is probably the most serious attempt to locate the place of the computer in modern society and assess its sociological impact. In addition, further attention is paid to the Third World, to American developments, the transfer of technology, appropriate technology, ecological systems and other recent pre-occupations. These features, in themselves, make the new edition worthwhile.

As for the refinement of his position, Ellul here pays attention to various questions arising out of his original position. First, there are more explicit references to his own sources and methodology as a social scientist. Still, he has never drawn these completely together, so that readers must continue to piece together the parts of his approach for themselves. Secondly, he answers questions about his tendency to consider technique as a thing in itself or to take dramatic flights that sound anthropomorphic. In a similar vein, he refutes charges that he is hostile to technique. At the same time, he reiterates even more forcefully that technique is not a neutral commodity whose assessment depends on its use for good or evil. Of course human actions are involved, but because of our corporate decisions, he maintains that all our choices are now *within* the technical system. That is the reason why the laws of technique have achieved autonomy. In these areas, he goes to some pains to clear up misconceptions that abound in reviews of the first book. Thirdly, he has tightened up certain arguments that, although always important, have previously been understated. For example, there is use of "concepts," the reality of the technical milieu, the absence of finality in a matter dealing exclusively with "means." The increased precision of terms comes from almost a quarter of a century of reflection on approach and content. (Although not essential, it certainly is a help to have read the other books, and novices may, therefore, find some sections frustrating.) None of the aspects I have mentioned is new: the reformulation is, nevertheless, enlightening.

There is no point in recapitulating *Le Système technicien.* Much of his primary argument remains intact. The most substantial innovation, as I have already indicated, is the emphasis on the technical system. In conversation, Ellul described the relationship between his earlier book and his most recent one. He began *The Technological Society* from a number of factual observations and ended up more or less with the idea of a system. In *Le Système technicien,* he now starts from the concept of a system in order to determine what more can be seen from that vantage point. Any changes are not radical in the sense that they mark a different direction or a repudiation. (That is why I have devoted considerable attention to the first edition in this article.) They are inherent in the first version—only not spelled out. We must now, therefore, address ourselves to what he means by a "system."

* * *

With the language of systems, Ellul moves most definitely into the area of professional and scientific usage—an area that, at first glance, may appear to laymen as either esoteric or jargonistic. Ellul himself denies that he is simply jumping on the latest bandwagon of the social sciences that is popular because it makes any study look more "scientific"! It is, in fact, the case that a close study demonstrates that the concept of a system delineates more scrupulously and expands on what he has previously implied by the word "ensemble." Because of the prevalence of systems analysis as a scientific tool, it is important to see Ellul's precise appropriation of it.

> The system is an ensemble of elements in relation to each other, so that every evolution of one part provokes an evolution of the whole and every modification of the whole ensemble has repercussions on each element.
>
> The second characteristic I shall retain is that the elements composing the system present a sort of preferential aptitude to combine more among themselves than to combine with external factors.
>
> The third characteristic is that a system that can be grasped at a moment in its composition is, nevertheless, dynamic: the interrelations of the parts are not of the same type as those that exist among the parts of a motor which act on each other and where one part is a function of the other, but which repeat the same action indefinitely: in a system, the active factors *modify* the other elements and the action is not repetitive but constantly innovative. The interrelations produce an evolution.
>
> The fourth characteristic is that this system existing as an entirety can enter into relationship with other systems, with other entireties.
>
> Finally, it is well known that one of the essential traits is the feedback or the "structures of referral" without becoming a system itself.[36]

These are the general characteristics for any system and Ellul asks us to bear them in mind for an analysis of the particular example of the technical system.

The basis of the book is the argument that the specifically technical system is constituted by the coming together of the technical phenomenon along with technical progress, so that the bulk of the book describes these two dimensions. The part entitled "The Characteristics of the Technical

Phenomenon'' corresponds very closely to sections of *The Technological Society,* where he devotes attention to the internal workings and laws of technique. There is little need here to repeat the arguments concerning the autonomy, the unity, the universality and the monism inherent in technique. More fully developed is Part III, ''The Characteristics of Technical Progress,'' where he stresses the unlimited expansion of technique. (Anyone familiar with Ellul will rightly assume that he does not mean ''progress'' as ''things getting better'' so much as ''more of the same.'') Each technical change brings in its wake a whole series of related changes (each of which can be absorbed only technically) and problems (each of which can be dealt with only technically). Therefore, technique develops in geometric proportions, with no apparent end in sight; that is to say, as the pace accelerates, there are no built-in checks on how far the process can go. Such a movement absorbs more and more fields that previously were removed from such a system. For example, he maintains that it is quite inaccurate to view leisure as an antidote to increasing technicization, as do so many authors. Not only are forms of recreation increasingly dependent on technical apparatus, but also, more important, leisure serves mainly to make the functioning of the system possible by giving the illusion of freedom from it. This is only a single instance of what Ellul means by the expansion of technique which contributes to the ''perfection'' of the overall system.

In short (and perhaps a bit simplistically, for the whole book is an examination of the thesis), again we see the double aspect of technique—its inner life and its expanding relationship with external factors and sectors of society. The two are no longer simply concomitant or working to change the face of social reality and to raise questions about liberty. By now they are so integrated and coordinated that they are more correctly seen as a system. The second version of Ellul's book follows how that evolution has taken place.

Cérézuelle, a former student of Ellul, has summed up the technical system most succinctly. To avoid the risk of plagiarizing what has already appeared in this publication, I shall merely quote what he has said.

> But while it establishes a new milieu that is as demanding as nature has been for human beings of other civilizations, modern technology tends to establish itself as a *system*. Like living systems, modern technology is characterized by an autonomous mode of growth which we can label as ''self-augmentation,'' each technical progress engendering new innovations. It is also characterized by a growing interdependence of technological elements, with each sector becoming more and more sensitive to the changes which affect other technological sectors. Such an interdependence which evolves into a rational integration that is progressively pressured by the different aspects of technology has clear social consequences. Technological society now tends to establish itself as a coherent, unified, and totalitarian system. However, Ellul modifies his model by saying that from a global viewpoint technology does not

establish itself as a complete system along the line of cybernetics or systems theory, because there is no coherent or effective "feedback" when technology interacts with another natural, human, or social milieu.

Thus the growth of technology can only multiply irrationalities and malfunctions (harmful effects, pollutions, etc.) to an increasingly global level. Ellul agrees with Bernard Charbonneau that technological society generates a true dialectic of system and chaos which leaves modern man with only the choice between progressive social and ecological disorder or totalitarian organization—if not both at the same time![37]

* * *

From this summary, it should follow that *Le Système technicien* in no way alters the general tenor of Ellul's analysis. As I have already indicated, however, it does add considerably to his corpus. In particular, the centrality of technique as system elucidates many of his other writings, even if it does not change their conclusions. In order to round out this section, I shall point to three areas in which this book raises certain questions and problems to a new level.

First, as seen in the quotation from Cérézuelle, the dangers attendant on the unrestrained growth of technique appear as even more grave. *The Technological Society* stressed that our adherence to technique involves a relinquishing of freedom in favor of security. One often got the impression that technique never malfunctioned and that its spread could well be a painless (albeit enslaving) one. Now he states (whereas previously he merely alluded to the possibility) that even this security is at stake. Because of the lack of feedback, the technical system is much less stable than most of us have envisaged. Earlier, Ellul has more than hinted that there is a tendency, constant enough almost to be formulated as a sociological law, for monolithic societies to collapse. Still, the *inherent* reason why *this one* might come to an end has never been self-evident even to those sympathetic to Ellul's extrapolation of present trends. Now, we have at least an introduction to the reasons why he sees a probable breakdown, and that possibility becomes much more pertinent. For more on this dimension of the dangers of technique, as well as the possibilities for avoiding them, we can only look forward to the publication of the next volume of *Le Système technicien*.

Secondly, much more light has been shed on the discrepancy between the apparent dynamism and rapid changes we all sense in our society and Ellul's contention that our society is increasingly closed and stagnant. Again, strong intimations of an explanation are scattered throughout his books but never pulled together. On the one hand, there is the recurring theme of illusion vs. reality. Ellul does not pose this dichotomy precisely in the manner of the Platonic tradition; nevertheless, it is of paramount concern for him—given the primacy of clarity and awareness and our proclivity for dangerous confusion and illusion. Doubtless, the pursuit of clarity permeates *Le Système technicien* and one of his prime targets is the spoken or unspoken assumption that growth equals development.

Focusing directly on the technical system, on the other hand, brings to the foreground the specific questions of openness and closedness, suppleness and rigidity in our society. The very nature of any system entails combinations, modifications, evolutions, changes, and Ellul even uses the word (perhaps loosely) innovations. In the technical system, these movements are strikingly dramatic for a number of reasons—the newness of our situation, the rapid, self-perpetrating acceleration of technical progress, the dominating efficiency at the core of the technical phenomenon, etc. It is no wonder that we perceive our society as dynamic and open to radical change! At the same time, all this activity either takes place *within* the technical phenomenon or results from the extension of the scope of the system. The system itself seems immune. A good example comes in the well-known and highly touted flexibility and tolerance of the modern world.

> What we have just said gives the impression of a very great rigidity, and in fact the imperatives are more and more numerous and demanding. But it seems that if such is the case on the level of the subsystems, the whole tends to function in a more supple way, and it seems that the force and stability of technique resides exactly in its ability to adapt.[38]
>
> The suppleness also allows the maintenance of cultural diversities where centralization has not yet taken place. Certainly, there is still a Khmer culture and a Sahel culture which are survivals perfectly tolerated by technique, but classified by it as remnants of the past, folklore, ethnographic museums, etc. These cultures are an appearance covering the reality of a technical system which is similar everywhere (but with varying degrees of advancement) and where there is any risk of conflict, the local culture disappears (for example, in the course of a war where technique presents its most brutal and pitiless side). But, in general, technique is flexible enough to be able to adapt to local conditions. . . . Thus, let us retain the position that cultural diversities are a witness to the suppleness of the system and are not proof that human realities are beyond it.[39]

These two quotations, verging on the ironical, show the double-barreled impact of technique. There is much activity, but it is all self-generated and self-perpetuating; it is all in the service of the technical system; it is a conformism contributing to the strength and stability of the whole system. The appearance of dynamism and change is accurate on one level and as far as it goes. Still, on the level of the whole, any such appearance dissipates into an illusion, for none of the changes affects the system itself—as a self-growth system. Thus, this book is in no way at odds with the thesis presented in *The Betrayal of the West*. For Ellul, it would be paradoxical only in appearance to call our society an increasingly closed dynamism. It can continue for some time under its own momentum, but, given the "cracks" in the system, that dynamism, unchecked, has a finite life expectancy.

Just in case all the talk of system and subsystems seems abstracted from the reality of everyday life, Ellul concludes with a section called "Man in the Technical System." Here, he formulates five propositions that put together the two sides of the technical coin—variety and the monolith. (i) "[M]an, aware of himself, finds technique as a 'given.' Technique constitutes for him a milieu into which he came or inserts himself."[40] This proposition involves the whole thesis that the technical milieu has supplanted the natural one. (ii) "All intellectual formation prepares people to enter the technical world in a positive and efficient manner."[41] In other words, the high premium put on education, whether formal or informal, and the proliferation of educational institutions do not serve to "lead people out" to genuine options and understandings. Rather, universal education universalizes adaptation to technique. (iii) "The mode of conditioning has produced a new psychological type. . . ."[42] We have become psychologically adapted to our milieu and all the diversions and entertainments and gadgets we have served to make us fit in and consider our life to be one of freedom and spontaneity—when the reverse is actually the case. (iv) "In our society, the exaltation of desire can only make us move further along the path of technique."[43] The emancipation of the passions, in every form, cannot realistically be seen as an andidote to the technical system, so much as going along with it hand in hand. The breadth of possible gratifications is no sign of an increased freedom. (iv) "Man in our society has no intellectual, moral or spiritual reference point from which he could judge and make a critique of technique."[44] In short, although there are many choices to make daily, they are all *within* the technical system and *conditioned by* the technical system. Ellul sees none of this activity as indicative of the genuine dialectical movement that makes for a healthy society.

Thirdly, once the language of system predominates, we are more firmly than ever in the realm of the monolith, the closedness that Ellul has always decried. *Le Système technicien* raises, in a slightly different way, those questions and problems. Can the system ever be complete? Is society totally technicizable? The few statements he does make in this regard are almost enigmatic. For example, there is the closing paragraph of the book.

> Of course, we have seen that man is not perfectly integrated, adapted into the system. But it will suffice to state here that it is not the presence of man that prevents Technique from constituting itself as a system: man as he acts and thinks today is not

> an independent subject in relation to a technical object, but he is inside the technical system—he is himself modified by the technical factor. Man who today makes use of technique is, in fact, the very one who serves it. And, conversely, only man who serves technique is truly fit to make use of it.[45]

At the same time, he refuses to say that society itself can become a machine.

> The malleability, the plasticity of the social organism is not indefinite, to the point where one is led inevitably to make choices between techniques, between innovations. . . .[46]

Or:

> But it is curious that some sociologists can accept the reduction of society to the state of a machine. However mechanistic or determinist one might be, it is evident that no society has ever functioned in this way. . . . But I believe that it is very dangerous to use an apocalyptic vision; it is in effect too easy to show by the facts that *our* society is not mechanized . . . [and] man in this society is not truly mechanized to the point of being only a cog-wheel.[47]

What exactly does Ellul want to indicate in these passages? Despite the compelling grandeur of his analysis, does he at the last minute fall into a kind of wishful thinking that anywhere else he derides? I think that there will be some disappointment that such questions are still left somewhat hanging. The lack of a complete answer is of a different order from his general refusal as a sociologist to spell out answers. The questions referred to here are perhaps more philosophical and, as I shall discuss in the last section, they are the issues Ellul tends not to discuss. To see clearly what he means by the technical system in relationship to society and individuals in it, the reader has to draw out what Ellul means by each of the terms "man," "society" and "system." To delineate specific answers would take at least another chapter and would move us beyond the field to which Ellul limits himself. By way of introduction to these ideas, I shall examine one rather illustrative passage.

> In reality we must not confuse the technical system and technical society. The system exists rigorously, but it exists in society, living both in it, from it and grafted onto it. There is a duality exactly as there is between Nature and the Machine—the latter functioning thanks to natural products: but it does not transform nature into a machine. Society is itself also a "natural product." At a certain level culture and nature interpenetrate, forming a society as a whole that becomes a nature for man. And it is in this complex that the technical system has situated itself as an invading and irreplaceable foreign body. It does not make society into a machine. It models society in accordance with its necessities, it utilizes society for support, it transforms certain structures, but there is always an unpredictable, incoherent, irreducible part of the social body.[48]

From this relatively brief reference, taken in the context of the rest of Ellul's basic stance, I have tentatively drawn certain conclusions.

The irreducible element is the human individual or the human beings that, of course, form the content of society. Ellul rejects out of hand any arguments concerning an eternal human nature and stresses how malleable the human species is. Nevertheless, he does not go so far as to say that this plasticity is infinite or that man is completely technicizable, for such a statement is still an open question and to draw conclusions would be to go beyond the evidence. In fact, he fights against trends to mold people into conformity and the tendencies to desire total integration. When it comes to society, to people coping with their environment through living in groups, he sees that there is more at stake than simply atomized individual responses. There are collective forces at work that go beyond any specific individual, even though these forces are fostered through the participation of each person in the society.

> To me the sociological does not consist of the addition and the combination of individual actions. I believe that there is a collective sociological reality, which is independent of the individual. As I see it, individual decisions are always made within the framework of this sociological reality, itself pre-existent and more or less determinative.[49]

As a sociologist, he can study these forces and make predictions of probability on the basis of the trends. Still he cannot and will not reduce them to a mere mechanism, for human beings, even in a technical society, are still at the heart of the matter. It is always possible for there to be either a quite different collective response or else so many problems emerging that the pattern will change. In short, the technical system can perhaps be seen as a mechanism (albeit not neatly as a machine, as we have seen), but it remains situated within the forces of a wider society which in turn is made up of human individuals who have never proved totally predictable. The complexity is compounded by Ellul's view of the whole as characterized by the relationships among the parts. This complexity, however, cannot be short-circuited by confusing the system, society and the individual.

The citation above from *Le Système technicien* also indicates the direction of Ellul's thinking as to how present society might become livable, a way different from the liberal view that technique can simply be used for good goals. Although he has not elaborated what he means, a hint does come in his allusion to the interpenetration of nature and culture to form society. Societies of the past worked out their particular *modus vivendi* through the development of a nature-culture relationship. Societies developed to assert and preserve man's collective existence over against the threatening and fearful aspects of the natural environment. That environment, at the same time, sustained and nurtured man and he was inextricably part of it. As we have seen, Ellul argues that we have moved from a natural to a technical milieu, so that we must develop a parallel technique-culture relationship from which society can live. The establishment of such a society would then be different from a technicized society, in the same way that cultures were established over against nature and not simply as an integration into it. Similarly, a technique-culture relationship would preclude the view that we can master the environment as a whole or turn it to the service of some outside goal or seek some different milieu. Societies in the past neither tried to escape from nature nor were indifferent to its demands, nor underestimated its power. Just as the whole fabric of social life flowed from taking a healthy cognizance of nature, so we must do the same in the new milieu of technique.

Although Ellul does not think that our age is necessarily worse than any other, he has suggested that we have been negligent in our collective responses. Because we *feel* we are in charge, we are in danger of giving ourselves over, lock, stock and barrel, to the forces of our milieu, with no critical distance from it. Thus, for example, in contrast to premodern religions, he sees modern "religious" movements as short-lived and destructive (cf. *The New Demons*). It is not a matter of controlling technique, so much as living within it in such a way as to minimize the crises and preserving genuine (as opposed to conformist) relationships to it. It is not utopianism, so much as the survival of human society within anything that might be called civilization. As to how or whether such developments might come about, Ellul does not speak. Nor does he articulate what they might look like. Nevertheless, he does anticipate that only through some such route can we develop an antidote to the encroaching monolith.

4. Ellul as a Social Scientist

In the course of reading *Le Système technicien,* we see again Ellul's massive contribution to our knowledge of modern society. The strengths (as well as weaknesses) of this book are those of his whole body of sociology. Apart from specific insights that come like flashes of light, he also does succeed on the broad level. That is to say, he does provide the reader with a framework focusing on the centrality of technique. Ellul's way of fitting together the apparently disparate phenomena we experience is quite different from either the rather inexplicit liberal one of pluralism or the Marxist alternative. It is not surprising that many young Americans find some version of Marxism appealing, for no other equally comprehensive account has been seriously presented to them. Ellul could provide such an option. Whether or not his insights are entirely original, whether or not his accuracy is open to challenge, we must thank him for reaffirming sociology as a discipline worthy of the name and, more important, of our attention. As North American sociology becomes increasingly boring and irrelevant, we must count Jacques Ellul as a blessing. While maintaining professional rigor, he displays deep passion for our common situation and engages the reader in it. He reinstates the life of the mind as crucial to a healthy society, instead of accepting the intellect as a joke.

In fact, what finally distinguishes Ellul is his refreshing clarity or clear-headedness or, I am tempted to say, his sanity. As we would say, "He calls 'em as he sees 'em," or, as my grandfather would say, "He knows how many beans make three!" In this undertaking, Ellul is assisted by his lucid French prose style, which comes through, albeit somewhat diminished, even in English translation. Ellul's ability to call a spade a spade, without any diminution of breadth or attention to detail, makes him a contender as the most important social analyst of the twentieth century.

Having made such an assessment, however, it is mandatory to take seriously those who have real difficulties with Ellul. No questioning, no matter how badly formulated, is entirely rootless. It should be obvious that I do not intend to fall into the trap of labeling him a pessimist, a fatalist, or a person consumed by his own fear or hostility toward technology. Ellul himself has more than adequately spoken to these charges. He calls his own writings a "critique" of society, but only in the sense of analyzing the components, and he refuses the negative connotations attached to the word "critique." Similarly I am not concerned with those who do not read him carefully, because they have, simply temperamentally, rejected him out of hand. (Granted, these two groups constitute the majority of his critics.) Nevertheless, a simplistic acceptance of Ellul would be contrary to his own unflinching quest for clarity. Therefore, I want to conclude by opening, very briefly, three areas of concern about Ellul's stance: his refusal to give solutions to the problems he so ably evokes, the relationship between his sociology and his theology, and his relationship to philosophy. The first is the most unsettling for concerned readers; the second is the most misconstrued by commentators; the third is the most neglected by Ellul, who considers it to be inconsequential.

* * *

> It is true that the message of the writers in question is not that of Macbeth. They scream that life is gutter. But one cannot sense that life is gutter if one has not sensed purity in the first place, and of this which is by nature sensed first, they say nothing, they convey nothing.[50]

These are strong words, and it would be wrong to imply that Ellul claims life is gutter. Still, in his refusal to suggest alternatives, the point is there. Up until now, he has replied that a person need not be able to prescribe a cure to know that he is ill and he has underscored that, even for a physician, diagnosis must precede cure. Again, he remains consistent, but is his answer completely adequate?

In the first place, it would hardly be an exaggeration to suggest that Ellul, in practice, pushes the distinction between judgments of probability and judgments of value about as far as it can go and sometimes even further! The belief, for example, that sociology should pose the problems of society, although one many of us would like to advocate, is a stretch of neutrality as it has been presented in the social sciences. The very vehemence of his "warnings" implies that Ellul does have some notions of health and/or cure (as does the physician seeking a diagnosis). One is left, sometimes, with the impression that he will go as far as possible along a certain path; nevertheless, at some point, the dictates of his profession and his distrust of the distortions of the discipline prevent him from taking the next (seemingly logical) step. Surely it is no imposition per se to move beyond the stages of separating illusion from reality and the raising of questions about false solutions, to include some account of what might be more nearly adequate. Why he stops where he stops is not always clear. For instance, to use words like "monolith," "enslaving," "betrayal," "shipwreck" is not exactly a "value-free" stance and many thoughtful readers have wondered why he would want to claim it as such.[51] Such words, used deliberately, do give every impression of another, more healthy possibility—even within the limits of sociological rather than strictly theological discourse. Of this, to paraphrase Leo Strauss, "he says nothing, he conveys nothing."

At other times, it appears that Ellul's scrupulous refusal to spell out answers, or even to indicate directions except in general terms, puts almost a cloud over the very clarity he seeks. By preserving the proper place for the individual and social consciousness and response, Ellul sometimes leaves the reader perplexed. The few times he does use examples of the search for alternatives, as in a brief section in *Autopsy of Revolution,* I, for one, sense that his explanations are greatly enhanced. Even though there are perhaps all too few illustrations upon which to draw, those available should be closely examined and possible instances such as the formation of "poles of tension" should be further pursued, so that what he does say is not absorbed into liberal rhetoric. In terms of this article, perhaps the best example comes in Ellul's argument that a technical milieu has replaced the former natural milieu. That position is brilliantly presented, so that it is open to good and serious discussion. When it comes to the question of the establishment of a livable society within the new milieu, however, readers are left hanging by his reticence. Does he mean a technique-culture relationship such as I have suggested? What could that direction possibly entail? (We have, sadly enough, all too many examples of how not to go about it.) Could one possibly talk about techniques for a technical milieu? These questions are not meant as a desire for blueprint answers or unwarranted optimism, so much as they are meant to fall into the category of the clarification of options.

The precise nature of these queries does not deal with what Ellul has said; rather, they ask for more. Perhaps asking for "more" is dangerous in the field of social analysis! On the one hand, Ellul has taught us much that discourages any forcing of sociological observations into preconceived patterns—especially if they stem from beliefs in the myths of history, science and human progress. On the other hand, after taking the dangers into account, there remains the responsibility to help all those caught up in social forces to engage in the struggle for options. It is not surprising that we would look for help to someone of Ellul's caliber!

At this point, we must return to Ellul's own undertaking, to his overall perspective into which his sociology fits, and consider his sociology vis-à-vis his theological stance.

* * *

The content of Ellul's theology is too vast to summarize easily. Suffice it to say that he can be characterized (if not caricatured) as a Bible-centered, Protestant Christian, deeply influenced by Søren Kierkegaard and Karl Barth. Very briefly, he reads the whole Bible as testifying to the radical contrast between God and fallen man. They share no common identity in terms of reasoning, understanding, feeling or activity. Rather, Scripture reveals a rupture, a chasm which man cannot bridge without God. At the same time, he also sees the Bible as pointing to the truth that God has preserved His creation and has refused to accept the alienation of man and human corruptibility where all relationships are concerned. God has maintained an enduring connection in which He wills the salvation of all people, no matter how intransigent they may be. From this dialectical perspective, finding its ultimate unity in the person of Jesus Christ, Ellul draws his biblical understanding. Although Ellul does not concentrate on dogmatics, he finds Karl Barth's emphasis on the dialectical nature of biblical revelation compelling, for it forces the Church to self-scrutiny and impels Christians to their task in the world.

What is important to examine here is the way in which he sees the two aspects of his work. We must note that Ellul stringently separates his sociological studies from his theological reflections. Sociologically, this distinction corresponds to and maintains the distinction between judgments of fact and judgments of value that we have just seen. Therefore, he is careful to confine his analysis of society to one set of writings and his understanding of the meaning of those trends and the source of ultimate response to them to his other set of writings.[52] Theologically, the separation of his studies is equally important, for he stresses the Barthian/Kierkegaardian belief in the "infinite-qualitative distinction" between God and man. Practically speaking, this belief means that it is impossible to move from knowledge of the world to knowledge of God. In other words, divine revelation is a one-way street. God's revelation sheds light on the world, but not vice versa. Therefore, theologically, one cannot draw an understanding merely from an observation of the world. To challenge Ellul's broad and intransigent position (which for many Christians, at any rate, is basically modern in its denial of natural law)[53] would not be particularly easy. It would be hard, if only because Ellul is one of the most sophisticated theologians of the century—even though he is not formally a theologian. For example, an attack on his attack on natural law would have to take into account the whole of the Western tradition.

For our purposes, we should look rather to the way in which he does see the connection, for finally he does *not* see theology and sociology in water-tight containers. Again, predictably, the key to his understanding comes in his emphasis on dialectics.

* * *

> We must seek the deepest possible sociological understanding of the world we live in, apply the best methods, refrain from tampering with the results of our research on the ground that they are "spiritually" embarrassing, maintain complete clarity and realism—all in order to find out, as precisely as may be, where we are and what we are doing, and also what lines of action are open to us. The Christian intellectual is called frankly to face the sociopolitical reality. This is one demand of the Christian intellectual. The other is that he must also develop and deepen his knowledge of the biblical and theological fields. But he must be aware of "inflecting" theology for the sake of the cultural. . . . The only thing that will be of any use is not synthesis of adaptation, but confrontation; that is, bringing face to face two factors that are contradictory or irreconcilable and at the same time inseparable. For [it is] only out of the *decision* he makes when he experiences this contradiction—never out of adherence to an integrated system—that the Christian will arrive at a practical position.[54]

Theological understanding is not sufficient: it demands a response. First, Christians must be concerned with the preservation of the whole world and that preoccupation alone gives them an interest in making society livable. Furthermore, the Church is also called to bring forth new life from the very center of human disobedience in the age of technique. He is convinced that there is no mechanical channel between what is learned in the Bible and what can be known about contemporary society. The demands of the two spheres are totally in conflict. Ellul hopes to remind Christians that they must live out the resulting tensions in challenge and love.

> Thus on the one hand the Christian is called upon to participate in the activities of the world in which he lives, to recognize not just justice before God but relative usefulness and normality. It is normal for the world to develop its own morality. But on the other hand, the Christian is called to live in all its dimensions the fullness which has been given him, without distinction of temporal and spiritual, private and public, etc. The Christian faith implies for him transformation in politics, in business, etc., which are directly referred to the person of Jesus Christ. Thus the tension is set up in one world (and not a separation), tension which does not mean rupture but not adherence either. It is a dialectical relationship which eliminates at one and the same time the possibility of the autonomy of the morality of the world and the possibility of the autonomy of revelation in relation to the world.[55]

Of course, opponents can argue that it is logically impossible to live simultaneously in two contradictory realms. For Ellul, however, this *is* the paradoxical commandment of Christ to be "in" the world but not "of" the world. In our time, it is the technological society as a whole which forms one pole in the never-ending confrontation within the human project (itself dialectical). The technological society is the concrete realm to which the word of the Bible is now directed and in which the Church must be prophetic.

One last question for Ellul's theology. Can the sociological books really stand alone when they form only one part of Ellul's complex structure? At times, one does feel that he has a hidden agenda and comes close to indulging in secret preaching. Certainly, he does use words like "freedom" a trifle ambiguously; certainly, many readers are left feeling hopeless and helpless, with no idea as to how to respond. Readers who feel stranded are not simply imagining their frustration! Some of these complaints come from Ellul's exuberance of style; others are answered in different writings (as, for example, the end of *The Political Illusion*); others result from the fact that he is a sociologist. Because Ellul is such an impassioned sociologist, again, perhaps we expect more from him than the discipline can deliver.

To conclude about problems people have with Ellul the sociologist in relation to Ellul the Christian believer, it is obvious that he is not beyond discussion. Nevertheless, I would caution against a facile repudiation or correction based on an underestimation of Ellul's grasp of the whole. He may well be wrong, but if he is, it is on a grand scale and it would take an equally grand scale to prove it.

* * *

The area most difficult to assess, mainly because of Ellul's refusal to enter the discussion, is his relationship with philosophy. He claims not to be a philosopher at all, only a describer, and thinks that the influence of philosophers on lived reality has been overrated. Again, his approach is rather modern. The problems concerning Ellul and philosophy arise on three levels. First, there are the usual grounds of the difference between philosophy and sociology; nevertheless, for Ellul, that question for Christians more properly belongs in the realm of theological reflections (i.e., the idea of philosophy as "the love of wisdom"), already alluded to above.

Secondly, comes the question of whether he has underestimated the change in political philosophy as a major source in the rise of technique itself. In the final analysis, the causes of the rise of technique are peripheral to Ellul's central task. His lack of preoccupation shows in the section in *The Technological Society* on the historical development of modern technique—probably the weakest one in the book and a topic not included in *Le Système technicien*.[56] The only significance in the roots lies in the quest for alternative suggestions. However much one might like Ellul to spell out the impact of belief factors or make a philosophical analysis of the emergence of our unique society, still it would not appear that such a tack can lead to a radical attack on what he has presented.

The third level, the most serious and complex one, involves the philosophical assumptions (assumptions he tends to deny or disregard) implicit throughout his own approach and description. In very general terms, they can be included in his acceptance of the modern social sciences. Specifically, one should examine the presuppositions he takes from Weber and Marx as leaders in the field. For example, he takes for granted Marx's belief that a thing is defined in terms of its relationships and his view of society as developing through relationships among opposing forces. Such a view may seem self-evident to many of us moderns, but it does rest on a philosophical basis quite different from anything that went before the modern era. To be able to uncover exactly what are the philosophical currents at work requires a mind of equal scope to Ellul's. For an introduction to the intricacies of the relationship between modern social science and philosophy, I can point only to Leo Strauss's *Natural Right and History,* with particular attention to the chapter "Natural Right and the Distinction Between Facts and Values."[57] There, Strauss demonstrates the affinity between Weber's account of social science and the Kantian account of the world. Although Ellul is not identical with Weber, they share enough in common to warrant further reflection on this issue. The outcome is not merely of "academic" interest, for such a study would revolve around the following question: To what extent does Ellul participate in the same modern project he wishes to expose?[58] If he does, to what extent does that cloud his analysis and blur the quest for options?

I am unable to answer either question in a sustained or systematic manner; therefore, I shall point only to two areas that indicate the specifically modern cast to Ellul's thought. First, let us consider his account of reason. Although he is careful *not* to equate rationality and reason (the former being only one form that reason can take), he does firmly subscribe to Kant's view of the instrumentality of reason as its sole "function." Unlike the ancients, Ellul denies that reason is capable of leading to any knowledge of the good. Rather, he is much closer to Hobbes's dictum that reason is the slave of the passions—and the overriding passions in man are the will-to-power and the desire for security. Reason itself, for Ellul, cannot alter the contents of those passions, even though they have obviously varied in different civilizations, as well as in different times in history. Connected to the restricted definition of reason is his acceptance of the modern denial of any fixed human nature, in favor of belief in the almost infinite malleability of man.[59] As a sociologist, therefore, there is no solid basis for confidence that people are not completely technicizable. Nor are there grounds for preferring, for example, a natural milieu to a technical milieu, as the human habitat. The only legitimate ground for concern, on a theoretical basis, is that rigid societies *tend* to collapse, but no fixed law on that score can be established. It is to Ellul's credit that he completely understands the implications and consistently accepts such limitations on any conclusions. Nevertheless, it is a somewhat parochial appreciation of what it means to be human—parochial in the sense of being limited to the last few centuries of specifically Western thought. We do not often think of it as being narrow, for it is the account that has captured the whole world. It is precisely the account that has engendered modern technique and has facilitated its success.

It is obvious, then, that Ellul's enterprise represents, on a very high level, a critique (in all senses of the word) or challenge of modernity from within. He speaks a language modern people can understand, for he speaks entirely from the center of what they profess to believe. It may be a dangerous game, but he plays it consummately well.

* * *

To question Ellul for sharing certain assumptions could imply only that he should have embarked on an entirely different enterprise. It does not necessarily undermine the power of his descriptions. One *is* led to wonder how he remains immune to the seductive power of technique and the myths that surround it like a halo. At this point, we come to the direct influence of his Christian faith on the doing of sociology. His Christianity does not inform his methodology nor its content: it does provide him with the outside source that enables him to look, without flinching, at things as they really are. He manages "to pull if off," because his ultimate faith does not depend on the vagaries of the present human condition. Thus, he can stand in the middle of it, with calmness and compassion, without making it into an absolute. In biblical terms, his refusal to worship false gods gives him an insight into the falseness of present idols. His belief in creation allows a reliance on common sense that Weber could never quite trust.[60] In traditional language, his trust in God supplies him with fortitude and prudence in the area of the social sciences.

NOTES

1. "From Jacques Ellul," *Katallagete* 2:3–4 (Winter/Spring 1970), p. 5.
2. Translations for this article are from the hardback English edition of *The Technological Society*, 1964. For *The Political Illusion* and *Propaganda*, I am using the English translation, paperback editions. For untranslated books, the English version is my own.
3. *Le Système technicien*, Calmann-Levy (Paris: 1977). To date, it remains untranslated.
4. *The Technological Society*, p. xxviii.
5. By occupation, Ellul is also a historian with several publications in the field. I leave them aside only because they have never entered into the controversy. Nor have they had the same public impact as his other writings. On the question of selection, it is important to note that Ellul has published a prolific number of short pieces, many of which are subsequently incorporated into his books. Finally, how is it possible to think of sociology at the University of Bordeaux and not think of Durkheim?
6. *The Technological Society*, p. xxv.
7. *Ibid.*, pp. 5–7. From the very beginning of the book, it is clear that technique is very similar to Max Weber's ideal type of rationality.
8. *Ibid.*, p. 146.
9. *Ibid.*, p. 41.
10. *Ibid.*, p. 58.
11. *Ibid.*, p. 97. Put differently, one could say that there are obviously goals in society, but these now all fall *within* technique and there is less possibility (if any) of a goal *for* technique itself.
12. *Metamorphose du bourgeois*, p. 20.
13. *The Technological Society*, pp. xxvii–xxviii.
14. *Ibid.*, pp. xxxi–xxxii.
15. *Ibid.*, p. xxxiii. Although this quotation alludes to Ephesians 5:14, it is not used here with specific theological implications.
16. For Ellul, a concept is not something made up, but itself is in the realm of fact. What he means by "concept" is further elaborated in *Le Système technicien*.
17. Ellul is still working on the impact of technique in various fields; for example, he lectures extensively on jurisprudence and in *Le Système technicien*, p. 113, he refers to a forthcoming book on art and technique.
18. *Autopsy of Revolution*, pp. 268–269. Here, the word "technique" is wrongly translated as "technology."
19. *The Betrayal of the West*, p. 129.
20. Two points should be made here. First, he says that the origin was the desire for security and control, but the autonomy of technique has gotten so out of hand that there is an increasing feeling of oppression and lack of security and no control, for which there seems to be no answer. Secondly, he argues the tight link between ends and means. If the means used are those of domination, then the end will be domination as well. Therefore, for example, he has said that he thinks Orwell's *1984* is probably more likely than Huxley's *Brave New World* with the benevolent Mustapha Mond. In the final analysis, it is impossible to predict that all will be so "happy" then.

21. *Propaganda*, p. xiii.

22. *Ibid.*, p. 255.

23. "De la Signification des relations publiques dans la société technicien," *L'Année Sociologique*, 1963. This publication is not, strictly speaking, a book, but it is an important example for our purposes.

24. Ellul recognizes special debts to both Weber and Marx (who had a firmer idea of seeing society as a whole and the relationship among the parts and forces). It would be wrong to say that Ellul is entirely opposed to Marx, for he believes that *qua* sociologist, Marx was one of the giants. This is not the place, however, to go into Ellul's intellectual debts.

In the interests of appreciating Ellul's approach, it is important to note the care he takes in trying to understand other thinkers in the way in which they have understood themselves. For example, in his outstanding lecture series on Marx, not once did he state his own opposition to Marxism and only in the closing lecture did he raise questions about Marx's account of the modern world. Despite his own forthright and, at times, almost aggressive style, he does go to great pains to hear what others are saying, both in the way of questions and answers about our society, and to draw from them.

25. For example, both regard the French Revolution as pivotal, albeit with different interpretations, and each gives little weight to the American Revolution.

26. In this book, he does link Marx with his followers, though in *The Betrayal of the West*, p. 144, he also maintains that the followers have grossly manipulated what Marx said and meant. This is not to say, as many would try to maintain, that Marx can be totally divorced from his followers.

27. By "phenomenological approach," he means a study of what can be observed from the outside of the religion and not necessarily an account from within. Although he does not fall into a definite school or philosophy, here he takes what he finds useful in modern phenomenology and incorporates it into his own work. He denies any judgment as to whether religion is an entirely human phenomenon, but deals with its human manifestations. He maintains that biblical revelation is not "religion," even though the practice of it easily and constantly succumbs to that temptation.

28. *The Betrayal of the West*, pp. 126–127. It is interesting that on p. 82 of that book, he distinguishes between the anarchist left and the socialist left.

29. *Ibid.*, pp. 136–137.

30. It should be remembered that both *The New Demons* and *The Betrayal of the West* are sociological books. According to Ellul, the points where they may sound theological are simply discussions of what Christians have said and done, historically and socially.

31. *The Betrayal of the West*, pp. 167–168.

32. *Ibid.*, p. 199.

33. *The Political Illusion*, pp. 222–223. (The Pierrelatte establishment refers to the French nuclear network.) Here, he can be accused of exaggerated language, for elsewhere he says that it cannot be known whether man can be reduced to a machine-like object. Nevertheless, the danger is real and he wants to warn that such a danger cannot be ruled out of court.

34. *Le Système technicien*, p. 7.

35. *Ibid.*, p. 361.

36. *Ibid.*, pp. 88–89. The term "feedback" is borrowed from the natural and "hard" sciences and it refers to the mechanism whereby inputs into the system are tested as to whether they will contribute to the proper functioning of that system. Feedback is a built-in self-correction device. It is not clear from this book whether Ellul considers nature itself to be a system, though there is no reason to doubt that he does. At the same time, it would be wrong to characterize Ellul as using a simple biological metaphor for technique. The concept of system is much wider. Finally, this passage shows, even more dramatically than before, that Ellul does not equate technique with machines.

37. *Research in Philosophy and Technology,* Vol. II, p. 64. It appears that Cérézuelle maintains Ellul's emphasis on dialectics, but it also appears that he is using the term here in a somewhat different context, one that is neither healthy nor fruitful.

38. *Le Système technicien,* p. 121.

39. *Ibid.,* p. 124. By suppleness or flexibility, he is including here the power of assimilation which was characteristic of the bourgeoisie and to which he devotes considerable attention in *Metamorphose du bourgeois.*

40. *Le Système technicien,* p. 344.

41. *Ibid.,* p. 345.

42. *Ibid.,* p. 347.

43. *Ibid.,* p. 351. Here, he refers the reader to a book by J. Brun, *Le Retour de Dionysos,* 1969, where he argues that technique is not the blind and cold machine, but a manifestation of the dance of Dionysius. One of the dangers Ellul sees is the merging of the Apollonian and the Dionysiac, rather than maintaining the two dialectically. (*Cf.* p. 225 of this article.)

44. *Le Système technicien,* p. 352.

45. *Ibid.,* p. 360.

46. *Ibid.,* p. 336.

47. *Ibid.,* p. 25.

48. *Ibid.,* p. 25. All of this reference, except the last sentence, is in italics in the original.

49. *The Technological Society,* p. xxviii.

50. Leo Strauss, *Liberalism Ancient and Modern,* p. 262. Even though Strauss makes this comment in a different context, it is not inappropriate here.

51. Two comments are perhaps in order here. Those who raise this kind of question are themselves not admirers of the separation of facts and values in the social sciences. It would not be the case that they applaud the tendencies to be caught up in the modern trends any more than does Ellul himself. Secondly, this question does not imply that value-free sociology is, in fact, immune from specific assumptions.

52. As a Christian, Ellul denies that Christianity is a "value" which is more of a sociological category. A detailed argument of what he means here would lead us too far afield.

53. Ellul's *The Theological Foundations of Law* is one of the most sustained and complete rejections of natural law theology around. Although such a rejection is modern for Christians, we must remember that throughout the Jewish tradition there has been a greater ambivalence toward natural law foundations and philosophy in general. Since Ellul often consults Jewish commentators for his studies, we cannot say that his rejection of natural law is self-evidently modern. Still, it is true that such a complete rejection is characteristic of modern thought.

54. Ellul, "Mirror of these 10 Years," *Christianity and Crisis* LXXXVII (18 February 1970), p. 201.

55. Ellul, *To Will and To Do,* p. 290.

56. In fact, Ellul does not exclude ideal factors from his analysis of the development of society and sees them in a dialectical relationship with material factors. Nevertheless, it seems accurate to say that he is not as compelling when he discusses philosophical change and changes in outlook in *The Technological Society.*

57. Strauss, *Natural Right and History,* pp. 35–80. I choose Strauss not only for the scope of his philosophical study, but also because he recognizes the greatness of Max Weber, as *the* social scientist, and it is on this basis that he undertakes his criticism.

58. See G. P. Grant, *English-Speaking Justice,* for a discussion of the links among Kantian philosophy, technique and political liberalism.

59. Generally, he denies the existence of any unchanging human nature, but he does leave the door open slightly, or one might say that he hedges slightly, as, for example, in *The Political Illusion,* p. 76.

60. By belief in creation, I am not here implying a full theological doctrine that informs Ellul's sociology, but rather something simpler. I mean the conviction that things actually do exist concretely and can be perceived. Theologically, Ellul does not think that we can know the world as it was created before the fall and its implications of the order of necessity. At the same time, he can state with confidence that "reality is reality" and that confidence, as the source for some form of reliance on common sense, is ultimately rooted in belief in creation.

NATURE, TECHNIQUE AND ARTIFICIALITY

Jacques Ellul

For the past two decades the question of the relation between Nature and Culture appears to have been examined from every angle. Yet I believe that despite everything I have read on the topic, the implications have not been fully drawn for a slightly different formulation of the question—viz., that between Nature and Artificiality. We remain implicitly convinced of the excellence of nature, and the word "artificial" always includes at least a hint of negative connotation and criticism. It is enough to think of the distinction so often made between "natural needs" (always legitimate and justified) and those "artificial needs" that are the butt of every criticism, particularly those of the consumer society. It must be clearly shown that the most natural needs are lived, expressed and satisfied in an artificial way; that the line between artificial and natural needs has always been impossible to draw; and, finally, that so-called artificial needs always have a natural basis, that it is impossible to create human needs out of nothing. We have to relearn Marx's lesson that man is a

Research in Philosophy & Technology, Volume 3, 1980, pages 263–283.
Copyright © 1980 by JAI Press Inc.
All rights of reproduction in any form reserved.
ISBN: 0-89232-102-4

unique being; as is not the case with animals, when man gratifies one need or a certain order of needs, a whole new order of needs immediately unfolds for him. Even if we know all this, spontaneously we still judge natural needs positively and artificial needs critically. This is simply one example of a rather general attitude relating to many components of our civilization.

One might say that this belief dates from the acceptance of the ideas of Rousseau, but in reality it is a very old attitude which, at least since 500 B.C. in the West, presents nature as a model, an excellence, a good, or a value. As a result, we consider law or morality in terms of the "demands" of nature; we set up a model of law in natural law; and we try to construct a natural theology. (This adjective legitimates the theological enterprise.) Now, the movement of technique, the formation of an almost totally artificial world, pushes us to the wall and forces a more fundamental reflection. At the same time, this development leads directly to a more extreme evaluation of nature. Nature is endangered by technique, it appears as the sole refuge, the only remedy, the only compensation vis-à-vis the excesses of the technical system. It is enough to think of the reaction of hundreds of thousands of young people in the world, the founding of communes, the search for revolutionary "spontaneity," the flight into an apparent nature during vacations as an escape from technical work. These are wholly disparate signs, but they spring from the same convictions.

With this in mind, let us try to start from some simple givens in order to see what consequences follow. The two basic givens are that man is essentially artificial and the producer of artifacts and that theorization on the basis of nature is an error.

I. ARTIFICIAL MAN

If indeed man comes from the animals, it is necessary to recognize that man does not reveal himself as such until he manifests an artificial creation. We must be precise. According to some thinkers, one can be sure that human beings are present only when a tool, however rudimentary, is produced. It is, then, the appearance of this artifice that characterizes the passage from animal to man or that would mark the appearance of man. For others, it is language that makes man. They mean a language that would be neither mechanical, nor a simple automatic transmitter of concrete data, but a language that brings into play, for the speaker and the hearer, the realm of the imaginary and something beyond mere reality. Consequently, it is a language totally different from the establishment of a code of communication such as we find among bees and ants. The debate on this subject is not closed. Some thinkers (e.g., Laborit) estimate that language is a recent development—that man as such existed long before language and that language refers only to the material elements (tools). On the other hand, a very fruitful school (e.g., Leroi-Gourhan) teaches that man and language are consubstantial—that man is *always* a talking man. In both cases, however, we should note that language is an artificial creation of man. It is not the same thing to say "man from the beginning is one who talks" and "dogs have four legs." Man is not endowed with language in itself. Language is instituted progressively and it does not change by natural evolutions, but by invention and imagination. There is no "natural" unfolding of a language. Man is no more a "language-er" than a "maker." The production of language is as artificial as the production of tools and the two obviously go together.

Furthermore, Marx (but following others!) is equally right to insist on the fact that only man is a worker. He fulfills his life through work. He changes his environment through work. Already (as a third example of the *same* view of artificiality) we have here a basic question of ecology. Each animal is integrated into its environment. We have all seen diagrams of the balance of nature showing that such and such an animal—nourishing itself in such and such a way, diverting the course of a stream, etc.—by its very *being* contributes to the balance of the environment through a sort of mechanical system. As soon as the aspect of work appears, there is a disturbance in the environment by a nonnatural intervention that cannot be integrated. That is to say, man as worker does not obey the laws of the environment and the ecological balance, but rather he obeys his own "law," his desire, his power, his interests—that is to say, a collection of voluntary and sometimes calculated factors (for example, the hunters or fishermen who respect the game in order to allow reproduction). It is no longer spontaneous adaptation, but the choice of balancing and successively making adjustments among three factors—what man desires, what the natural environment gives, and the means man has available (the instruments of action). Thus, when a balance is re-established, it is no longer a fact of nature, but of choice and artifice (for example, the birth control practiced in the Polynesian islands in terms of what is possible for subsistence).

Linked to this first set of artificial givens which are coextensive with humanity, we find a second grouping with problems that can be called morality. For some people, man is man only after he poses a norm different from what his spontaneity leads him to do, and this norm is essentially: "Thou shall not kill." Not to kill is nothing innate in the nature of man.[1] Even if you cannot see yourself as a carnivore pure and simple, a killer by nature, you still have to recognize that the *formulation* of the norm (which is very general) is an a-natural intervention. When man gives himself rules of conduct or family organization, he performs an artificial act. Let us understand this properly: ethnology shows us certain correspondences between animal behavior and human behavior. The unique factor, however, is that this behavior is made explicitly normative. The rule establishes a behavior as obligatory or else it opposes that behavior and prohibits it. This is what is human and, at the same time, artificial.

We can consider two examples. If we accept Konrad Lorenz's thesis about the natural restraints on animal aggression against animals of the same species (namely, that the restraints are more powerful in proportion to the harm the animal can do), then on man the "natural" restraints, the physiological ones, are obviously very weak, since the natural human resources for agression (nails, teeth) are very weak. A shift takes place when man invents artificial resources for aggression (weapons); from this moment, he is dangerous. But his physiological restraints are fixed. In order to counterbalance the power of these formidable new resources, for himself as well as for his neighbor, he has to come up with commandments, prohibitions, a morality which is also both essential to man and, at the same time, artificial. The other example concerns organization. A complex system of taboos in relationships, prohibitions (for example, against incest), etc., is neither the sign of primitiveness nor the expression of crude interests.[2] On the contrary, it is the sign that man is not a simple animal. None of these things are "inscribed" on his "nature"—neither the prohibition against incest, nor the durability of sexual relationships transformed within marriage, nor the length of the educational relationship between the mother and the child (filiation). All these things were willed; the whole system is an artificial institution from which man finally issues. It did not just appear. Man had to establish it, to conquer it, and it is in self-restraint that man was wrought.[3] On this level, the problem is not to find out whether some morality is worth more than some other one (as to content), but rather it is the coexistence of man with a system of norms (moral, religious, political, juridical) which he imposes on himself, making himself in the process of self-regulation.

In these two cases, we see that the limitation and the coercion do not come from outside, nor are they imposed by necessity, nor do they express an internal spontaneity (any more than putting a handle on a piece of flint is spontaneous). Once more there is the interplay among the external constraints, the possibilities for action (here the possibility is the norm itself!), and the choices (and it is hard to say whether they come from a sense of utility, efficiency, conservation, etc.—a matter of small importance). Thus we are constantly led back to the decisive character of the artificial.

Basically, we find exactly the same thing in the Bible. Nowhere do we see a human nature affirmed there. On the contrary, we are shown the aspect of human initiative and innovation willed by God. Man is made of clay (hence the natural factors), but God breathes His spirit on him, and this is a spirit as free and as initiating as the Spirit of God. God charges man "to dress it and to keep it" (Genesis 2:15), that is, to enhance Eden in an artificial way, by human "art" and not simply to watch growing whatever is growing. In addition, man, endowed with the word of God, does not have a language set in advance. The story of the animals is very characteristic. "God brought them unto Adam *to see* what he would call them" (Genesis 2:19). It is impossible to state more clearly the initiative left to man and the artificial, fabricated character of this language. Biblically, then, man is an "uncertain" being who is not endowed with a fixed nature, a purely mechanical movement; he is not a figure of repetition but of history. Even if it is said that man is made in the image of God, it is not made at all definite *in what this image of God* consists. It is a mistake to say that this image is *a nature for man,* for this would mean that God is a nature! Thus, the Bible presents man's constant invention of the artificial as his "vocation," his role. This description comes "before the fall," but it remains the same "after the fall," only with a double transformation—the natural environment has become hostile to man and man is going to express the artificial, above all, in technique.[4] Therefore, the artificial aspect of techniques is not abnormal and the invention of techniques, the production of an artificial environment, is not evil. But neither is it purely and simply the product of necessity where man finds himself taking this route in order to survive. (Certain interpretations of Marx describe this development far too mechanically.) If, biblically, techniques are inscribed in the world of the fall from the beginning until today, it does not mean at all that, as such, they are the product of sin. Man is shown as artificer before as well as after the separation from God. Thus, the problem raised by modern techniques is not at all their artificiality which would be judged by the standard of nature[5]—an artificial evil in the face of a natural good. We shall see what the real problem is, and that it is something completely different. Let us remember simply that man is the artificial producer and that, from the beginning, he has asserted himself *over and against* nature.

II. THE ERROR OF THEORIZING ON THE BASIS OF NATURE

I am leaving aside the question of the existence of a fundamental human nature. It seems to me that the staggering variety among peoples, as shown to us by historians, ethnologists, sociologists, etc., makes the question of a human nature at least problematic. Quite the contrary, I see the decisive importance of the artificial both in the appearance of man as well as in his cultural development. I have sketched some givens concerning man. Brought into existence by means of the artificial and himself the producer of this art, man is the "artificer." If this is so, then we cannot adopt the usual simplistic assessment that says behavior is good if it conforms to nature, that it must obey the "laws" of nature, that it must obey a natural model. The Greeks invented this idea which has pursued us relentlessly and we still live with it. "Conformity with nature" is normal; "nonconformity" is abnormal. Right from the start, we meet a formal difficulty, one that Hans Kelsen shed light on some time ago. Nature is not an animated being, so that, even if we admit that its "laws" are the good which man should obey, it is never nature "who" tells us we ought to obey. Nature does not give commands; it never says "you ought." Consequently, it is left to man's freedom to impose on himself the duty to observe nature, but in this case, it is an artificial obligation, one that comes from a human choice. He prescribes the law of nature to himself as law, exactly as he could have chosen others, or else he must assume the existence of some transcendent which imposes on man, from the outside, a specific commandment to observe the laws of nature.

In any case, this obligation is not at all biblical. We do not find it anywhere in the Bible.[6] God does not ask us to observe what nature does. On the contrary, He declares that His Law is not at all comparable to anything found among other peoples. It is not a natural given, but a declaration of a separation. As for the formula that God will place His law in the heart of man and will transform these hearts of stone into hearts of flesh, this is a prophecy of the end of time; this is a promise through which we are assured of some future accomplishment, but it in no way refers to the present. In fact, it demonstrates just the opposite. If the prophet announces it as a final miracle, then it is not the case now! Just as the law of God is exterior to man, written on tablets of stone, so the word of the Beatitudes is not at all what we find in human nature (if one exists). It is even the exact opposite. Let us think of a single formula, "Blessed are the meek: for they shall inherit the earth" (Matthew 5:5), an expression that is clearly contrary to reality! And again, "My thoughts are not your thoughts, neither are your ways my ways, saith the Lord" (Isaiah 55:8). What better way to say that whatever is found spontaneously in man has no common ground with the will of God? The will of God, then, is not natural. Nowhere do we come across a commandment to obey nature. In other words, nature has its laws, its structures, in which man is situated, but there is no natural morality with a commandment to obedience, no natural law which would be valuable to man in itself.

We come then to this necessary double formulation: *nature is not a model,* and *nature is not a limitation on human action.*

Nature is not a *model,* for man entirely within nature obeys other models (the mythical, the rational, etc.), and from the beginning his relationship to nature has been at least twofold. It is the environment from which man derives the necessities of life, yet at the same time, nature endangers him and threatens to obliterate him. Facing this situation, man is always the one, as I have said, who disturbs and upsets the natural order. Human artifice means that the natural cycles can no longer recur indefinitely without changing. Man substitutes history for cycles. Clearly man cannot be disconnected from that nature which provides the entire substratum of his existence; he cannot claim to lead a life itself totally artificial (and in reality this is our present drama, although it was nobody's intention); he cannot ignore the "laws" of nature. On the contrary, it is through discovering them that he uses them in order to make something other than what seems to be the natural direction. He cannot claim to live without breathing, but the fact that he uses nature also involves the non-natural in life. We do not have to reject what he does, for that would be to deny man the benefit of his environment. The observance of natural laws does not at all express the Good which man should be content to imitate and to which he should adapt. On the contrary, the observance of nature is the efficient means for changing things, for making them useful. It follows that we do not have to look to what happens in nature to try to find a model of goodness and right behavior. Those who believe that they are coming back to the good by returning to nature present another problem altogether. They continue to live in the realm of the artificial, to cultivate, to modify the countryside, to use it. They just lead healthier lives. They respect nature. They have happier work. They find a happiness other than the consumption of industrial goods. All this is fine, but they are not at all "imitating" nature. They do not lead a life *of* nature. To take up nature as a model appears as a false temptation for ecologists and others as well.

Two examples: Much has been said about the natural balances and feedback loops within natural systems. They seem excellent and a good to be imitated—as if the existence of these mechanisms could be worthwhile for the human group as a model for human action. Now, we must make clear distinctions. The processes of feedback and balance are indispensable for the natural environment to maintain itself. Without it, man clearly would no longer exist. To disturb the processes, to prevent nature from functioning, is to run a considerable risk, but it does not mean that human behavior should be modeled exclusively on nature, and that it should be copied, taking, for example, feedback as a model. Feedback (which is a process of constantly returning to an antecedent balance, or a reintroduction into some cycle of new elements destined to be reused so that the cycle can recur) cannot be a model either for human action or for human organization. When we speak of feedback in human systems, it is a question of something totally artificial, having nothing in common with the natural model. Of course, it is possible to argue that this utilitarian attitude is not moral, that to relate nature to whatever is necessary for human life is an untenable anthropomorphism. My reply is to ask what "nature" would be without man. An empty sky. An empty earth that could continue its cycles without any meaning. It is man who gives meaning to nature by modifying it, by not limiting himself to copying it. Man has to respect this natural environment in order to continue his own work which is precisely the invention of something other than this environment.

The other example is the instincts. We see the spread of proclamations on the excellence of instincts which we must obey as opposed to the abnormal and perverse character of law, morality, etc., and, particularly nowadays, the liberation of the sexual instincts from prohibitions and taboos. Instincts are good because they are natural. It is most remarkable to note, however, how much the human "instincts" in the area of sexuality are different from animal instincts. We know that animals mate about three months during the year. Since man has no cyclical limitations, human beings make love anytime and not in terms of their needs, but in terms of a desire totally divorced from the natural and the procreative. This sexual "instinct" of man is perfectly anarchic and explosive and it can be tolerated only if it is transformed, reworked, modified. This change can take place in an aesthetic manner and then it will be eroticism (which unfortunately is confused with the frantic liberation of the sexual instinct although it is a perfectly artificial construction). More frequently, and fortunately for the survival of human groups, the change takes place with the elaboration of rigorous behavior, taboos and prohibitions that finally make the anarchic sexual behavior of man tolerable for all. It is, however, the same problem as with violence—to the extent that for human beings there are no fixed and determining laws of behavior, man must freely and artificially give himself some laws of peculiar behavior which are not the expression of his nature nor the copy of a natural model.

Secondly, nature is not a limit in itself. That is to say, man has always set himself up as an enemy of nature, as an exploiter, as a more or less respectful transformer. Furthermore, to the extent that nature has always been his enemy, it does not present any absolute and impassable limit to him. There is no natural limitation on his action—that is, no other limitation than the fact that if he transgresses the "laws" directly he dies, but this is a problem we have already discussed above. There is a point at which, if he wants to stay alive, man should choose not to go on. Nature, however, does not dictate any conduct, does not impose any limit on him: he can go on just the same and commit suicide. Surely man has an instinct for life, for self-preservation that will lead him to obey the laws of nature in order to survive? Yes, but it seems to me obvious that there is also a death wish (despite recent denials by philosophers and psychoanalysts who claim it did not exist for Freud) which leads to the passion for suicide and absolute risk. Even if it were not possible to speak of it on an individual plane, I find it indubitably present in the history of societies. There are periods when a society wants to self-destruct completely.

We encounter the same ambivalence with reason. For a very long time, it has been asserted that reason is the nature of man—that by obeying nature man behaves reasonably, and that reason leads to a behavior in conformity with nature. It seems to me that when we consider human reality, we are now in the presence of a construction that is both ethical (and it is desirable that it be so) and metaphysical (if pushed, and despite appearances, I would affirm that this is the case). Yet, there is nothing "natural" here. Reason is a human conquest. It does not seem that "primitive" man had reasonable, let alone rational, behavior. It seems that progressively and in different parts of humanity reason has been affirmed and when people pass to the stage of rationality, they clearly have entered into the realm of pure artificiality. When reason draws reasonable limitations on human action, it is a deliberate act, a choice which is not at all dictated by a natural impulse. Reason imposes a mode of behavior on man—one that is not derived from nature, for the nature of man is just as irrational as it is rational. There are, then, no limits in themselves within nature, for nothing is irrefutable or insurmountable. There have been only limits that man could not yet go beyond, simply because he did not have the means. When these were found, it was possible to transgress everything. We see then another shift in the problem—not the existence of natural limits in themselves, but rather the technical possibility or impossibility of going beyond them.

Finally, in order to confirm this certainty in another area (the certainty that it is an error to want to take nature as a model and limit, and that we must not theorize on the basis of nature), I shall go over two theological aspects.[7] First, contrary to classical theology, which tried, at any price, to find a continuity between nature and God's intervention, if we respect the biblical message, we are obliged to say that grace is not a super-nature, but the antithesis of nature, that grace is the contrary of nature. The Law in the Old Testament was an obstacle, a check, a canalization of the natural tendencies: grace is a pure contradiction. We merely have to read the parables of the Kingdom of Heaven to see at what point the activity of grace is the contrary to what nature would command (not to uproot the tares, not to reject the prodigal son, to pay the workers from the last hour the same as the others, to invite beggars and vagrants to the banquet of the king). Also, the behavior of Jesus (for example, the washing of feet: it is because I am the Lord that I act as a servant) is contrary both to "natural" nature, to human nature and to social nature. Grace which expresses the fullness of love contradicts the processes of nature which can be governed by "laws," but not at all by love. There is no love within nature; there is no "natural" love. Love as *Eros* is a human invention which is superimposed on the sexual instinct, and *Agape* is found only in God. Grace is a refusal of the natural.

The second point to underline is that the Kingdom of God is not the outcome, the fulfillment of nature, but the assumption and the recapitulation of history. As I have often discussed, the Kingdom of God is in no way a return to a previous state of nature (Eden) with the annulment of every human invention, nor is it the establishment of a natural balance. It is not the triumph of nature. It is the fulfillment of the will of God. Let God be all in all. Let us be able to see face to face and know even as we have been known. This has nothing at all to do with any kind of nature however reintegrated in its purity, its splendor, etc. The Kingdom of God is the irremissible communion between God and man, but with the fullness of man; that is, with all he has been, all he had made, all he has lived, his history. Put differently, we can put forward the scandalous proposition that *Christianity is anti-nature*. It does not refer us to a natural model. It is not a question of a fulfillment or re-establishment of nature and, if God loves all His creation, it is in terms of man and not in itself. Genesis is very clear in this respect. It is man who responds in love and not the stars: it is man who manifests the love of God in nature and not a nature by its own existence. Finally, within Christianity, nature is not superior to man, and the laws which operate within nature are not the school he should attend. He has to listen to the word of God, to know the will of God, to listen to Him and to love Him. That is the one and only school. There is no identity between the will of God and the laws of nature. A simple comparison: a technician who fabricates computers sets up mechanisms capable of operating according to some program. The technician is married; he loves his wife. Obviously he does not expect his wife to obey the computer program. It would not be a proof of the wife's love to function as a computer or to repeat the program.

When I say that Christianity is anti-nature, I do not mean that it is anti-creation. If man has to love, respect, admire and use creation with moderation, if he has to keep it and make it fruitful with his care, it is not because it is nature, but rather because it is the creation of God. Because he loves God, he has to love God's handiwork and to admire God in His creation and to praise God for this creation. The Bible teaches us to read creation as a mirror of God, but it sends us to Him through it. To be anti-nature is to challenge—to refuse to accept that nature is a value in itself, that it exists for itself, that its intrinsic laws are good. Christianity challenges the project brought into being especially since the eighteenth century in the West, a project which corresponds exactly to the religious transformation of nature brought about in antiquity—the adoration of the moon or the sun so strongly attacked by the prophets. To glorify the natural, as do the present-day nudists or all those who justify themselves by saying "because it is natural," is precisely to practice the idolatry condemned in the Bible and is, in reality, antihuman.

III. NECESSITY—FREEDOM

It is a quite traditional commonplace that life within nature is a life of freedom, while social life is a life of constraint and obligations. I myself have often shown that technique is a destiny, a fatality, an anti-freedom. Here, I should like to specify further something that is not of the order of the technological society. Beyond the shadow of a doubt, man enclosed within nature is in no way free. He is strictly conditioned by the environment and, as has been shown so often, he is constantly threatened by environmental forces whether these be animals, poisons, the danger of famine, illness, cold, or lack of water. Because of these things, he is tightly locked into a struggle for his survival that was surely no freedom.

Furthermore, nature is a combination of necessities. Everything science discovers has been summed up in the terms "chance" and "necessity." Anything that is not in the order of necessity is in the realm of chance. And chance does not correspond to freedom. Chance too is an order of necessity. Everything we know about heredity tends to show that man emerges in a web of necessities and of conditioning. Genetics does not give even a glimpse of freedom, nor can we look to culture for the source of freedom. For some years now, I have insisted on cultural conditionings—the fact that language is a yoke, that the social environment ensures "reproduction," and that culture becomes a second nature for man. As a result, everything in the natural order is of the order of necessity. This is what the nineteenth-century scientists were implicitly saying when they were searching to formulate laws.

Thus we see the extraordinary and astonishing error on the part of "liberals" when they figure they can guarantee freedom by a policy of "laissez faire" (nonintervention). Nonintervention in what? In economic or sociological laws? Not to intervene in the economy, so that wages and prices are settled all by themselves through the interplay of the economic laws of the marketplace? This approach, however, does not take into account the pain, the disasters, the uprootedness, the despair that this free play of natural laws is going to produce. The freedom in question is not human freedom; it is the freedom of economic laws and that's all there is to it. It's just like saying, "The laws of gravity, heat, etc., want this avalanche to fall here. There is nothing to do except let it fall and if there is a village in the way, that's too bad." There are as many good things as bad things in nature, but "good" and "bad" are anthropomorphic judgments. And, in fact, it is a question of good and bad things *for man*. When we assess that we ought to follow the laws of nature, this "ought" is clearly directed to man! Then the argument does become relevant. *Essentially nature represents the locus of human necessities, implacable fatalities and, in reality, man's destiny.* Man asserts his freedom only by putting checks on the apparently inevitable chains of nature, only by using these "laws" in order to bring about other results, only by opposing certain regularities to accomplish something other than the copying of the natural cycles—in other words, by involving himself in the artificial. The artificial (and technique) are the expression of human freedom and the road to this freedom. Man is free only in his own invention, when he builds an environment for himself (society) and gives himself the tools for it (knowledge, religion, art, and technique). It is here and not in nature that he is free, that he frees himself from natural servitude, and that he revives freedom. The natural realm is the world of necessity; the artificial realm is the expression of his freedom.

But it is necessary here to be careful not to transform this historical and sociological statement of fact into a philosophical, metaphysical, eternal truth. I am not saying that nature is THE DESTINY, nor that technique is THE FREEDOM. For me, there is neither absolute destiny nor absolute freedom. There are situations and circumstances more or less oriented in one direction or the other. Similarly, I am not saying that they are stabilized and established conditions. In the next section we will have to see the reversal that our society experiences in this area. For now I am trying to determine what has been the historic route man has taken and also the way in which man lived these things and these situations, more perhaps than the reality itself. For concrete man, the formidable natural environment was the center of fatal forces, and the means with which he had to struggle against it constituted his freedom. If this is the case, then we must accept an undeniable consequence: namely, that there is no irreparable contradiction between technique and Christianity. Coming from me, this conclusion might seem astonishing, to the extent that people have often interpreted what I have written as being based on this contradiction. If, however, on the one hand, technique cannot be declared evil because it is artificial and anti-nature (as would be the case if you drew from nature a theory, a generalization, a good and protective diversion); and if, on the other hand, as I have indicated, Christianity is anti-nature; then there is no opposition. If man becomes man through his declaration of an ethical intention and also by making tools; if man wins his freedom through artificial means, in the realm of the artificial, and by dominating natural determinations; and if, at the same time, the grace which is in Christ is the activity which truly liberates from what is inscribed in human nature (namely, sin); if grace, as opposed to nature, is the covenant and reconciliation of whatever by nature has become hostile; then there is even a coincidence between technique and Christian faith. Both are equally artificial and artful.

From these statements, however, we should not draw any unwarranted corollaries. In particular, I am not saying that one springs from the other. Technique is not brought forth from Christianity, nor does Christianity appear because of a certain level of technique. Both are the expression of a self-same human movement; both are situated in the same relationship to nature. No more. Similarly, in this coincidence, we should not see a permanent generality. This is why I have so often criticized the arguments of philosophers (Heidegger, Habermas, Mounier) and theologians (texts from the World Council of Churches, Harvey Cox). They present universal arguments, whether starting from a conceptual analysis in order to apply it to our times or from a concrete study of our times in order to derive a rule for every age. In other words, some commit Hegel's idealist error and others commit Marx's materialist error. They are missing (and this is the meaning of what I have been trying to shed light on from the beginning) the radical newness of our age—the fact that technique today has nothing at all in common with previous forms of technique; that the modern state in no way follows from royal, feudal, etc., political power; and that the ideas of ancient philosophers are of no use to us. With regard to Christianity, traditional interpretations as well as modernist adaptations are false and useless.[8] We must clearly understand that if man remains artificial and the artificer, then the relationship between nature and the artificial has been reversed, has been thrown into disorder, and we have to situate ourselves in relationship to this reversal. The considerable difficulty then is to become capable of thinking the reverse of what man has lived for thousands of years, to overcome our schemata and our stereotypes, to retrace our steps. That is to say, we now have to consider that what used to be means to freedom has become a condition of slavery and what has always been the mechanism of necessity has become the occasion for a possible freedom. This is the main difficulty for ecology which should not be a romantic dream or a medieval pastoral, but which should be correctly thought in terms of a precise analysis of the technical world. Again, we must ask ourselves how a similar reversal could be carried out.

IV. THE REVERSAL

History has changed course. Two phenomena have provoked this unbelievable reversal. First (something that could not be imagined before), it seems that technique carried within it immoderation, pride, unlimited power. In ancient times it was impossible to understand the strange relationship in the Bible between Cain and techniques, while technique was simply a modest means for survival and freedom. Today we can see the prophetic meaning of the passage. It is within this meaning that technique is essentially in contradiction to the limitlessness of God, the grace and love of God. It is arrogance competing with the biblical God; it is the concrete and unchained power against the symbolic and self-bound power of the God of Abraham, Isaac, and Jacob. It is the victorious demon against the humility of Jesus. Already technique contained all the possibilities for expansion, but it needed certain economic, social, and cultural conditions for it to blow up. Now we are immersed in this path of power, in this delirious expansion that no one can any longer control or measure or canalize. Everything is subordinated to its growth and everybody participates by his choices, his work, his consumption—the evidence of progress. As I have often said, it is no longer a matter of man's good will nor of man's ability to come up with a good use for technique. The only problem is the unlimitedness of the growth and the unlimitedness of the power. We can search out all the other material elements in human cultures and see that only technique contained, in itself, these implications. This possibility was included from the beginning, but it was unknown.

The second factor in the reversal involves the danger itself of freedom. Surprisingly, it too carried a sort of potential for indefinite growth. For man, it was a question of a hard-won freedom against the natural environment and then it was the struggle against social oppressions. As soon as freedom is secured, however, intolerable obstacles appear again. When nature is dominated, we are irritated that it does not always provide more possibilities for utilization. When space is dominated (by rapid transportation) we feel terribly bullied if vehicles are slowed down a bit, either by traffic problems or by administrative decisions. Obstacles to "freedom"! The more freedom there is, the more the anger against the little that remains of limitations. In the social realm there is exactly the same attitude; we have overthrown the tyrannical power, we have annihilated the taboos or collective moral ruses, but the few that remain seem like a monstrous tyranny. The few remaining limits on pornography, family or social constraints, or work obligations appear in turn as the negation of freedom. When there were genuine constraints, some genuine lack of freedom, the struggle that took place day by day, one step at a time in order to win freedom, was the struggle of hope and the affirmation of man, with heavy sacrifices and a vocation. Today, when we live in an incredible laxity, nobody wants to endure any more sacrifices for freedom. It is mistaken for laziness, negligence, stupid incoherence, infinite unreasonableness, etc. Actually, the change has been made greater on a material level. As long as man had few instruments for action, freedom for him was centered on body-to-body combat. He had to be involved completely in a contest where he knew the real obstacles that put him to the test. From the moment when he had powerful enough instruments to overcome all obstacles, then freedom changed its meaning and its significance. It is clear that freedom was the very life of man as long as he did not have the absolute means to put it into practice, as long as it was a difficult game to be played against real oppressions. It became a danger and a lie when it was matched up with unlimited means.

This double factor has brought about two results, the tragedy of which we are living out. First, the mortal danger of the exclusively artificial. Man is an artificer, *but in a natural environment* and in relationship to it. Everything is changed by the complete victory of the artificial which involves a negation and an elimination of the natural. When man replaces everything given him in nature with the artificial, then he makes an environment for himself that is totally unliveable. All the natural products replaced by plastic, all natural foods replaced by new chemical substances, to the extreme that the image promised us is that of the living conditions of the astronauts. This is even the image presented as the ideal in animated cartoons for children. We forget that man still remains a creature of flesh and blood and also of passion and freedom and that it is not human to be reduced to the stage of a guinea-pig living in a totally artificial world. The elimination of the natural does not coincide with what is still man. When I speak in this way, I am not referring to a human nature, but to the long march over thousands of years toward our present destination. That is, the destruction of the natural in favor of the artificial leads to the negation of what successive generations have desired and searched for throughout five thousand years. It is the negation of the human model which has been long elaborated throughout the course of history. At this point, there is a total negation: artificial man becomes the negation of man the artificer. He is himself subjected to the artificial that has become independent of him. In turn, this destruction of the natural environment removes any meaning for this human work, this art, these artifacts, for they had meaning and value only to the extent that they referred to an existing natural environment. Nature was the basis, the condition, the substance, the possibility for making the artifices that let man live. Human life was ensured to be meaningful in the interrelationship between the natural and the artificial. If the first is destroyed or allegedly replaced, the second no longer has any meaning or value. This reduction and transgression of nature is no longer the victory of freedom, but on the contrary, the destruction of freedom. In effect freedom exists only in the recognition of necessity. Where there is no longer any necessity, there is no longer any freedom. We know freedom only when we have to struggle against a destiny. Without that, we are in a weightless universe. It is very significant that beyond the pull of gravity, it is impossible to move. The weight which keeps us on earth is also the force which allows us to distinguish high and low, to situate ourselves and, even more, allows us to walk, to move forward. Without this weight, we gesticulate in a perfectly ridiculous manner coming to nothing. Without this heaviness, we are much freer than with it, for, freed from this constraint, we can certainly float in the air, but without having a hold on anything. Here we put our finger on the folly of those who have always dreamed of escaping the constraints of this earth and of being *free* because of being no longer connected to the ground. We discover then that this "freedom" is only an impossibility of doing whatever and an impossibility of being. It is a freedom of perfect indifference in which right and left or high and low are identical and meaningless. It is a freedom that "man" at this moment shares with things, since an object thrown into this weightlessness floats in just the same manner. Man can no longer intervene at all; he has been reduced exactly to the state of an object. In effect, this is what happens to us globally, collectively. In return for having made a totally artificial universe, having eliminated the natural constraints, we have placed ourselves in a world where freedom is no longer possible *at all.*

We come then to the last implication of this exclusively artificial world, to what Bernard Charbonneau has called "the lie of liberty."[9] This lie seems to me to be twofold. First, it consists in having transformed into an ideal, an absolute, a principle, a declaration, or a system what really is the order of experience, the concrete, the lived, or the everyday. The lie consists of changing freedoms into Freedom, of trusting in the program of Freedom when we are deprived of all practical freedoms, of claiming to be free through the realization of new institutions. Secondly, there is this transformation that has taken place: that which was the instrument of freedom has become our slavery. Technique, as I have said, used to ensure a certain freedom for man, but now everything has changed. Technique has ceased to be an instrument and become instead the law of our being. It has its own determinations, it is autonomous and it develops independently of the will of man. It is a fatality that conditions man in everything. Following his path slothfully, man maintains that since technique has increased, his freedom has increased as well. Yet this argument is based on an assimilation of our situation with that of the past. Nothing is harder than to understand this reversal of roles and that the liberator of yesterday has become today an even more authoritarian master than the one he succeeded in conquering and dominating.

* * *

In our situation it is truly impossible to go back in time, to recover nature, to move back to a natural level, and to yield again to the laws of nature. We cannot forget our history. I have also said that it is inconceivable to take the laws of nature as a model. We can take only ourselves as a model! But under what constraint? Heraclitus wrote, "The sun itself could not transgress its limits without the Erinyes catching it in the act" (fr. 94). This formulation has often been wrongly interpreted by those who believe in the noncontravenable laws of nature. Heraclitus does not say that "the sun cannot . . ." but that if it does do it, then "the Erinyes. . . ." Put differently, transgression involves the vengeance of the Furies. This is exactly what we now know. Let us not say that "nature takes vengeance" for that would imply an anthropomorphism which I have criticized above. Nevertheless, by the excess of technique and liberty, we ourselves have set in motion a destructive reaction. We are confronted with a decisive choice which must be conscious. The choice is between whatever is going to allow the continuation of life and whatever is going to end it. It is between life and death, bearing in mind that life for the human species is both biochemical and spiritual—the two have to go together. It is not only a question of death through atomic destruction, but of everything that produces the extinction of that which has characterized the human being.

Now what avenues are open to us? A return to the past or to nature is excluded, at least in the order of voluntary choice, although it is not impossible that, after an atomic catastrophe, a remainder of humanity would find itself back in "natural" conditions—a classic theme in science fiction that we cannot go into here. Yet it is time to proceed to the voluntary choices. Once again we have to choose the artificial. But what kind? It is a question of what I shall call the Second Level of the artificial. In what would this Second Level consist? One temptation open to us is thinking that technique will resolve technical problems, a solution that is correct for anything involving *individualized and fragmentary problems,* but radically incorrect for the central problem I have alluded to here. No technique is able to respond to nor is it permitted to dominate the complex phenomenon of technical pride and incoherent freedom. Some techniques will be useful in responding to such and such an inconvenience piecemeal (for example, it is possible to mitigate pollution, to reduce the effects of an immoderate consumption of water or energy, etc.), but it is not a collection of specific problems that constitutes the situation we are in. Technical solutions (necessarily more and more technocratic) will not suffice. The classic formula—that it is not by grace of less technique but by more technique that the situation will be altered—is thus at once correct and incorrect.

Also, we must not think that the Second Level of the artificial will be made up of the information processing system perfecting and superimposed on the technological system. Elsewhere I have explained that modern technique implies an ensemble of techniques of the second degree, situated on the informational level. This second stage of information processing is not, however, the second level of the artificial.

Finally, this second level cannot be formed by science. We see more and more scientists thinking that science is the only way open that will allow us to overcome all our problems—"more science." In France, there are Monod, Jacob, Laborit, etc. from the natural sciences, but also E. Morin and many others from the social sciences. In the "committed" world, the Marxists think that Marx provides the true science and thanks to this science all our problems will be resolved. Now I do not have to express my opinion on the validity of such and such a science (namely, whether Marxism or the social sciences are really sciences), but I proceed from one statement of fact. Man has always succeeded in rising to the test, in meeting the challenge presented to him by means other than those which are the source of the challenge. To respond to the danger of nature, man did not remain on the level or within the framework of nature. He created the means and then the universe of the artificial. Here we must rigorously draw out the implications of the fact that technique has established the new environment for human life, completely replacing nature as environment. One of the consequences is that science is essentially the same as the new environment. It is not at all the means for acting on the environment in question because it both produces this environment and is produced by it—the cause and effect of this environment without being in the least dissociated from the world. *Science is a reduplication of the technical environment.* It cannot provide any escape because it can only confirm what already is. To have recourse to science, *in our day,* would be to do exactly what the "animal-man" in the first ages would have done had he buried himself in the natural, not differentiating himself at all from the environment, copying it and identifying himself with it—that is to say, "refusing" to become man by refusing the route of the difference, the distance, hence the artificial.

The question is whether, with respect to the new environment, we are going to be capable of reproducing what specifically determined man with respect to his first environment. Are we going to differentiate ourselves from it, to use it, to master it, to reorganize it, just as man, through the first level of the artificial, was able to do to the natural environment? But what orientation do we choose now? Perhaps we have one indication from the analysis itself of the artificial on the first level. We have seen that it was characterized by two operations instituted by man—tools and the formulation of rules of behavior not imposed by reflex and adaptation, but chosen in order to dominate nature. With regard to the technical environment, we already have some tools for domination (and it is here that I place the techniques of compensation and information that certainly are not negligible), but we have no new orientation for behavior. Put differently, taking into consideration the profound reality of human history, the specification of man, on the one hand, and, on the other, the reality of the new environment, I am led, not by an arbitrary choice nor by an ideological preference but by the situation itself, to think that the second level of the artificial is ethical. With respect to this environment, it is a question of reproducing the decision of man formulated in the commandment, "Thou shall not kill." Of course, there can be no simple return to a traditional morality, since this involves taking a position toward the natural environment. It is indeed an ethical question, but one to be constructed and produced in order to discover not what conduct to follow (for this path would come back to the belief that it is enough to change how technique is used), but the *being* to be assumed. Who should we be in order to respond to the challenge and place ourselves on the second level? The experience of the first level teaches us, in any case, that if there is one impossible way, it is the path of adaptation. All those who preach a better adaptation of man to technique are the grave-diggers of the human species. On the contrary, it is by refusing the technical conditionings, by distancing ourselves in relationship to the conditions of life we have made, by allowing ourselves (through a radical critique of the new environment) to put it in front of us as an object—thereby denying its fatality. Even if it takes a "magical" operation comparable to the one carried out by primitive man on nature, it is by this path that we can discover which artificial will allow us to master the artificial. There is no other escape and no other possible reference.

In this area I shall distinguish two orientations: recognition and limits. Recognition is a form of complex research; it is recognition of the real, of our past which remains inviolable, of the multiple aspects of reality, recognition of the *other* as well as recognition of *nature*. This task takes place only within a constellation made up of contemplation, communion, respect, meditation (but not in itself, or centered on the Infinite or the Absolute, but on the finite and the contingent!), consideration. This framework for recognition is basically new in relationship to the usual grouping of Western ethical orientations. Further, it is a question of the recognition of simple things, of elementary "principles" (as against the present intellectual sophistication that can be good for nothing and makes us more and more impotent toward our environment; I am thinking of the whole Parisian school—Lacan, Derrida, Guattari, but also Foucault, etc.). It is the recognition which is a movement of discernment and reclassification of the essentials. Recognition as opposed to desire. Recognition in the sense of joy and gratitude for what we have, for what we are and for what we have been given. Recognition instead of demands (as against the whole school influenced by Marxism which makes class struggle the beginning and the end of every explanation and all behavior).

The second major orientation is the attempt to find and set voluntary limits on action. Now man has to limit his consumption and his demography. By himself, he should set limits on his powers. This idea is totally different from that of a "good use of means," but on the contrary, it moves exactly in the direction of man's search for a restraint to complete the insufficient physiological restraints—the search, as we have seen, that allowed man to become man. To set limits, to give up his own power, not to do everything it is possible to do, to reject the "always more" (larger, faster). To refuse both the temptation to unlimitedness (the conquest of space!) and the identification of freedom with the disappearance of limits. To learn, on the contrary, that it is the formulation of limits on his own action that is the act of freedom *par excellence*. The refusal of fatality and dogmas, through the critique of power, becomes the imperative for setting limits, ones that are genuine because they are coherent and not fleeting unto utopia. *This, above all, is the supreme test.*

What I have just said is not new? I know, but what is important to me is whether it is accurate. What I mean to show is that this way is the *one and only possibility* for man to resolve the problem posed. It is not at all idealistic, but given the situation, it comes from a rigorous realism. (The disconnected idealists are the politicians living in their imagined reality.) My suggestion is not a gratuitous choice founded on personal preferences, but is the result of a rigorous analysis both of "becoming man" and the modern situation in which the artificial of the first degree has placed us. The artificial of the second degree can be only the invention of an ethics allowing man to situate himself and live in the world he has made for himself. All the rest is futile talk. Of course, and this is the big question, an ethic cannot be made artificially: it cannot be the product of one person. It is a question of a collective elaboration (which can be formalized by one person or group, but no more) and a common adoption of a certain way of being. Nothing guarantees us that people, all together, will enter the path for which I am laying down some signposts. At the same time, nothing allows us to say that they will not do so. Presently there are some small indications that did not exist twenty years ago. There is a certain awareness. There are some attempts at new living. Probably the elaboration of human morality and society commenced in the same manner. The second level of the artificial does not imply that we are supermen nor that everyone will adopt this new way of being without any violations. ("Thou shall not kill" was also transgressed; yet it is what allowed the species to exist and man to be constructed as man.) The second level of the artificial does imply that this new way of being will be laid down as the human aim, not abandoning responsibility either in institutional change (politics) or in utopian evasion (religion).

Translated by Katherine Temple;
edited by Carl Mitcham and Jim Grote.

NOTES

1. Clearly, we must not generalize, as is done so often, from the fact that ethnologists have discovered here and there some tribe, some small human group that is peaceful, nonwarlike, nonaggresive. Such a situation is an accident and not the general prehistoric condition.

2. This is why I agree only partially with Lévi-Strauss's explanations of the elementary primitive structures of parenthood. He shows their artificial character clearly, but he is wrong to structure them too rigidly in terms of exchanges between groups by the bartering of women.

3. Here we see the absurd character of the present attitude that is destructive of taboos and moral norms: Man himself is destroyed in the name of a pseudo-Freudianism. Freud saw more clearly when he spoke of sublimation.

4. Jacques Ellul, "La Technique et les premiers chapitres de la Genèse," *Foi et Vie* 59, no. 2 (1960): 97–113. See also my *The Meaning of the City* (Grand Rapids, Mich.: William B. Eerdmans, 1970).

5. I would like to underline the general misunderstanding at the time of the publication of my first book on technique in France, when people thought that I was "condemning" technique from the standard of nature and the excellence of natural man. (See, for example, Duverger, "La Langue d'Esope," *Le Monde*, Spring 1954.)

6. Except in certain texts of Proverbs (e.g., Proverbs 6:6—"Go to the ant, thou sluggard; consider her ways, and be wise").

7. It is beyond the scope of this article to state the radical difference between the concepts of creation and nature, which are perfectly antithetical. The Bible does not know the idea of nature.

8. I am pointing to three tendencies: a) the *adaptation of Christianity* to what some people believe to be the new social reality (e.g., abundance, rapid "change," decolonization, socialism, etc.); b) the *following of an ideology* which some people hold to as the actual truth to which Christianity must be adapted—I mean those who are convinced of the truth of Marxism, communism, materialism, and who want a materialist reading or interpretation of Christianity in terms of class struggle; and finally, c) the *view of scientists* convinced that rigorous methods allow a more accurate understanding of the Bible—for example, structuralism or linguistics. Actually, these honest theologians radically by-pass the problem that the modern world presents to man and they render Christianity completely useless.

9. This is the title of an admirable book by Bernard Charbonneau, in my opinion the most profound analysis of our age and our society—but one which has never been able to find a publisher.

PART IV:
SYMPOSIUM ON CRISES IN TECHNOLOGICAL PRAXIS

SHOULD MONTANA SHARE ITS COAL?

TECHNOLOGY AND PUBLIC POLICY[1]

Albert Borgmann

In this paper I want to explore the question how the philosophy of technology can illuminate issues of public policy. There is a growing uneasiness about the force of technology in modern society, and the recent eminence of the term "public policy" is itself an indication of that concern. Developments that formerly seemed to be well directed by the mechanisms of the market and by private initiative now appear to threaten the commonweal. They call for collective responsibility and action. It is the entry of new concerns into the realm of public discourse and decision that has given public policy its present position of prominence.

Yet the universe of public discourse is poorly equipped for the consideration of our deep-seated uneasiness. Discourse on these matters fails

Research in Philosophy & Technology, Volume 3, 1980, pages 287–311.
Copyright © 1980 by JAI Press Inc.
All rights of reproduction in any form reserved.
ISBN: 0-89232-102-4

more often than it succeeds.[2] The difficulties are so profound that it is not even clear what precisely is at issue. There is something like agreement on two issues, however. First, it is the conflict of technology and the environment that is often at the center of these newly searching debates. Second, the greatest difficulty in these debates stems from our inability to deal in a principled manner with certain values, variously called soft, fragile, humane, qualitative, intangible, or higher.[3] The second agreement is of course little more than a confession of perplexity, one that Dorfman has put as follows: "The challenge is . . . that our skill in attaining objectives has outrun our skill in determining or even articulating them. One might say that science has done its work but philosophy has not, and the gap has become so egregious that even the scientists have noticed."[4]

I agree that the dominant perplexity is a challenge to philosophy. But in what follows I want to suggest that even the tentative and general terms in which this challenge is brought to the fore already mislead our deepest concerns and aspirations. More precisely, the means-ends distinction that Dorfman invokes to make his point is far from innocent, and so is the value talk commonly employed. I want to argue that these parlances reflect features of a framework that is normally taken for granted and that the success in our attempts to do justice to our profoundest concerns depends on whether we can bring out and consider this framework in its crucial features. I believe that this is possible and that from the delineation of this framework we can go on to show how the framework is commonly understood and what kind of allegiance it commands.

I share the common agreement that the conflict of technology and the environment is a vortex of our uneasiness. Hence the following considerations depart from such a conflict. Montana's coal policy is of course an enormously complex phenomenon. It lies within still wider and more difficult issues, those of national energy policy, of the standard of living, and of the common good. It comes to a narrower focus in the question whether Montana's coal should be exported only or converted into and exported as electricity. This question again has, within the last five years, turned on the specific issue of whether two 700-megawatt generating plants should be built at Colstrip, Montana (commonly called "Colstrip 3 and 4"). It is obviously impossible to give even a rough survey of these issues within the space of an essay. I will simply draw on these materials for guidance to the essay's central considerations, and I will return to some of the issues to see what illuminating force those considerations may have.

To begin with the title question, the issue is not whether Montana should begin to share its coal. More than 27 million tons were mined in Montana in 1977, and most of it went out of state.[5] The initial question is rather on what basis decisions to mine and export coal have been made in

Montana. It seems obvious at first that the answer should depend on whether or not Montana's coal is needed for our energy requirements. The pivotal significance of need is recognized in the Montana Major Facility Siting Act.[6] It provides "that no power or energy conversion facility shall hereafter be constructed or operated within this state without a certificate of environmental compatibility and public need."[7] But what is need? There is no lack of investigations. But their findings, taken together, stress nothing so heavily as the elusiveness or subjectivity of needs.[8] One may try to objectify the presence of needs in two ways. The first is by polling. In regard to Colstrip 3 and 4, a state-wide poll was taken, and a plurality of the respondents saw no need for the electricity from these plants.[9] But there are two problems with this finding. First, the poll was not representative. The people who take the trouble of responding to such surveys belong to the "opinion leaders" or to the "attentive public."[10] They constitute the politically active population. Two-thirds of the voting-age population surrender the shaping of politics to the remaining third, and one-third or one-half refrains from participation altogether.[11] The second question regards the quality of the response. It may have been incompatible with other views of the respondents.[12] Or their information might have been deficient. Alternatively, one might objectify the need for electricity by equating it with actual and projected consumption. But past consumption might have been wasteful. Even if it was not, there remains the problem of how to provide a projection of future consumption or need. In a coordinate system the data of past consumption yield a scattering of points which accommodates very different projection curves.[13]

The determination of public need for generating plants in Montana is made by the Board of Natural Resources. It relies in part on the research of the Department of Natural Resources. The notion of need is one of two decisive concepts in the Major Facility Siting Act. But it does seem prohibitively elusive, and the staff of the Department gave little help to the Board in capturing its meaning. The *Statement,* prepared by the staff, concedes that "there is probably no criterion of need that is universally acceptable." Hence "need itself is not defined," and the "difference between need and consumption is left to the reader to decide."[14]

Yet the scruples that thoughtful analysts of need avow contrast with the ease and assurance with which proponents of new generating plants use the term. The *Environmental Analysis* of Colstrip 3 and 4, prepared for the proponents, puts it this way: "Many factors account for the anticipated increase in power requirements which make the Colstrip plant necessary. They include higher standards of living, pollution control, waste treatment and industrial expansion."[15] The presidents of the Montana Power Company have been more emphatic. One foresaw a "real catastrophe'' in future years with massive power shortages and closure of state industries by 1979 or 1980 if Colstrip 3 and 4 were not allowed to be built.[16] His successor warned: ''Without Colstrip 3 and 4 we can forget about economic growth, about creating new jobs, about expanding agricultural production, about all the sound, healthy economic development Montana needs so desperately.''[17]

The warnings of doom aside, the proponents of expanding power generation appear to be in agreement with empirical findings regarding the hopes and fears of the people in this country. A high and rising standard of living seems to be among their most important concerns.[18] Although the notion of the standard of living seems vague and far removed from genuine need, closer inspection shows it to provide a fruitful approach to the framework within which people orient themselves.

A rising standard of living provides increased material well-being. The latter, in turn, is thought to be the basis for the realization of most of the great human values. A rise in the standard of living is in fact regarded as very nearly the same as an increase of freedom. In a simulation of the political shaping of the future, one group of policy makers was asked to maximize the gross national product, the other to enhance freedom. The shape of the two resulting worlds was very similar. Freedom was, however, more sought in both worlds by the representatives of the economically less powerful members.[19] Accordingly, other studies have found that poverty is experienced in this country as a lack of freedom. To escape poverty is to become free of debts, free from dirty surroundings, free to take long vacations in distant places, free to go to films, to concerts, to the races, and to eat out.[20] Since energy is the life blood of affluence, an energy crisis is a crisis of freedom.[21] On this view, Montana, in sharing its coal, would aid the cause of freedom.

The bond between values other than freedom and growing wealth is less tight. Still, even so thoughtful and radical a critic as Feyerabend expects that an increase in the number of the devices that technology has given us will make possible the rise of ''the freedom, the charm, the beauty of Greece'' for all of us.[22] Rescher believes that ''there seems little doubt that in the better-taught and more leisure-oriented America of the year 2000, artistic creativity and the arts (creative and performing) generally will occupy a far more important and influential place than they do today.''[23] More broadly, the wealth that technology has given us is said to constitute ''an Age of Opportunity.'' ''Our modern civilization,'' says de Jouvenel, ''is far superior to any other in terms of power, and thereby we enjoy undreamed-of possibilities to foster the good life not for a tiny minority, as in earlier societies, but for the multitude. This is a wonderful privilege. . . . It is right to rejoice in the greatness of these means: it is proper to feel responsible for their personal and optional use.''[24]

And sometimes these opportunities are claimed to be so extensive and radically new that we are in danger of lapsing into shock.[25] If a high standard of living is so fundamental and fruitful for a creative and vigorous life, then the only consistent alternative to the pursuit of a higher standard of living seems to be stagnation, resignation, or a world-denying asceticism.

These views of the significance of the rising standard of living can be condensed into the following points: (1) A high standard of living provides rich opportunities for a great variety of human achievements. A rise in the standard of living will further enlarge the range of opportunities and achievements. (2) A strong majority of the people in this country are deeply committed to a rising standard of living even though they are overwhelmed by the novelty of the coming opportunities. These claims, it seems to me, reflect a serious misreading of the contemporary world. I want to argue to the contrary the following theses: (1) The present high standard of living is not the basis of a variety of achievements but is itself an achievement of a definite style with a well-defined pattern. The outline of the pattern has become clearer over the last century and a half and it will be further sharpened. But this development is devoid of radical novelty. (2) In a profoundly implicit way, this pattern is well understood by people. They have implicitly agreed to take up with reality according to that paradigm. But their commitment to this agreement is increasingly ambivalent.

This paradigm provides the framework within which people normally orient themselves. The appropriation of reality within this framework or according to this paradigm is what I want to call technology. Thus technology, formally defined, is the characteristic way in which contemporary man takes up with reality. Obviously this definition is too narrow and too wide in relation to well-established uses of the term. It is not my aim to capture all or even the major senses of "technology."[26] I regard the proposed use of the word as justified if the word in this usage, first, embodies one good sense of "technology" and, second, has from its various common uses more helpful connotations for its intended denotatum than other possible titles for the present enterprise. To establish the first of the two theses above we must give substance to the formal definition of technology and show that there is in fact a definite and distinctive pattern to the way in which man today takes up with reality. Let us begin with a tentative and summary thesis regarding the goal of contemporary commerce with reality. It is the procurement and enjoyment of availability. Something is available if it has been rendered instantaneous, ubiquitous, safe, and easy.[27] Warmth, for example, is now available in this sense. We get a first glimpse of the distinctiveness of availability when we remind ourselves that warmth was not available in Montana a hundred years ago. It was not instantaneous because in the morning a fire first had to be built in the stove or the fireplace. And before it could be built, trees had to be felled, logs had to be sawed and split, the wood had to be hauled and stacked. Warmth was not ubiquitous because some rooms remained unheated and none was heated evenly. The coaches and sleighs were not heated, nor were the boardwalks or all of the shops and stores. It was not entirely safe because one could get burned or set the house on fire. It was not easy because heavy work, some skills, and attention were required to build and sustain a fire.

Such observations, however, are not sufficient to establish the distinctiveness of availability. In the common view, technological progress is seen as a more or less gradual and straightforward succession of lesser by better implements.[28] The wood-burning stove yields to the coal-fired central plant with heat distribution by convection which in turn gives way to a plant fueled by natural gas and heating through forced air, and so on.[29] To bring the distinctiveness of availability into relief we must turn to the distinction between things and devices. A thing, in the sense in which I want to use the word here, is inseparable from its context, namely, its world, and from our commerce with the thing and its world, namely, engagement. The experience of a thing is always and also a bodily and social engagement with the thing's world. In calling forth a manifold engagement, a thing necessarily provides more than one commodity. Thus a stove used to furnish more than a mere warmth. It was a *focus,* a hearth, a place which gathered the work and leisure of a family and gave the house a center. Its coldness marked the morning, and the spreading of its warmth the beginning of the day. It assigned to the different family members tasks which defined their place in the household. The mother built the fire, the children kept the firebox filled, and the father cut the firewood. It provided for the entire family a regular and bodily engagement with the rhythm of the seasons that was woven together of the threat of cold and the solace of warmth, the smell of woodsmoke, the exertion of sawing and of carrying, the teaching of skills, and the fidelity to daily tasks. These features of physical engagement and of family relations are only first indications of the full dimensions of a thing's world. Physical engagement is not simply physical contact, but the experience of the world through the manifold sensibility of the body.[30] That sensibility is sharpened and strengthened in skill. Skill is intensive and refined world engagement. Skill, in turn, is bound up with social engagement. It molds the person and gives him character.[31] Any one person's limited skill confines his primary engagement with his world to a small area. With the other areas he is mediately engaged through his acquaintance with the characteristic demeanor and habits of the practitioner of the other skills. That acquaintance is importantly enriched through one's use of their products and the observation of their working. Work again is only one example of the social context that sustains and comes to be focused in a thing. If we broaden the meaning of "thing" to include events, we can see similar social contexts in entertainment, in meals, in the celebration of the great events of birth, marriage, and death. And in these wider horizons of social engagement we can see how the cultural and natural dimensions of the world open up.

We have now sketched a background against which we can outline a specific notion of the device. We have seen that a thing such as a fireplace provides warmth, but it inevitably provides those many other elements that compose the world of the fireplace. We are inclined to think of these additional elements as burdensome, and they were undoubtedly often so experienced. A device such as a central heating plant procures mere warmth and disburdens us of all other elements. These are taken over by the machinery of the device. The machinery makes no demands on our skill, strength, or attention, and it is less demanding the less it makes its presence felt. In the progress of technology, the machinery of a device has therefore a tendency to become concealed or to shrink. Of all the physical properties of a device those alone are crucial and prominent which constitute the commodity that the device procures. The remaining properties are changed, normally on the basis of scientific insight and engineering ingenuity, to make the commodity more available. Hence every device has functional equivalents, and equivalent devices may be physically and structurally very dissimilar from one another.

The distinction in the device between its machinery and its function is a specific instance of the means-ends distinction. In agreement with the general distinction, the machinery, the means, is subservient to and validated by the function, the end. The technological distinction of means and ends differs from the general notion in two respects. In the general case, it is very questionable how clearly and radically means and ends can be distinguished without doing violence to the phenomena.[32] In the case of the technological device, however, the machinery can be changed radically without threat to the identity and familiarity of the function of the device. No one is confused when he is invited to replace his watch, powered by a spring, regulated by a balance wheel, displaying time with a dial and pointers with a watch that is powered electrically, is regulated by a quartz crystal, and displays time digitally. This concomitance of radical variability of means and relative stability of ends is the first distinguishing feature. The second, closely tied to the first, is the concealment and unfamiliarity of the means and the simultaneous prominence and availability of the ends.[33]

The concealment of the machinery and the disburdening character of the device go hand in hand. If the machinery were forcefully present, it would *eo ipso* make claims on our faculties. If such claims are felt to be onerous and are therefore removed, then so is the machinery. A commodity is truly available when it can be enjoyed as a mere end, unencumbered by means. It must be noted that the disburdenment which rests on a feudal household is ever incomplete. The lord and the lady must always reckon with the moods, the insubordination, and the frailty of the servants.[34] The device provides social disburdenment, i.e., anonymity. The absence of the master-servant relation is of course only one instance of social anonymity. The starkness of social anonymity in the technological universe can be gauged only against a carefully drawn picture of the social relations in a world of things.[35] Such a picture would also show that social anonymity necessarily shades off into one of region, culture, and history. Only those social, regional, cultural, and historical elements remain apparent in technology which are themselves procured as commodities. But such procurement only aggravates the dissolution of the context and rootedness of the original world.

Even this analysis of the distinctiveness of the device is deficient, and again the deficiency can be brought into relief through two objections. Is not, one may ask, the concealment of the machinery and hence the specific character of the technological means-ends distinction due to widespread scientific, economic, and technical illiteracy? And quite apart from one's level of education, is not everyone in his work directly and explicitly engaged with the machinery of devices?

We can approach the first point through one of its companion phenomena, people's alleged unwillingness and inability to maintain and repair technological devices.[36] How well founded is this allegation? One way in which commodities are made available is that of making them discardable. It is not just unnecessary but impossible to maintian and repair paper napkins, cans, Bic ball points or any of the other one-way or one-time devices. Another way to availability is that of making products care-free. Stainless-steel tableware requires no polishing, plastic dishes need not be handled carefully. In other cases maintenance and repair become impossible because of the sophistication of the product. Microcomputers are becoming increasingly common and influential as devices that free us of the tasks of allocation, record keeping, and control. The theories and technical processes that underlie the production of microcircuits are too complicated and too much in flux to be known in detail by more than a handful of people. And the microcircuits themselves are realized at a functional level so minute and dense that it does not permit the intrusions necessary for repairs even if structure and functions are fully understood.[37]

Still, education in engineering and in the natural and social sciences would make much of the machinery, i.e., of the context, of technological devices perspicuous. But even if such education were to become more common, the context of functions and commodities would remain different from the world of things for two reasons. First, the presence of that context would remain entirely cerebral since it increasingly resists, as we have seen, appropriation through care, repair, the exercise of skill, and bodily engagement. Second, the context would remain anonymous in the senses indicated above. The machinery of a device does not of itself disclose the skill and character of the inventor and producer; it does not reveal a region and its particular orientation within nature and culture. In sum, the machinery of devices, unlike the context of things, is either entirely occluded or only cerebrally and anonymously present. It is in this sense necessarily unfamiliar.

The function of the device, on the other hand, and the commodity which it provides are available and enjoyed in consumption. The peculiar presence of the end of the device is made possible by the means of the device and its concealment. Everyone understands that the former rests on the latter and everyone understands as well that the enjoyment of ends requires attention to the means. Only in magic are ends literally independent of means. The inevitable explicit concern with the machinery takes place in labor. But labor does not in general lift the veil of unfamiliarity from the machinery of devices. The labor process is itself transformed according to the paradigm of the device. A first indication is the use of devices in the production process where every device has its own unfamiliar machinery. The fundamental character of this transformation derives from the endeavor to remove from the process of production the constraints of time, skill, and risk. Production itself, as far as possible, is to be instantaneous, easy, and assured. For the skilled worker, such transformations constitute the heartrending loss of his work world. The laborer who first moves into the labor force is largely unacquainted with the exercise of skill. It has already been made superfluous in the area of leisure and enjoyment. The laborer is not suprised that he is not expected to acquire a skill. Labor remains toil, of course, and although technological labor as a rule is lighter and safer than work in a world of things, labor is more draining and brutal than skilled work.[38] To the narrow engagement in leisure which results in disburdenment there corresponds a mutilated engagement in the labor process which results in numbing repetitiveness. To the mindlessness of consumption there corresponds the mindlessness of labor. Most everyone tolerates the demeaning character of labor since it is after all a dealing with the machinery of devices, with the means which no one can expect to be its own reward. Technological progress can ameliorate the conditions of labor by making them safer, more pleasant, and shorter in time, and above all, it can vastly increase the end and purpose of labor, the consumption of commodities. This is what the labor movement fought for once it had accepted technological progress.[39]

It has scarcely, since then, endeavored to restore labor to the fullness of work.[40]

The transformations that resulted in the loss of skill were themselves the result of great skill on the part of scientists and engineers. But that sort of skill is essentially different from the skill of craftsmanship. The skills of technological innovation are not embodied skills but largely theoretical, intellectual, and intuitive. They are not common skills in the sense that the mechanization of bread baking, for instance, need not be invented by every bread factory.[41] If the skill of mechanization has been exercised successfully once, it has been exercised for all. There is a highly skilled elite at the leading edge of technological innovation. But every one of its successes makes a skill obsolete. Hence the circle of skill is contracted more and more and the army of essentially unskilled laborers grows accordingly.[42]

The consideration of the skill that is proper to technology leads us to two final features of the device paradigm. The people at the leading edge of technology are engaged in enlarging the universe of availability through the procurement of more and better devices. Their number comprises not only scientists and engineers but also businessmen, lawyers, administrators, and all the persons who make up Galbraith's technostructure.[43] They are pioneers and conquerors; their leading role in giving this era its character fills them with pride and satisfaction. The growing domination that they establish over the world and the gratification that flows from it is shared by the other members of the technological society vicariously and secondarily. They experience domination by enjoying its fruits. There is a peculiar restlessness about domination that comes to the fore when it is contrasted with mastery. In mastery man has through discipline attained the skill that allows him to be equal to some great thing. Art and athletics are areas where mastery can be found. A thing that has been mastered, such as a sonata or a slalom course, has not been destroyed but rather has been heightened in its significance. It remains a challenge because in every great performance man risks himself; he goes to the limits of his ability in the face of possible failure. Domination, however, in procuring availability, overcomes the challenge of things by replacing them with instantaneous, ubiquitous, safe, and easy commodities. The act of procurement, achieved by the technostructure, is a triumph which is echoed in the thrill of the consumer who for the first time allows himself to be gratified by a device when the claims of the thing whose place is taken by the device are still vivid. But an existing device, by its nature, is devoid of challenges. The use of devices becomes more boring the more their novelty wears off. If the paradigm of availability is to hold the interest of the technostructure and the consumer, it must not only rule but conquer.

Closely related to domination is diversion, which also serves to combat the endemic boredom of availability. Diversion is entertainment which does not engage. Television and tourism are prominent examples. Diversion uses culture and history for its primary resources and it is therefore less voracious than domination as regards things and physical resources.

We now have a reasonably clear outline of the pattern of technology. It prescribes a commerce with reality which is devoted to the procurement of devices and which is characterized by domination, disburdenment, and diversion.[44] We make over in the image of the device not only things, but also social relations. Organizations and techniques of interpersonal communication are as much devices as machines. Since the objects and relations that surround us have increasingly device character, our universe is acquiring a profound and pervasive sameness and definiteness. The novelty of the technological universe is parasitic. It depends on the continued destruction of things at its frontier. Within its boundaries an essential uniformity increases as the standard of living rises.

I now turn to the problem of the terms to which people in this country have come with the pattern of technology. A natural starting point is the question why the significance of the standard of living appears so differently in the common view sketched earlier. It is due to the characterization of the universe of devices in terms which were appropriate to the world of things and at least plausible for the transition from the latter to the former. The phenomena of domination, disburdenment, and diversion provide illustrations of this point.

Life in the pretechnological world of things is often seen as a bitter and precarious struggle for survival. This surely is a misleading use of evolutionary perspectives. Biologically, the place of the human species has been secure on this planet for at least a million years. And as soon and wherever we have detailed knowledge of human communities, we normally see them prevail with a grace and resourcefulness that goes beyond bare survival and issues in games, songs, and celebrations. Though all this is globally true, there were always numerous local hardships and catastrophes. And quite generally, the world before modern technology was a more resistant material. When tools and instruments began to take on device character, that resistance was increasingly overcome. The first experience of this success was legitimately one of conquest and triumph. But the pattern that has emerged in the two centuries since those beginnings does not support the view that technology in its modern phase is used by man to turn "toward the material world and dominate it in a new active way, making it a medium for the spirit's creative *self-expression* and *self-realization*."[45] Technological domination is procurement, not mastery, and it leaves in its wake not the sort of challenge to which creativity would be the appropriate response, but availability which is taken up in consumption.

Similarly, the early conquests and triumphs of technology certainly freed man in many instances from toil, disease, hunger, and ignorance of the world at large. Yet again it is a mistake to believe that modern technology is taking a course such that the "energy thus liberated can be diverted upward into his [man's] various higher and more characteristically human levels of activity."[46] Technological liberation has more and more clearly exhibited the character of disburdenment which in turn is becoming aggravated to the point of disengagement from things and persons. Finally, it is undoubtedly true that technological progress has made people's lives richer with instruments, books, athletic equipment, and other implements that allow one to become a vigorous and articulate citizen of the world. But here too it is increasingly obvious that technological enrichment is no counterforce to disburdenment and so has turned into diversion which is exacerbated to distraction, a debilitating scattering of one's attention and character.[47] Whenever through technology a limitation of human life was overcome and a new device was procured for man, the question was raised how this new development would affect modern man. This was so at the beginning of radio, television, paperback publication; and so one now asks how the impending flood of home video devices will enrich our lives.[48] The result will be the same as in the past. The new video developments will greatly increase our disengagement and distraction.

If that prospect disturbs us and if we want to act constructively on our dismay, we must try to delineate yet more incisively the terms on which people are dealing with technology. Surely the current state of affairs is not simply due to the misleading vocables that we use in characterizing technology. One may in fact challenge the thesis that technology has essentially ceased to be liberating and enriching. Is it not plainly true that it is up to the consumer whether or not he wants to buy the new video equipment and, if he does so, whether he uses it toward enlightenment or stupor? Is it not a matter of the consumer's values whether technology is liberating and enriching or disengaging and distracting?[49]

In answer to the first of these questions it must be noted that there is indeed a possibility of choice, one that is exercised against the prevailing paradigm of technology by those who practice voluntary simplicity.[50] But the fact that, at least now, they constitute a weak minority must give the practitioners and the theoretician pause. If a critic of the life style of domination, disburdenment, and diversion assumes that the adherents of that style have adopted it on the basis of something like an explicit decision, he must conclude that about 150 million people in this country have simply and gravely gone wrong.[51] This is surely too simple a conclusion and one that lacks compassion. The theoretician and critic is after all the contemporary, neighbor, friend, and relative of the people who deal with reality largely in agreement with the device paradigm, and he knows those people to struggle earnestly with their daily problems. Yet the paradigm of availability is not the object of these struggles but their basis. To be sure, the basis is affirmed in the daily endeavors, but this is a deeply implicit affirmation. The paradigm is one in Kuhn's sense, the unspoken and unquestioned model or framework for normal activity.[52] Its profound novelty appears only when we hold it in its mature form against the world of things.

Another reason for the concealment of the technological paradigm is its resourcefulness and flexibility. A consideration of the value question will highlight these features. While the common agreement with technology is implicit, it is also uneasy. There is a literature of discontent with technology which springs from this deep-seated uneasiness.[53] It is often thought that we can radically clarify our relation to technology by raising the question of values.[54] But a treatment of technology in those terms serves only to strengthen the reign of the paradigm of availability. To see this, let us follow Baier and distinguish between the value possessed by things and the values held by people.[55] "The value of something," Baier says, ". . . is a certain sort of property of it."[56] More precisely, it "is the thing's capacity to confer a benefit on someone, to make a favorable difference to his life."[57] We can see here that the discourse of values does not recognize things in their entire depth where every discernible property is significant and an essential tie to the world of the thing. Such discourse rather presupposes the means-ends distinction that comes into its own in the machinery and function of the device. Accordingly, Baier defines values held by people as the "*tendencies* of people to devote their resources (time, energy, money [the means of all means]) to the attainment of certain ends."[58] The relative stability of ends and the radical variability of means which again comes to fruition in the device is likewise congenial to value talk and stressed by Clarke: "The essential principle of education involved here [i.e., in the endeavor to control technological progress], it seems to me, is a shift of emphasis from means to ends, from teaching customs or *ways* of doing things—so quickly obsolete or irrelevant today—to teaching basic values or *goals* to be aimed at steadily through the flux of changing ways and means."[59] The affinity of the discourse of values with the paradigm of availability is palpable when Baier talks about ways of assessing the value of things:

> A book, a lecture course, an invention, a suggestion has greater value than another if it satisfies more fully than the other thing the listed desiderata; if it is generally easier to satisfy the conditions under which it can or will play its characteristic causal role (being available for reading, attending, etc.); if it more reliably brings about or is a

> greater help in bringing about certain intended changes in the lives of people, or brings them about in more lives, or in a larger proportion of lives to be improved; and if the changes brought about constitute a greater improvement.[60]

No matter how the question of values is raised and settled, the pattern of technology itself is never in question. Technology comes into play as the indispensable and unequaled procurement of the means that allow us to realize our preferred values.

One can, of course, extend value talk to encompass the discourse of things by defining the values of objects and subjects in such a way that the values are satisfied or instantiated only by things. But value talk then becomes awkward and misleading. It is not true in the end that some values are hard and measurable and others soft, fragile, and elusive. Rather, there are certain goals that are consonant with the paradigm of technology, and these appear as hard and measurable in value talk. But when value talk is about things, it falters, and the object of discourse slips from our grasp. Discourse that is appropriate to things must in its crucial occurrences abandon the means-ends distinction. It must be open to and guided by the fullness of the thing in its world, and it can communicate the thing only through testimony and appeal. The nature of principled discourse of things requires an investigation of its own. Here I only want to set it off against value talk and to expose the futility of value talk for a radical analysis of our relation to technology and the environment. In spite of its shortcomings one should, as a matter of prudence and pedagogy, encourage discussions that raise the value question. Without this familiar if inadequate approach, a fundamental analysis of technology remains forbidding. Moreover, values will remain indispensable as ways of summarizing, recollecting, and preparing for our experiences with things. It is the fundamental status of values that must be rejected.

Thus the common ways in which we speak about technology reflect the unquestioned and deeply implicit manner in which technology reigns. Technology is understood. It is presumed not to require discussion. But as noted before, uneasiness is beginning to color this understanding of technology. To the scholarly discontent with technology there correspond popular sentiments of a more muted kind. They are expressed toward the standard of living and toward technology. It has been found that a general rise in the standard of living is not generally experienced as an increase in happiness.[61] And the desire for a rise in the standard of living, though still prominent among personal aspirations, seems to abate.[62] Sometimes the disenchantment with the increase in our amenities that technological progress yields shows itself in the plainly erroneous denial that there has been such an increase. Still, people are uncertain whether they want to return to earlier ways of living.[63] Similarly, when people are questioned about technology, they are ambivalent between acceptance and apprehension, and this ambivalence seems to be growing.[64] But once more the power and implicitness of the paradigm of technology make themselves felt in an ambiguity of this ambivalence. It is never clear whether people's discontent is within or about the paradigm. Are people dissatisfied because the world has not yet been rendered available entirely and as a whole? Or are they disturbed by the rapid spreading of disburdenment and distraction and by the concomitant eclipse of the forces that used to engage us in their own right?

One must accordingly ask what is at issue in the energy crisis. Is it the embarrassing realization that we have failed to procure the energy basis for the procurement of availability? Or are we beginning to question procurement as the dominant manner of taking up with reality? The circumstances under which the energy crisis is coming to pass in Montana may allow it to be questioned at the more radical level. The advent of strip mining and power generation in eastern Montana is in some respects the brutal irruption of technology into a world of things. The common recognition of what we are losing is as implicit as the understanding of the ascending paradigm. When an explicit characterization of the thing character of Montana's world is required, Toole calls that character haltingly enough a "something" that Montana has (and not Montana alone, of course).[65] But that an implicit recognition can be strong nonetheless is clear from Toole's detailed and forceful picture of how Montana engages its people.[66] The strength of this understanding is also clear from the crucial tie that Toole sees between the fate of a place like Montana and the understanding that this country has of itself. The conquest of this continent is perhaps the most prominent case in our time of the ambivalence of domination. If the entire land is made available as a gigantic device for a life of disburdenment and diversion, then the resourcefulness and courage of the pioneer will give way to the disengagement and distraction of the citizen in the technological society, and our capacity to respond vigorously to our world will atrophy. If places where things prevail in their own right are respected and saved, we can experience first-hand what it is to be challenged in all of one's powers, and we can hope to learn new ways of engagement.[67]

In Montana, the ranchers are among the people who are most deeply engaged with their world. They speak eloquently of this engagement which is one of bitter struggles and grudging respect as much as one of joy and gratitude.[68] This engagement is of one piece with a social context of mutual trust and cooperation and of forbearance and discretion. Strip mining moves into this world as a particularly brutal mode of domination. In the first phase, it uses sophisticated and unintelligible legal devices to disrupt the social context and secure access to the coal deposits.[69] In the second phase, it breaks up the ground in gigantic explosions and removes the coal with machines that are nearly 400 feet tall and scoop up 75 cubic yards of earth per minute.[70] Power plants are similarly awesome devices. Moreover, the labor force that is needed for their construction and for the mining exhibits and suffers the thoughtlessness and brutality of labor as a mere means. These newcomers, without evil intention or personal guilt, destroy the social setting and overtax the public facilities of the old communities. They are in turn unable to form a new community. Normal levels of medical care, educational facilities, shopping, and entertainment are unavailble to them.[71] They tolerate these discomforts for the promise of high wages. In all this the prominence of the means toward a higher standard of living overshadows the realization of its end.

This clash has forced the ranchers to break their accord with the movement that is represented by the big corporations.[72] The ranchers have become more radical and have formed alliances with the environmentalists.[73] And still the resourcefulness of the paradigm of availability threatens to overtake the ranchers. The disruptive appearance of technology has made them realize that in more subtle ways technology has been a transformative force among them for some time and it has weakened their social fabric.[74] At times they accept the paradigm and argue their case from within it. They accept the need for more energy but urge that it be generated where mining and generating cause less disruption, and that eastern Montana be reserved for food production.[75]

In the state at large the energy crisis is most often cast into the terms of energy and economics vs. the environment. In the prevailing view, one would like to make both abundant energy and an unspoiled environment available. But physical restrictions force us to agree on trade-offs. This is the paradigmatic framework of opinion surveys, and it restricts severely the scope of responses.[76] However, if there is a distrust of technological progress, it would surely manifest itself in responses favoring the environment which are indeed in the majority. [77] Similarly, when a plurality of the respondents holds that a major energy conservation program would have a positive effect on the standard of living, one might ask whether that does not indicate a preference for a less wasteful and more engaging style of life.[78]

A political focus of the energy crisis in Montana is the Major Facility Siting Act. As mentioned before, it requires the Board of Natural Resources to issue a certificate of environmental compatibility and public need for power-generating plants. Environmental compatibility is more nearly within the ken of an enlightened pursuit of availability and of technical expertise. The concept of need is more searching. We saw in the first part of this paper that there is a strong tendency to locate it within the paradigm of availability. To counter the boredom that comes from the disburdenment and disengagement of availability, technology must, as argued above, constantly expand the universe of availability and procure new devices. This is the motive force of the ever rising standard of living. On this view more energy is always needed. To be sure, the need for Colstrip 3 and 4 is sometimes given a more humanitarian rationale by its proponents. The energy from those plants, so they say, will help to feed the starving and to provide work for the unemployed.[79]

This raises the fundamental question of how social justice fares within the pattern of technology. I do not believe that the highly productive technological society requires and is the result of a systematic and conscious exploitation of the workers by capitalists.[80] As I have tried to show above, the pattern of technology is fundamentally understood and accepted, uneasily perhaps, by almost all. The technostructure, of course, has the satisfaction of working at the frontier of technology. For the majority of the people work is an unloved means, a necessary evil.[81] But as long as the technostructure can count on the overall acceptance of the pattern of technology, it can count on an acceptance of that necessity. It can count moreover on the fundamental support of the people in the enterprise of advancing availability. It is the confidence in such acceptance and support that sustains the proponents of Colstrip 3 and 4. But the support may be weakening and the confidence also. Thus when Toole pointed out to the president of the Montana Power Company that the firm had been "under almost ceaseless attack from almost all quarters," he received this reply: "I guess I have had no problem during the years that I have had a certain amount of executive authority that has haunted me worse, given more hours of concern, or found me as totally helpless to maybe determine a solution; and if it is determined, to get an execution of it."[82]

On the other hand, I do not believe either that the paradigm of availability is nearly synonymous with democracy, understood as "a set of institutions which do aim to make everything available to everybody."[83] It seems to me that technology is socially indifferent though not inconsequential. There is some tendency to let the scope of availability expand and to let more and more people have its benefits. But the paradigm of availability becomes socially active only where it is threatened by social unrest. In such instances measures are taken to transform the negatively engaging social relations into commodities, that is, into mutual services that are safely and easily available. The development of labor relations illustrates this pattern.[84] On the other hand, where human misery borders on the universe of technological affluence without disturbing it, that misery is callously neglected. Surely the cultivation of social disburdenment in technology informs this neglect.

In our technological society, the distribution of availability is unequal with a rather continuous and regular rise from the lowest to the highest level. As the case of technologically advanced socialist societies shows, this unequality is not a necessary trait of the universe of devices though it favors the advancement and stability of the reign of technology. The unequal levels of availability represent a synchronic display of the stages of affluence that many people can hope to pass through. What the middle class has today the lower class will have tomorrow, while the middle class aspires to what the rich have now. The goals of tomorrow do not consist of vague conceptions and promises; they are realized and lived by those above my standing in the economic order. Everyone identifies his lot and aspirations with those above him. Thus the pervasive relative deprivation fuels the motor of technological advancement. The aspirations of the people are to some extent fulfilled. I do attain eventually what my wealthier neighbor has today, but not until he is wealthier yet. The rising tide lifts every ship, but it does not equalize their sizes. Social justice, understood as an equal level of affluence, will not be advanced merely by technological progress. A slowing or decline of general affluence, on the other hand, increases injustice. The scale of wealth provides a principle of selecting those who will pay for economic declines. In letting the poor and powerless suffer primarily, the allegiance of the more prosperous to the status quo can be secured. And since the slope of inequality is continuous and fairly linear, the boundary of the favored can be shifted without creating or revealing clearly demarcated social classes.[85] Class formation is also prevented by the upward orientation of most people and by their refusal to make common cause with those below them against those above them. I believe that it is this shifting and blurred outline of social injustice which profoundly discourages so many of the poor and less educated from participation in politics and gives rise to the limited politically active and attentive public. And of course quite a few of the people at the middle and upper levels of wealth and education feel disburdened from genuine citizenship because they are certain of the stability of the political framework and content with the benefits they derive from it.[86]

Naturally, no position is taken on these matters when the Major Facility Siting Act speaks of the need for energy. The Act is remarkable, however, in requiring thoughtful deliberation before new generating plants are approved. If someone is asked to show need, he is held to more stringent and essential criteria of justification than if he were invited to list his demands and wants. *Need* is the recurring and dominant term of the Act; *demand* and *want* are not admitted as benchmarks for the discussion of energy availability. The Act also recalls "the constitutionally declared policy of this state to maintain and improve a clean and healthful environment for present and future generations."[87] It requires a consideration of "promotional activities of the utility which may have given rise to the need for this facility'' and of ''conservation activities which could reduce the need for more energy.''[88] Finally, it entrusts the decision of whether or not a certificate of need is to be granted to a lay board. If the determination of need were a technical matter, the calculation of a definite magnitude within an unchallenged framework, it should be in the hands of technical experts. But if the consideration of need involves the question how we are to orient ourselves in this time and in this country and how we are to take responsibility for that orientation, then the issue must be moved beyond the scope of technical expertise.

What decision did the Board of Natural Resources reach? To begin with, it is probably fair to say that the governors' appointments to the Board did not in general encourage thoughtful reflection. Moreover, the hearing process which serves as a basis for the decision turned out to be so massive and prolonged that its conduct had to be turned over to a hearing examiner while most Board members were absent most of the time. The record of the proceedings amounted to 17,000 pages and thousands of exhibits. The machinery of the decision-making device was allowed to assume its paradigmatic inaccessbility, and the Board allowed itself to be overpowered by the paradigm of technology. It quickly gave conditional approval to Colstrip 3 and 4. But the insistence of the authors of the Siting Act that there be thoughtful consideration prevailed in District Court.[89] As of this writing the matter is pending before the State Supreme Court. The project has also become entangled at the Federal level. The Environmental Protection Agency has refused to grant a construction permit because it found that the plants would impair the pristine air quality on the Northern Cheyenne Reservation nearby. That decision too is being appealed.

Environmentalists sometimes think of these elaborate review processes as mere detours through which the inevitable despoliation of the environment will come to pass eventually. Business representatives, on the other side, regard them as irresponsible measures of disruption and delay.[90] I agree with Stone, however, that they do not have to be either; they can engender thoughtfulness and deliberation.[91] Indeed, it seems to me that the review process is the way in which we make room for ourselves to raise the question whether we want to confirm or revoke our agreement with technology.[92]

The question is being raised tentatively, haltingly, and often misleadingly. It is possible, of course, that it will never come fully into the open. At times it surfaces for an instant. A newspaper account of a hearing on Colstrip 3 and 4 began thus: ''The modern world is becoming disenchanted in its worship of technology, and Montana, with its Major Facility Siting Act, is in the forefront of that change, a lawyer for opponents of a major power development told the Montana Supreme

Court. . . ."[93] But eventually, the present crisis of energy and the environment may turn out to have been merely a temporary obstacle to the transformation of the earth into one smoothly and reliably functioning device. The alternative to this trend can surely not be a return to nature. Today the alternative to disburdenment cannot be toil, it must be engagement, to diversion it is not confinement but joy, and to domination it is not passivity but mastery. No one can procure a world of things; that endeavor would be essentially and internally inconsistent. Public policy can procure it least of all. But it can make room for the engagement with things. Engagement must begin by personal example. But unless public policy clears the ground for it and thus affirms it collectively if tentatively, engagement soon comes up against insuperable barriers. A conservative energy policy would make room for thoughtful deliberation and perhaps for a new world of things.

NOTES

1. A brief and early version of this paper appeared in *Montana Trails*, no. 2 (1977), 27–33. This paper has benefited from conversations and correspondence with Will Griffis.

2. Cf. Robert H. Socolow, "Failures of Discourse: Obstacles to the Integration of Environmental Values into Natural Resource Policy," *When Values Conflict: Essays on Environmental Analysis, Discourse, and Decision*, ed. Laurence H. Tribe, Corinne S. Schelling, and John Voss (Cambridge, Mass., 1976), pp. 1–33.

3. The anthology mentioned in the preceding note is devoted to the consideration of this issue.

4. Robert Dorfman, "An Afterword: Humane Values and Environmental Decision," *When Values Conflict*, p. 160.

5. See *Missoulian*, April 6, 1978, p. 22. In 1975, more than 22 million tons were shipped out of state; the projection for 1980 is more than 50 million. See Paul E. Polzin, "Transporting Coal from Montana," *Montana Business Quarterly*, XV, no. 2 (1977): 19–31.

6. See *Revised Codes of Montana* (1947): 70–801 through 70–829.

7. *Ibid.*, section 802.

8. Cf. William Leiss, *The Limits to Satisfaction* (Toronto, 1976), pp. 49–71.

9. See Energy Planning Division, Montana State Department of Natural Resources and Conservation, *Draft Environmental Impact Statement on Colstrip Electric Generating Units 3 and 4. Volume One. Summary* (Helena, 1974), p. 183.

10. See *ibid.*, pp. 174–75; the "attentive public" has been defined by Donald J. Devine, *The Attentive Public* (Chicago, 1970). Todd R. LaPorte and Daniel Metlay similarly speak of the "potential public" in "Technology Observed: Attitudes of a Wary Public," *Science*, CLXXXVIII (April 11, 1975): 125. On the attentiveness of Montana's public in particular, see Jerry W. Calvert, "Representation, Reform, and Accountability: A Look at the Montana Legislature," *Montana Business Quarterly*, XIV, no. 2 (Spring, 1976): 10–11.

11. See "Why Voters Don't Vote," *Newsweek*, October 30, 1978, p. 33.

12. The respondents of the *Impact Statement* clearly oppose the generation of electricity in Montana for export (p. 182). Respondents to another survey support such generation even more strongly. (One should note though the misleadingly brief question put to the respondents.) See Budget and Program Planning, Office of the Governor, *Montana Futures: A Survey of Citizen Choices* (n.p., 1977), pp. 40 and 63. This latter view also seems inconsistent with the concern for the environment expressed by the same respondents (*ibid.*, pp. 39 and 63).

13. See *Impact Statement,* pp. 12–34.

14. *Ibid.,* p. 12.

15. Westinghouse Environmental Systems, *Colstrip Generation and Transmission Project, Applicant's Environmental Analysis* (n.p., 1973), pp. 1–3; cf. pp. 1–6.

16. *Missoulian,* March 5, 1975, p. 12.

17. *Missoulian,* March 5, 1978, p. 1.

18. See Albert H. Cantril and Charles W. Roll, Jr., *Hopes and Fears of the American People* (New York, 1971), pp. 19 and 23. Regarding expectations for the future standard of living, see Nicholas Rescher, "A Questionnaire Study of American Values by 2000 A.D.," in Kurt Baier and Rescher, eds., *Values and the Future* (New York, 1969), p. 136.

19. See Alvin Toffler, "Value Impact Forecaster—A Profession of the Future," in Baier and Rescher, pp. 9–12.

20. See Milton Rokeach, *The Nature of Human Values* (New York, 1973), pp. 59–72; Lee Rainwater, *What Money Buys: Inequality and the Social Meanings of Income* (New York, 1974), pp. 41–135; Irene Taviss, "A Survey of Popular Attitudes toward Technology," *Technology and Culture,* XIII (1972): 621. Lillian Breslow Rubin, *Worlds of Pain: Life in the Working Class Family* (New York, 1976).

21. That view is implicit in the contention of the president of the Montana Power Company that the energy crisis constitutes "the greatest threat to this nation since World War II" (*Missoulian,* May 14, 1975, p. 1). Similarly President Carter called the effort to meet the energy crisis "the moral equivalent of war," and our failure to make this effort, he said, would lead to "an economic, social, and political crisis that will threaten our free institutions" (*Missoulian,* April 19, 1977, p. 1). In a more circumspect and qualified way Robert L. Heilbroner makes the same point in *An Inquiry into the Human Prospect* (New York, 1975).

22. See Paul K. Feyerabend, "On the Improvement of the Sciences and the Arts," in Robert S. Cohen and Marx W. Wartofsky, eds., *Boston Studies in the Philosophy of Science,* vol. 3 (Dordrecht, Holland; 1967), p. 414.

23. "What Is Value Change?" in Baier and Rescher, p. 85.

24. Bertrand de Jouvenel, "Technology as a Means," in Baier and Rescher, p. 217.

25. This is Alvin Toffler's thesis in *Future Shock* (New York, 1970).

26. For a comprehensive and helpful survey of the meanings and types of technology see Carl Mitcham, "Types of Technology," *Research in Philosophy and Technology,* I (1978), pp. 229–294.

27. Earlier developments of this notion of technology can be found in "Technology and Reality," *Man and World,* IV (1971), 59–69; "Orientation in Technology," *Philosophy Today,* XVI (1972); 135–147; "The Explanation of Technology," *Research in Philosophy and Technology,* I (1978), pp. 99–118. Daniel J. Boorstin similarly describes the character of everyday America in terms of availability and its constituents. See his *Democracy and Its Discontents* (New York, 1975).

28. Cf. Emmanuel G. Mesthene, *Technological Change* (New York, 1970), p. 28.

29. Cf. Melvin M. Rotsch, "The Home Environment," in Melvin Kranzberg and Carroll W. Pursell, Jr., eds., *Technology in Western Civilization* (New York, 1967), II, pp. 226–228. For the development of the kitchen stove (the other branch into which the original fireplace or stove developed) see Siegfried Giedion, *Mechanization Takes Command* (New York, 1969 [first published in 1948]), pp. 527–547.

30. See my "Mind, Body, and World," *The Philosophical Forum,* VIII (1976): 68–86.

31. See George Sturt's description of the sawyers in *The Wheelwright's Shop* (Cambridge, Mass., 1974 [first published in 1923]), pp. 32–40.

32. Cf. Morton Kaplan, "Means/Ends Rationality," *Ethics,* LXXXVII (1976): 61–65, and Laurence H. Tribe, "Technology Assessment and the Fourth Discontinuity: The Limits of Instrumental Rationality," *Southern California Law Review,* XLVI (1973): 617–660. On the means-ends relation in technology see Langdon Winner, *Autonomous Technology* (Cambridge, Mass., 1977), pp. 226–236 and 238–251. I think Winner overstresses the shift in

guidance from ends to means. More importantly, the emergence and the radically novel character of the technological means-ends distinction are more remarkable and consequential than the shifts in emphasis that occurred once the distinction had become established. I agree with Winner's main point that contrary to widely held beliefs technology is not an instrument in the service of politics; this essay argues the same thesis. There are, however, two noteworthy differences in approach and orientation. (1) The present argument tries to establish its central thesis on the basis of an ontological analysis. Winner's analysis is primarily one of political science and philosophy. (2) The crucial thesis of the present argument is summarized and centered in the notion of the technological paradigm. Winner is attracted by the notion of autonomous technology, but he is too circumspect to embrace it. Hence his positive notion of technology remains a collection of traits or tendencies without a clear center. These differences naturally have some significant consequences.

33. Martin Heidegger gives a careful account of the interpenetration of means and ends in the pretechnological disclosure of reality. But when he turns to the technological disclosure of being (*das Gestell*) and to the device in particular (*das Gerät*), he never points out the peculiar technological diremption of means and ends though he does mention the instability of the machine within technology. Heidegger's emphasis is perhaps due to his concern to show that technology as a whole is not a means or an instrument. See his "The Question Concerning Technology," in *The Question Concerning Technology and Other Essays,* tr. H. William Lovitt (New York, 1977), pp. 3–35, pp. 6–12 and 17 in particular.

34. It also turns out that a generally rising standard of living makes personal services disproportionately expensive. See Steffan B. Linder, *The Harried Leisure Class* (New York, 1970), pp. 34–37.

35. Sturt's book provides such a picture. See note 31 above.

36. Robert M. Pirsig describes this aversion to technology and contends that we can find wholeness at the center of technology if we begin to understand, maintain, and care for our devices. See his *Zen and the Art of Motorcycle Maintenance* (New York, 1974), pp. 11–35, 49–50, 97–106, 276, 290–292, 300–326.

37. Joseph Weizenbaum argues that certain computer programs have altogether escaped comprehensibility. See his *Computer Power and Human Reason* (San Francisco, 1976), pp. 228–257. On the unintelligibility and the consequent ignorance of technology see also Winner, *Autonomous Technology,* pp. 282–295. While ignorance (of the machinery) is to be admitted and stressed, one must add that this ignorance goes hand in hand with an understanding (discussed below) of the overall pattern of technology. As long as one condones the rule of the pattern and can count on general obedience to that rule, it is legitimate and safe to acquiesce in this ignorance.

38. Cf. Sturt, pp. 16–17.

39. Cf. J. B. Schneewind, "Technology, Ways of Living, and Values in 19th-Century England," in Baier and Rescher, pp. 125–132.

40. This distinction between work and labor should not be confused with Hannah Arendt's in *The Human Condition* (Chicago, 1958).

41. This story is told by Giedion, pp. 169–208, and it provides, as does the entire book, an example of how technology replaces work with labor.

42. See Harry Braverman, *Labor and Monopoly Capital: The Degradation of Work in the Twentieth Century* (New York, 1974).

43. See John Kenneth Galbraith, *The New Industrial State,* 2nd ed. (Boston, 1972), pp. 59–71.

44. Heidegger describes everyday existence as characterized by disburdenment and distraction (and related features such as ambiguity, curiosity, and others). See his *Being and Time,* trans. John Macquarrie and Edward Robinson (New York, 1962), pp. 163–168, 210–224, 396–400. In the tradition of the transcendental approach Heidegger mistook this mode of being for a universal and necessary feature of human existence. His later writings

disavow the transcendental standpoint and show that he had really been characterizing existence in a technological society. Heidegger is symmetrically mistaken in his description of the pretechnological engagement with the work world. See *Being and Time*, pp. 95–122.

45. See W. Norris Clarke, "Technology and Man: A Christian Vision," in Carl Mitcham and Robert Mackey, eds., *Philosophy and Technology* (New York, 1972), p. 249. Similar views are expressed by Mesthene in "Technology and Wisdom," in Mitcham and Mackey, pp. 109–110.

46. Clarke, *loc. cit.;* similarly Mesthene, *ibid.*, p. 111. Clarke's reluctance to take an incisive view of modern technology has not surprisingly led him to wonder "what detailed or precise contributions philosophers as such can make to the problem of technology." See his "Reflections on the XVth World Congress of Philosophy," *International Philosophical Quarterly,* XIV (1974): 117.

47. This point has similarly been made by Boorstin, pp. 118–120. Its empirical underpinnings have been presented in studies of how people value and spend their leisure time. See, e.g., Sebastian de Grazia, *Of Time, Work, and Leisure* (Garden City, N.Y., 1964), pp. 57–130; Tibor Scitovsky, *The Joyless Economy* (Oxford, 1976), pp. 149–247; John P. Robinson, *How Americans Use Time* (New York, 1977). See also Linder's book referred to in note 30 above, and Rubin's book, pp. 185–203 et passim, mentioned in note 17.

48. See Merrill Sheils, "TV of Tomorrow," *Newsweek,* July 3, 1978: 62–74.

49. Cf. Clarke, "Technology and Man," pp. 249–251 and 257–258.

50. Cf. Duane Elgin and Arnold Mitchell, "Voluntary Simplicity," *The CoEvolution Quarterly,* Summer, 1977: 4–19.

51. For this estimate see Elgin and Mitchell, p. 12.

52. See Thomas S. Kuhn, *The Structure of Scientific Revolutions,* 2nd ed. (Chicago, 1970), pp. 10–51.

53. Samuel C. Florman has given a survey of that literature in "In Praise of Technology," *Harper's Magazine,* CCLI, no. 1506 (November, 1975): 53–72. Florman capitalizes, unfairly perhaps, on the unresolved ambivalence toward the pattern of technology which characterizes much of that literature. The ambivalence is described below.

54. See Clarke, *ibid.*, pp. 249–250 and 257–258. Baier's and Rescher's anthology departs from that assumption.

55. See his "What Is Value? An Analysis of the Concept," in Baier and Rescher, p. 40.

56. *Ibid.*, p. 38.

57. *Ibid.*, p. 40.

58. *Loc. cit.*

59. Clarke, *ibid.*, p. 257.

60. Baier, p. 48.

61. See Richard A. Easterlin, "Does Money Buy Happiness?" *The Public Interest,* no. 30 (1973): 3–10, and Scitovsky, pp. 133–145.

62. See Cantril, p. 19.

63. See Rainwater, pp. 88–93; Taviss, p. 607; La Porte and Metlay, p. 123.

64. Taviss, p. 609; cf. La Porte and Metlay, pp. 124 and 126.

65. K. Ross Toole, *The Rape of the Great Plains* (Boston, 1976), pp. 233–235.

66. *Ibid.*, pp. 227–248 and the references in note 68 below.

67. Cf. Roderick Nash, *Wilderness and the American Mind,* 2nd ed. (New Haven, 1973) and Mark Sagoff, "On Preserving the Natural Environment," *The Yale Law Journal,* LXXXIV (1974): 205–267. Sagoff in the end is unable to resolve his ambivalence about technology and this is ably pointed out by Tribe in "From Environmental Foundations to Constitutional Structures: Learning from Nature's Future," *The Yale Law Journal,* LXXXIV (1975): 545–556.

68. Toole, pp. 78–79, 140–143; Institute for Social Science Research, University of Montana (Raymond L. Gold, Project Director), *A Comparative Case Study of the Impact of*

Coal Development on the Way of Life of People in the Coal Areas of Eastern Montana and Northeastern Wyoming, 2nd ed. (Missoula, 1974), pp. 90–93.

69. Toole, pp. 52–53, 106–108, 204–205 et passim; *Comparative Case Study,* pp. 27–35.

70. Toole, pp. 3–4 and 247.

71. The cluster of these and other social disruptions has been called the "Gillette syndrome." See Toole, pp. 96, 103, 108, et passim. See also *Comparative Case Study,* pp. 57–75, 94. At Colstrip, some of these disruptions may now have been ameliorated.

72. Toole, pp. 83 and 212; *Comparative Case Study,* pp. 27–35.

73. Toole, pp. 124–125, 208–226; *Comparative Case Study,* pp. 37–40, 59.

74. Toole, pp. 116–117; *Comparative Case Study,* pp. 87–88, 91, 136–137.

75. Toole, pp. 196–202; *Comparative Case Study,* pp. 28, 90 note 1, 99.

76. See *Impact Statement,* pp. 174–210 and *Montana Futures,* pp. 36–48. The same framework is found in surveys of other states. See Carl F. Hummel, Lynn Levitt, and Ross J. Loomis, "Perceptions of the Energy Crisis. Who Is Blamed and How Do Citizens React to Environment-Lifestyle Trade-Offs?" *Environment and Behavior,* X (1978): 37–88.

77. *Montana Futures,* p. 39; cf. *Impact Statement,* p. 180.

78. *Montana Futures,* p. 38. A Harris poll of December 4, 1975 is said to indicate that 77 percent of Americans are willing to simplify their lifestyles. See Kay Mills, "Simplify Your Lifestyle," *Missoulian,* January 4, 1976, p. 25. A similar Harris poll is quoted in *The CoEvolution Quarterly,* Summer, 1977: 34.

79. See the statement of the president of the Montana Power Company in the *Missoulian,* March 5, 1978, p. 1 and an editorial in the *Miles City Star,* January 29, 1975, p. 4.

80. Cf. Braverman, *Labor and Monopoly Capital* and G. William Domhoff, *The Higher Circles* (New York, 1971).

81. See Studs Terkel, *Working* (New York, 1974); Rubin, *Worlds of Pain,* pp. 155–184; Robinson, *How Americans Use Time,* pp. 116–132. How deeply ingrained is the view of labor as a means, appears from the finding of the University of Michigan's Survey Research Center that three quarters or more of the workers would prefer a ten percent wage increase to "more interesting work, more freedom, a more comfortable workplace, safer or healthier conditions, more job security or less tiring work." The same study reports that "65 percent agreed at least somewhat that their pay was good." (Reported in an AP news release of December 17, 1978.)

82. See Toole, p. 211.

83. Boorstin, p. 28; cf. pp. 29, 26, 102.

84. See Galbraith, pp. 263–282.

85. On the shifting line of poverty see Rubin, *Worlds of Pain,* pp. 8, 29–31, 47–48, 169–170, 206. In a socialist country, almost everyone is affected by an economic decline. Hence, Heilbroner to the contrary, it seems that capitalist countries are structurally better able, in the short run at least, to cope with economic and energy crises. See Heilbroner, pp. 61–95.

86. See LaPorte and Metlay, p. 125 and the *Newsweek* article mentioned in note 11. It appears from the work of Norman H. Nie, Sidney Verba, and John R. Petrovik in *The Changing American Voter* (Cambridge, Mass., 1976) that within the past two decades the people in this country have become politically more discerning and consistent in their views. The general level of political concern and activity is still deplorably low, and what improvement there has been fits entirely within the pattern of technology. On the other hand it is clear that a more penetrating view within the ruling paradigm can aid toward an examination of the paradigm.

87. Section 802.

88. Section 816.

89. *Missoulian,* March 5, 1978, pp. 1 and 2.

90. The vice president of the Montana Power Company said of the Department of Natural

Resources: "They're employing every possible delaying tactic that any lawyer has ever heard of." The president said: "I don't think I'm revealing any news here when I say we've known for a year that the department has been spending our $1.2 million to find a reason to turn us down." *Missoulian,* March 5, 1975, p. 12.

91. See Christopher D. Stone, *Should Trees Have Standing?* (Los Altos, 1974), p. 37.

92. An important and unique contribution to the broadening of the universe of public discourse so that this sort of question can be raised has been made by the Montana Committee for the Humanities which distributes funds from the National Endowment for the Humanities.

93. *Missoulian,* October 25, 1978, p. 16.

TECHNOLOGY, PUBLIC POLICY, AND THE PRICE-ANDERSON ACT

K. S. Shrader-Frechette

I. INTRODUCTION

Despite more than three decades of government-subsidized promotion and regulation of nuclear fission technology, a number of questions remain problematic. Some of these are economic, a few are purely scientific, while others are largely political. Nearly all of them raise troubling ethical considerations. From a philosophical point of view, one of the most interesting of these questions is whether the U.S. government ought to continue to protect the nuclear industry by imposing a limit on liability in the event of a reactor accident. Answering this query demands not only that one analyze the ethical presuppositions underlying assessment and regulation of nuclear technology, but also that, in general, one evaluate government guardianship of the utility industry. This latter task was urged long ago by Juvenal, when he pondered: "*Quis custodiet ipsos custodes?*" ("Who shall guard the guardians themselves?")[1]

Research in Philosophy & Technology, Volume 3, 1980, pages 313–342.
Copyright © 1980 by JAI Press Inc.
All rights of reproduction in any form reserved.
ISBN: 0-89232-102-4

Society has always feared governments of bureau administrators; their power typically has led to corruption and to rule by men instead of laws.[2] Federal regulation of nuclear energy presents a particularly difficult and important subject for analysis, both because the complexity of the technology virtually necessitates control by a regulatory elite rather than directly by the body politic, and because abuses within government and industry have potential for widespread destruction. In this essay I will investigate the philosophical presuppositions implicit in the government guardianship of nuclear technology. Specifically I will evaluate the logical, methodological, and ethical theories underlying the Price-Anderson Act. All too often the philosophical theories underlying such decision making have gone uncriticized because their proponents claimed their views were not theory-laden, but merely commonsensical or "just business."[3] Hence the philosophy implicit in technological guardianship has sometimes enjoyed an undeserved power. As a consequence, fundamental principles of freedom, justice, and security often have been compromised. As Albert Schweitzer put it:

> So little did philosophy philosophize about civilization that she did not even notice that she herself and the age along with her were losing more and more of it. In the hour of peril the watchman who ought to have kept us awake was himself asleep and the result was that we put up no fight at all on behalf of our civilization.[4]

Although the debate over nuclear energy is a many-sided one, the basic issue is the adequacy of Federal guardianship, since the government has assumed almost absolute authority for making the trade-offs between energy demands and the radiation risks of nuclear power.[5] If the thesis of this paper is correct, however, there are strong philosophical reasons both for the people, rather than the Federal government, to play the role of the Solomon who must weigh the goods and evils arising from nuclear power, and for the balance to shift in the character of these assessments.

II. THE PRICE-ANDERSON ACT

To understand the significance of the Price-Anderson Act, one must know something of the history of nuclear energy. Roentgen's discovery of the X ray in 1895 and Becquerel's discovery of natural radioactivity in 1896 mark two of the beginning steps in the development of radiation-related technology later exploited in medicine, industry, and warfare. The nuclear energy industry, however, dates to the 1940s, when the United States spent five years and $2 billion to develop the first atomic bombs during World War II. Thereafter the government took twenty years and more than $100 billion in subsidies to develop the first power reactors used to generate electricity; the reason for beginning to develop reactors in the forties was that the military wanted bombs and Eisenhower wanted "some kind of nonwarlike excuse for continuing the development of nuclear energy" and hence for obtaining the weapons-grade plutonium as a reactor by-product.[6] One legacy of this arms race is that U. S. nuclear technology is built around a type of reactor much more susceptible to catastrophic accident.[7] Moreover, despite the fact that the United States now has more than enough weapons material, it has sponsored a proliferation of a technology without ever assessing whether it is a desirable means to a quite different end, viz., energy rather than bombs.[8]

In one sense, industry did evaluate the desirability of employing nuclear energy to generate electricity; power companies refused initially to invest their funds in such a dangerous technology, since the legal consequences of a reactor accident could destroy the assets of a utility many times over. Because private companies would not provide more than a small fraction of insurance coverage, the government was forced to protect the industry itself. In 1954, Congress passed the Atomic Energy Act, whose object was "to promote the private development of atomic energy."[9] Three years later the Price-Anderson Act, as section 170 of the Atomic Energy Act, became law. Under its provisions, (1) the nuclear power plant owner was required to obtain $60 million of insurance, viz., all that private insurers would allow; (2) the U.S. government provided an additional $500 million indemnity protection for each nuclear "incident"; and (3) liability of all persons, in the event of an "incident," was limited to $560 million; the Act specifically insured that the law will "hold harmless the licensee and other persons indemnified . . . from public liability arising from nuclear incidents in excess of the level of financial protection required of the licensee."[10]

The unwillingness of private insurers to provide sufficient coverage, together with the nuclear industry's continuing insistence that a government-guaranteed liability limit is a necessary condition for their remaining in the business of nuclear generation of electricity, have had an unforeseen consequence.[11] Congress has been forced to extend the Price-Anderson Act, intended merely as a temporary measure, beyond its original ten-year period.[12] In 1965, the Act was extended for another ten years.[13]

Even though mounting evidence indicated that the government was not protecting the public adequately against the threat of nuclear accidents,[14] the Price-Anderson Act was extended again in 1975 to cover the period up to August 1, 1987. At this time Congress also passed two major amendments to the act. It provided a "deferred premium plan" according to which, after several more decades have passed, there will be a gradual substitution of industry-financed insurance for government-financed coverage of nuclear liability.[15] A second amendment stipulates that, if at least twenty-six more power plants are built, and if the maximum ($5 million per licensee) premium is assessed, then the nuclear industry's liability might rise (at most) at a rate of $5 million for each plant built after the next twenty-six.[16]

III. A LOGICAL ANALYSIS OF THE PRICE-ANDERSON ACT

From the preceding account of the Price-Anderson Act, it is easy to understand why this law has been called "the single most important legislative contribution to the development of the nuclear industry."[17] Moreover despite recent drastic cutbacks in orders of nuclear reactors, since 8 percent of this country's electric capacity is now provided by atomic plants and others continue to be built,[18] it is imperative that the philosophical presuppositions of the Price-Anderson Act be analyzed.

On a purely logical level, there is question whether the Price-Anderson Act is consistent, both with certain presuppositions underlying other government laws and with accepted factual and scientific assumptions. The most obvious sense in which this is the case has to do with the Brookhaven Report, WASH 740. On the one hand, "The Price-Anderson Act was designed to meet two basic objectives: (1) to . . . satisfy public liability claims [of up to $560 million] in the event of a catastrophic nuclear accident and (2) to remove the deterrent to private industrial activity in atomic energy presented by the threat of enormous liability claims."[19] On the other hand, the government document, WASH 740, concludes that *property damage* alone, in the event of a nuclear accident, could go as high as $280 billion, apart from the thousands of deaths, and the hundreds of thousands of injuries and cancers predicted.[20] Either the Price-Anderson Act is to "satisfy public liability claims" as the Atomic Energy Commission says, or it is not. If it is to meet these claims, then its liability limits must be consistent with governmental damage estimates, and not merely 0.2 percent (i.e., $560 million divided by $280 billion) of them. If the Act is not to compensate for damages, then Price-Anderson need not be amended, but government officials must retract the claim that it will "satisfy public liability claims." In a later section of the paper I will argue why the most ethical decision is to raise the liability limits of Price-Anderson.

A second disturbing fact is that the government estimates of the nuclear risk, presupposed in the Price-Anderson Act, are inconsistent with the estimates given by private insurers. Clearly, either the $435 million government insurance subsidy (provided to the nuclear industry above the $125 million now available through private insurance companies) represents a risk the government ought to take with the taxpayers' money, or it does not. If such a risk is economically desirable, then a profit can very likely be made on nuclear insurance and the taxpayers' money probably will not be lost. In this case, private industry ought to be ready to invest heavily in insuring the nuclear industry. However, facts do not bear this out. After more than twenty years during which nuclear reactors have generated power, the maximum insurance obtainable for them through private insurance pools is $125 million. This is especially interesting, since private insurance groups have provided, for example, up to $500 million in coverage for Pan American Airlines.[21] If nuclear plants are safe, and if private insurers are *incorrect* in failing to provide adequate coverage for them, then why is it reasonable for the government to limit liability for a reactor accident? On the other hand, if nuclear plants are not safe, and if private insurance companies are *correct* in their unwillingness to take such a risk,[22] then clearly citizens ought not to be limited in their right to collect damages on their property.

Several facts will indicate precisely why the assumptions of the insurance industry are incompatible with the actuarial presuppositions underlying the Price-Anderson Act. The premiums currently paid to private insurance pools for $125 million worth of coverage are based on the assumption that the chance of a serious accident is about one in twenty.[23] Price-Anderson (government) statistics, on the other hand, are based on the assumption that the probability of a serious accident is about one in 17,000.[24] Likewise the "deferred premium," established through the Price-Anderson Act, has been set by the government at a maximum of $5 million per reactor, whereas private insurers maintain that, to cover all damages resulting from a serious nuclear accident, an annual premium of at least $23.5 million per plant per year would have to be assessed.[25] Hence, at best, the actuarial presuppositions underlying the Price-Anderson liability limitation are inconsistent with fairly reliable insurance data. In a later section of the paper I will argue why, for methodological reasons, the statistics from private insurers are probably more reliable than the government figures.

Another apparent inconsistency arises, if one examines the reasons usually employed to justify the Price-Anderson Act. According to its proponents, the liability limitation will never actually deprive a person of the right to sue for property damages because a serious nuclear accident will never occur. However, industry refused to enter the nuclear field until a liability limit was established through the Price-Anderson Act.[26] Now either nuclear power is safe and catastrophic accidents are impossible, in which case no limit on liability is needed to protect the nuclear industry from bankruptcy; or, on the other hand, nuclear power is not safe and catastrophic accidents are possible, in which case a limit on liability is needed to protect the nuclear industry from bankruptcy. If the limitation is needed, it can only be so because successful claims can be made against the industry. But successful claims can be made against the industry only when injury can be shown to be the result of a nuclear accident. And if this can be shown, nuclear power is not safe. Hence one cannot argue consistently both that there is a need for a limit on nuclear liability and that nuclear reactors are safe.[27] Perhaps this logical difficulty is the reason why some proponents of nuclear energy have argued that the industry needs billions, not millions, of dollars of coverage.[28]

A final respect in which the Price-Anderson Act may be said to be questionable on logical grounds is in its relationship to the Energy Reorganization Act of 1974. In this legislation, Congress declared that "all possible sources of energy be developed."[29] Yet, in the very next year, Congress extended the Price-Anderson Act and reaffirmed its subsidy of nuclear power. As several Congressmen have noted, this continuation of the Act served the "effect of concentrating Federal funds on one energy option, namely, nuclear."[30] More than $100 billion already has been invested in nuclear power, through subsidies of the Price-Anderson Act, reprocessing, waste storage, research and development, enrichment, and site restoration.[31] Eighty-three percent of Federal energy monies have been spent for nuclear fission research, and all the other energy options have shared the remaining 17 percent of funds.[32] Although there is no formal inconsistency between the Energy Reorganization Act and the Price-Anderson Act, it does not seem logical both to continue massive subsidies of the most heavily supported energy alternative and at the same time, to urge development of all energy options.

IV. METHODOLOGICAL ANALYSIS OF THE PRICE-ANDERSON ACT

In addition to the purely logical questions that can be raised about the presuppositions of the Price-Anderson Act, relative to other theses supported by government, industry, and private insurers, this legislation also raises a number of methodological issues. Perhaps the most obvious premise upon which the liability limitation is based is that the likelihood of a catastrophic nuclear accident is so remote as to be almost impossible.[33] This thesis, however, which may be called (A), depends upon a number of other propositions which are debatable on methodological grounds. Some of the more important of these are the following: (A_1) The calculation of accident probabilities in the Rasmussen Report is based upon reasonable mathematical assumptions. (A_2) No data relevant to computation of these accident probabilities were suppressed. (A_3) The emergency core-cooling system is reliable. (A_4) Even if a core melt did occur, most people could be evacuated from the area surrounding a nuclear power plant. In the following paragraphs I will outline my reasons for arguing why the scientific methodology upon which (A_1) . . . (A_4) is based is doubtful, and consequently why thesis (A) and therefore the liability limitations of the Price-Anderson Act are open to question.

Thesis (A_1) has been used, perhaps more than any other premise, to support public confidence, both in this nation's nuclear fission technology and in the liability provisions of the Price-Anderson Act.[34] There are a number of facts, however, that suggest that (A_1) is not the case. For one thing, both the EPA and the American Physical Society, the most prestigious group of physical scientists in the world, conducted reviews of WASH 1400, the Rasmussen Report, and came to the conclusion that "the calculational methods used in both the research program and the regulatory function [are] fairly unsatisfactory. . . . The consequences of an accident involving major radioactive releases have been underestimated."[35]

The general problem with the mathematical methodology used in WASH 1400 is that its main technique is "fault free analysis," developed many years ago by the Department of Defense and NASA, both of whom consider it erroneous, "obsolete," and merely "educated guessing."[36] Its difficulties lie in the fact that the method provides no way to account for all parameters contributing to an accident. For example, the probabilities of a serious accident were calculated without taking account of the aging and deterioration of the plant, especially the reactor vessel, and the effects of terrorism and sabotage. The APS also criticized, for example, the assumption "that persons downwind of a reactor accident would receive radiation only through the first day following the accident," and noted that correction of this error alone would increase the number of cancers and genetic damages resulting from an accident by a factor of 25. This deficiency in mathematical methodology, however, is just one of many. The APS noted that WASH 1400 has hundreds of such errors, e.g., the neglect of damages resulting from resource contamination through land and water, with consequent effects on five or six generations into the future; the neglect of effects of "the selective and much larger irradiation of special tissues, notably the lungs and thyroid." For these and other reasons, the damages predicted by the APS represent an increase over the Rasmussen statistics by a factor sometimes as great as 50. For example, WASH 1400 said there would be 310 fatal cancers, apart from immediate deaths and genetic damage, as a result of a reactor accident. The same prediction made by the APS was for between 10,000 and 15,000 fatal cancers.[37] Moreover, as other groups of scientists have pointed out, the Rasmussen figures are too low because they exclude delayed deaths and injuries, such as cancer and genetic damage appearing in later years or in subsequent generations; this omission is significant because delayed nuclear fatalities are thought to exceed prompt ones by a figure of 100 or more.[38]

Even where the mathematical methodology of the Rasmussen Report is not erroneous, it is misleading because of the incompleteness of the analysis. For example, the report estimates that the per year, per reactor probability of a core-melt accident is one in 17,000 (see footnote 34). While correct, this figure is not as meaningful as the probability that, in the next thirty years, there will be a core-melt accident in one of the plants now operating or currently under construction. This probability is one in five.[39] In the light of a more complete mathematical analysis, it is doubtful whether (A_1) is the case and hence questionable that (A) the likelihood of a catastrophic nuclear accident is so remote as to be impossible. Therefore, if an accident cannot be said to be highly improbable, there are economic, political, and ethical reasons to question the desirability of the Price-Anderson liability limitation.[40]

Of course, if data relevant to the Rasmussen calculations were suppressed, or if the WASH 1400 was not developed in an objective and disinterested manner, than it would be easy to understand the discrepancies between statistical conclusions of government and the scientific community. Although it is not possible to determine (A_2) with certainty, that no reliable data relevant to computation of these accident probabilities were suppressed, a number of facts are relevant to making this determination. First, the Energy Reorganization Act of 1975 was adopted specifically to counteract the numerous problems arising because of the pro-nuclear industry bias of the Atomic Energy Commission. The Rasmussen Report was commissioned by the AEC and finished in 1974, just at the time that abuses in the AEC were beginning to come to light. Secondly, the government suppressed the results of the Brookhaven Report (WASH 740, completed in 1957) for almost twenty years until it was forced to release them under the Freedom of Information Act.[41] Both of these facts, along with suppression of data related to nuclear weapons testing in the fifties,[42] indicate that there is strong reason to doubt the truth of (A_2).

Other considerations also suggest the likelihood that pro-industry bias has resulted in suppression of important material. For example, the Atomic Energy Commission, later the Nuclear Regulatory Commission, has given repeated permission, not only for site preparation and construction activities prior to any public hearing or construction permit for a nuclear plant,[43] but also for required environmental impact statements to be ignored.[44] Government regulators allowed a fast breeder reactor to be constructed, even though its Advisory Committee on Reactor Safeguards opposed the permit, and then "concealed that fact from Congress and the public."[45] They are well known for the facts that they will finance only speakers who are proponents of nuclear fission,[46] that they have "punished" employees who have taken the safety aspects of their job too seriously,[47] and that they have censored government research reports critical of nuclear power.[48] For all these reasons, there is at least a strong possibility to doubt (A_2) that no reliable data relevant to computation of accident probabilities were suppressed, and hence to question thesis (A).

Another reason to question (A), from the purely empirical point of view of the practicing scientist, is that statistical assumptions (about the low probability of a serious reactor accident) are based on an empirical thesis (A_3) whose validity has never been tested. (A_3) is highly doubtful, on grounds of scientific methodology, simply because no full-scale empirical test of the emergency core-cooling system has ever taken place.[49] Moreover all of the small, scale-model tests, six out of six, have failed.[50] Instead, Nuclear Regulatory Commission officials have sanctioned computerized calculations of the probability that the system will work. All problematic possibilities obviously cannot be accounted for in a small physical model, and even fewer risk factors can be treated in a purely mathematical model. In fact, the mathematical model has been used, not because scientists are convinced of its merits, but rather because the full-scale tests would be "possibly very destructive" and would involve a great "risk."[51] The logical weakness of this reason for failing to test the emergency core-cooling system is obvious. If the system is effective, then a test of it would not be catastrophic. On the other hand, the only reason why a test would be destructive is if the system is not effective, which is prima facie evidence that it ought to be tested. Hence there is strong reason to doubt thesis (A_3) and therefore strong reason to question whether a core melt has a low probability.

Even if a core melt did occur, however, one of the crucial presuppositions undergirding (A), the thesis of the low probability of a disastrous accident, is (A_4), that 90 percent of the people could be evacuated from the area surrounding the nuclear power plant.[52] (A_4) is doubtful on methodological grounds because almost no evacuation plans have been tested, and there are no solid, empirical data to support it. Where the evacuation plans have actually been tested, almost all have failed, as in the "emergency response test" of the reactor in Toms River, N.J. recently.[53] Moreover, in the 1975 Browns Ferry incident, when both units had emergency core-cooling system failures as a result of a fire which burned out of control for hours, officials who should have been preparing to evacuate the area were not told of the danger.[54] In the partial core melt in the Detroit fast breeder reactor in 1966, much the same happened. No public alert was given, not because there was no danger, but instead because Detroit Edison feared "mass panic."[55] Hence there seems to be reason to doubt (A_4) not only because most evacuation plans have not been tested, and because those that have been tried have failed, but also because it is possible that they will not even be considered. For all these reasons it is not surprising that the American Physical Society criticized thesis (A_4). Moreover, if their criticism is well founded, then there is reason to question (A), as well as the liability limits set by the Price-Anderson Act.

V. ETHICAL ANALYSIS OF THE PRICE-ANDERSON ACT

Besides the logical and methodological problems inherent in the Price-Anderson Act, there are a number of ethical difficulties with the legislation. These are crucial to the debate over atomic energy, because both proponents and opponents of nuclear power admit that it is a very dangerous technology. The real difference between them focuses not only on (1) *how much* of a danger it poses, but also on (2) *how acceptable* this danger is when the risks are weighed against the benefits. Even if the preceding discussion about (1) revealed only sound methodological presuppositions supporting the liability limit, it would still be necessary to examine (2), the acceptability of assuming an admittedly low risk. This examination is necessary, not only because, in the past, humankind has had to bear the consequences of "impossible" accidents (such as the sinking of the *Titanic*, the falling of the Tacoma Narrows Bridge, the colliding of two planes over the Grand Canyon, and the bomber running into the Empire State Building), or because statistically an accident is just as likely to happen "sooner" as "later." Rather, it is necessary to examine the acceptability of the risk, because the failure to do so will only increase the probability that society will not see nuclear power as an ethical issue. If the greatest sin is to be conscious of none, then the greatest error of "nuclear energy morality" is to believe there is none, in other words, to assume that atomic power raises only technological but not ethical questions.

To some extent the formulation and governmental application of the Price-Anderson Act is built exactly on this presupposition, viz., that questions related to nuclear power are technological and not ethical ones. There is evidence that (A) and (A_3) have been taken, by government regulators of the nuclear industry, as sufficient justification for the Price-Anderson Act, and hence that they view the law as requiring only technological substantiation. So far as I know, no arguments have been made, either in the AEC Staff Study of the Price-Anderson Act, or in the Rasmussen Report (WASH 1400), to show that the Price-Anderson Act is *ethically* justifiable. Rather, the authors repeatedly used theses (A) and

(A_3) to "justify" the liability limit,[56] even though (A) and (A_3) are purely technological or scientific statements.

When citizens brought lawsuits and complained that the risk of nuclear power was unfair, that thesis (A_3) was doubtful because the emergency core-cooling system had not been tested, and consequently that thesis (A), regarding the impossibility of catastrophic damages, was questionable, they raised an ethical question. This was, of course, whether there ought to be a limitation on liability. The response of the courts has been to "defer" to the "expertise" of the Nuclear Regulatory Commission to assess the probability of an accident in which liability would exceed the Price-Anderson limit.[57] This response has been predicated on the assumption that the safety issue and hence the liability issue were merely technological, and not ethical, problems. Otherwise there would have been no reason to defer to the NRC. The NRC, however, is no more competent than the individual citizen to decide if a risk is worth it. The technical question is what the risk is, not whether it should be taken, and the latter issue is the one raised by citizens in numerous "moratorium" lawsuits based on the limits of the Price-Anderson Act. Moreover, deference to the agency which is the major proponent of both nuclear power and the Price-Anderson Act virtually denies the public a hearing on ethical issues they have raised.

Generic hearings provide another example of how government regulators view presuppositions about safety, used to justify the Price-Anderson Act, as purely technical issues. Generic issues are policy questions which arise in every licensing case, but which are not completely covered by regulation. Since such questions are not clearly decidable, the Atomic Energy Commission, and later the Nuclear Regulatory Commission, decided to remove them from individual licensing cases and treat them instead at a single, national, generic hearing, e.g., on the effectiveness of the emergency core-cooling system. In this way, explains the NRC, the government need not treat "wide-ranging discussion of the nation's energy policy."[58] Instead it can "eliminate where possible from the public debate over nuclear energy extraneous arguments which cloud and make meaningful dialogue impossible."[59] Questions of safety, and whether a risk ought to be taken, are now disallowed at individual hearings regarding licensing a power plant. For all practical purposes, the ethical issues have been removed from grass-roots-level discussions and deferred to a national forum where only technical experts are allowed to speak. Thus the value question of whether a certain group of people ought to take the chance that the emergency core-cooling system will prevent nuclear damages from exceeding the Price-Anderson limit is thereby changed into a technical question, viz., whether the emergency core-cooling system will work. This means that "the naturalistic fallacy" has been committed on two counts.[60] It arises, first, when questions relevant to ethics are disallowed at the local level and said to be "generic." This means that only purely technical questions will be treated at the individual plant-licensing hearing which determines whether a group of people will have a nuclear reactor or not. The second commission of the fallacy occurs when, at the national generic hearings on a particular safety topic, only technical experts are allowed to take part. As has been pointed out, citizens are able to participate only in the technical aspects of the hearings.[61] Hence theses such as (A_3), that the emergency core-cooling system is reliable, and other purely technical propositions, are taken as sufficient grounds for asserting (A) and for accepting the Price-Anderson Act. Since "there is no forum for the public to express its views on the policy questions involved,"[62] policy or ethical questions are simply ignored.

The naturalistic fallacy is evident not only in the types of evidence acceptable in a court or hearing relevant to the liability limit, but also in the evidence cited in government studies. For example, the authors of the only allegedly complete government study on nuclear power (WASH 1400, the Rasmussen Report) explain that a nuclear accident would cause 5,000 deaths, in addition to cancers and genetic damage. Next, instead of arguing why this risk is moral, they write: "however, such injuries would be insignificant compared to the eight million injuries caused annually by other accidents . . . the small increases in these diseases would not be detected."[63] They also argue that accidents such as fires, air crashes, etc., are more probable than nuclear accidents.[64] Likewise, in a recent government publication, ERDA argued for the thesis that nuclear plants were safe (and therefore that the insurance limitation was reasonable) by claiming: "Predicted nuclear accident risks are small compared to other possible causes of fatal injuries."[65] The purpose of their arguments has been to show why extensive damages from nuclear accidents are unlikely, and hence why the Price-Anderson Act is an acceptable law. Showing that damages resulting from taking a risk are low, undetectable, or less probable than those accruing from other risks, however, does not provide reasons why the risk is moral or ought to be taken. As a number of contemporary philosophers have pointed out, one cannot deduce a normative or ethical statement from those which are solely descriptive or scientific, without committing the naturalistic fallacy.[66] As Albert Einstein put it: "Scientific statements of facts and relations, indeed, cannot produce ethical directives."[67]

If it were possible to produce ethical directives solely on the basis of scientific facts, then it would be moral to do whatever it was scientifically possible to do. For one thing, this might violate considerations of equity. For example, if something (generation of electricity by means of nuclear energy) *benefits* a given set of people, why is it equitable for only a subset of that group to bear the *costs* (inadequate compensation for damages caused by a nuclear accident) of that activity? Considerations of equity demand that the rights of the minority cannot be ignored in order to benefit the majority. If they are forgotten, then the very concept of right becomes meaningless. What is operative instead is a concept of utility.[68]

While it is realistic to recognize that costs and benefits of technology can never be divided completely equitably, ethics demands that this be done as far as possible. Hence to deny, for reasons of financial expediency, the rights of a minority to collect damages owed them is inequitable, unjust, and arbitrary. Such a liability limit is arbitrary because it discriminates against people unfortunate enough to be living in the environs of a nuclear plant. The Fifth Amendment, on the other hand, applies to all persons in the United States, and does not discriminate on the basis of geographical considerations. It forbids depriving "any person of life, liberty, or property, without due process of law." What the Price-Anderson Act allows, by virtue of the limit on liability, is the situation where the public can be damaged, yet where the public cannot recover damages. For this reason it is contrary to equal justice, due process, and equal protection under the law.[69]

To the preceding arguments regarding the utilitarian emphasis of the Price-Anderson Act's limit on liability, many government regulators have responded that the limitation does not preclude the possibility that, were there a disaster, Congress would extend funds to the victims in excess of the limit specified.[70] The ethical problem with this response is, of course, that it does not guarantee rights to property, including rights to collect damages. If the distinction between "utility" and "right" means anything, it is not simply that harm will *in fact* be compensated, but that harm will *in principle* be conpensated. Moreover, the pragmatic difficulty with this response is that, unless it is required by law or obtained through the courts (which Price-Anderson excludes), compensation for damages is almost certain not to attain 100 percent.

A second ethical question regarding the Price-Anderson Act is whether it is permissible for the government (and therefore the taxpayers) to pay for the negligence of a utility operating a plant causing a disaster. Is it wise for government to subsidize the nuclear industry through insurance, and thereby remove the deterrent effect of liability as it affects industry decisions regarding cost-cutting measures, siting, and larger capacity? As Charles Frankel put it, a decision is responsible when the group that makes it has to answer for it to those who are directly or indirectly affected by it.[71]

Government subsidies are not merely dangerous, however, because they encourage the nuclear industry to be less responsible than they might be if they bore complete financial liability for their decisions. The basic problem with subsidies, such as those provided in the Price-Anderson Act, is that they mask the real cost of nuclear power. In providing indemnity, where private insurers will not do so, and in providing it at a much lower cost than that which would be assessed commercially,[72] the government allows an opportunity for misrepresentation of "the good" and for miscalculation of costs and benefits of atomic energy. Nuclear power is only said to be a "profitable" way of producing electricity because it is so heavily subsidized. Without the $100 billion in subsidies, including those provided through the Price-Anderson Act, it could never be calculated as inexpensive.[73] Because of this masking of real costs, the Price-Anderson subsidies lead to the false conception that nuclear power can provide unlimited, low-cost energy. "Cheap energy" then acts as an opiate which dulls the moral sense of the public and provides an apparently painless incentive to consume more and more; otherwise persons might be forced to consume less. If conservation does not seem economically desirable, then many people will not attempt to use less electricity. For this reason the "cheap energy" conception allows consumers to live under the illusion that we need not live simpler or less wasteful lives. It allows us to refrain from asking hard questions, such as whether it is just that North America have 6 percent of the world's population but consume 30 percent of the world's energy. Likewise, with no economic incentive to consume less, we have not seriously asked whether it is ethical for the United States to consume two to three times as much energy, on an average, per capita, kilowatt basis, as do other developed nations in the West.[74]

If nuclear energy did not protect us from such troubling queries, then we might examine the ethics of consumption.[75] Once it became financially expedient to distinguish our needs from our wants, we might question our current "more is better" ethic. We might also ask about the moral desirability of enabling a public resource, "developed at the cost of billions of dollars of public money," to be turned over to the exploitation of a monopolistic segment of private industry.[76] Why have public funds continued to be poured into private enterprise, despite the fact that the public has little real control over nuclear power? One reason might be that the public has not analyzed all the social and political costs of pursuing accelerating energy consumption. The nuclear electric power industry already enjoys antitrust immunity and the elimination of competition. If these benefits are of questionable ethical value, then so are the subsidies of the Price-Anderson Act. They increase the utility's power, which from the point of view of the consumer means that it is less likely that the industry can ever be challenged in its behavior or policies.

Besides helping to create a false economic security within which important ethical questions, related to conservation and control of public policy, do not naturally arise, Price-Anderson subsidies also contribute to other undesirable consequences. For one thing, they preclude a fair assessment of the merits of one technology with respect to another. So much money has been poured into subsidizing nuclear fission that the safer, environmentally sound technologies of fusion and solar power have taken a second seat because of lack of funding. As a result, the most ethical, equitable and justifiable energy technologies might not be chosen to supply our energy.[77] For example, the United States might be "locked into" a breeder technology, not because it is desirable on grounds of equity and safety, but because it is needed to provide fuel for the light water reactors. Obtaining fuel might be seen as necessary, if it is the only way to get a "return" on enormous capital investments (subsidies) in fission technology.

What all the above arguments come down to is that government subsidies through the Price-Anderson Act, although not necessarily wrong in themselves, provide an ethical context in which it is increasingly tempting to "sell our souls" in a Faustian bargain for cheap energy. What is wrong is not only that in selling our souls we ignore such questions as the ethics of consumption, and the ethics of leaving a legacy of radioactive waste for generations to come. What is also wrong is that, like every Faustian bargain, what we get is not really cheap. We receive only the illusion of cheap power because government has indirectly picked up the tab. Such illusions are dangerous, and if Price-Anderson leads to them, it is also dangerous. The long-term consequence of these illusions is that it is more likely that Huxley's *Brave New World* will really come to pass. We will, because of a false economic security, continue to build a world in which individual rights and freedoms are sacrificed for economic gain and material comfort.

The provisions of the Price-Anderson Act threaten authentic morality, however, for reasons other than the fact that its liability limits exchange a system of due process, property rights, and equal rights for a system of utility, and the fact that its subsidies skew authentic energy decision-making. For one thing, the Price-Anderson Act provides that damages are collectable *only when* the Nuclear Regulatory Commission certifies that an "extraordinary nuclear occurrence" has taken place. Since there is no such thing as a radiation threshold below which there is no deleterious biological effect, and therefore no safe dose of radiation,[78] use of such a "threshold rule" virtually necessitates that some cancer victims of a nuclear accident will go uncompensated.[79] What might occur is precisely what has happened in the past. When the partial core melt of the Fermi fast-breeder reactor caused a major release of radiation in 1966 in Detroit, the public was simply told that the heightened radiation levels presented no health threat.[80] In 1961, when the explosion and partial core melt of the Idaho Falls reactor killed three persons, the radiation alarm sounded and signaled releases of radiation into the atmosphere outside the building,[81] but government officials said that these releases were not harmful to the public. Likewise, in the 1957 Windscale (England) nuclear accident, a burst uranium fuel element caused a uranium fire and a massive release of radiation into the atmosphere. As a consequence, radioactive iodine deposited on pastures and foliage posed a threat to infants and children in the area. Radiation levels were so high that cows' milk from an area covering 200 square miles was confiscated. Fallout from the accident reached London, which was 300 miles away. Yet workers at the reactor and people from the surrounding area were told that there was "no danger to the public." Moreover, government officials lied when they claimed that the harmful radiation had been carried out to sea.[82] Likewise, when the 1975 nuclear accident occurred at Brown's Ferry, Alabama, the public was assured that there had been no release of harmful levels of radiation. The difficulty with this claim, however, is that the radiation monitors in the stacks were not working.[83]

Despite the fact that higher levels of radiation releases were documented in all these nuclear accidents, except the one at Brown's Ferry, the Federal government has continued to maintain that there have been "no nuclear accidents to date."[84] Hence if future accidents may be defined as "nonaccidents," and if abnormally high releases of radiation may be called "safe," then there is no assurance that the NRC will accurately define an "extraordinary nuclear occurrence." And, of course, without this definition no damages may be collected under Price-Anderson provisions. Hence there is reason to believe that the "threshold provision" of the Act would function as a utilitarian principle according to which some victims of nuclear injuries might not be compensated. Moreover, because the Act requires payment to be given only "upon proof of the causal relationship between the occurrence and the injury . . . and upon proof of damage,"[85] it is likely that some injured persons will be unable to prove this and hence likely that they will remain uncompensated. Although proof of injury is not difficult for the more serious and immediate types of radiation deaths and injuries, it is almost impossible in cases of cancers and genetic effects whose symptoms are manifested only years later. How could one *prove*, for example, that a cancer (appearing fifteen years after a person's exposure to radiation released in a nuclear accident) was truly caused by the accident?[86] If 25 percent of the population will contract cancer, apart from nuclear reactor hazards, how would one distinguish these incidences from the malignancies resulting from a nuclear accident? The ethical dilemma here is not just that some people who deserve to be compensated will not be, because of the "proof" and "threshold" clauses of the Act, but also that, *in principle,* there is no way to prove whether or not radiation from a particular plant has caused a cancer. Hence, retaining these clauses forces government and the courts to employ an *argumentum ad ignorantiam,* an argument from ignorance. They will have to conclude that an accident did not cause a particular cancer whenever the victim is unable to prove that it did. This means that employment of the Price-Anderson Act puts a double ethical burden on citizens. It has not only disallowed due process and equal protection/justice under the law, but also (because of the very nature of the radiation hazard) made it impossible, *in principle,* to adjudicate claims involving equal justice under law. For all these reasons, acceptance of the Price-Anderson Act wrongly encourages us to accept a utilitarian calculus, to ignore the constraints of equal justice, to forget the consequences of government subsidy, and to allow a situation to continue in which most of the costs of an accident will be borne by its victims rather than by its perpetrators.

VI. AN AGENDA FOR THE FUTURE

Admittedly the problems with the Price-Anderson Act do not allow one to conclude that this country ought not to use nuclear fission technology to generate electricity. At best, this analysis indicates only that meeting constraints of due process and equal justice/protection are necessary conditions for the ethical employment of any technology, and that in the case of nuclear energy, these conditions are not currently being met. A more complete examination of the philosophical presuppositions underlying nuclear technology, however, would have to address a number of questions not raised in this paper. Several of these are the following.

1. How ought the government to restrict civil liberties in order to guard against sabotage, terrorist attack, illicit fabrication of weapons, or radioactive contamination during the lengthy nuclear fuel cycle?[87]
2. What health costs ought the public to pay for its energy demands?[88]
3. Will the greater good of future generations be served by continuing to deplete the world's finite supply of fossil fuels or by leaving a legacy of nuclear technology, with attendant radioactive wastes, genetic damage, and increased cancer rates?[89]
4. Does the consumer have the right to receive the cheapest form of electricity available, and if so, is this generated by nuclear fission?[90]
5. Will nuclear energy render the U.S. energy-independent and therefore better able to exercise moral leadership in the world community?[91]
6. Is it morally and economically desirable to employ nuclear technol-

ogy in order to encourage increasing levels of U.S. energy consumption?[92]

7. Ought the nuclear industry to equate "energy demand" with "energy need"?[93]
8. Is it ethically responsible to employ a technology before a means of storing its lethal wastes has been devised?[94]
9. Are financial and economic criteria sufficient to justify current atmospheric releases of carcinogenic, low-level radiation from nuclear reactors?[95]

In answering these questions, as in evaluating this analysis of the Price-Anderson Act, we will be forced to make a perennial distinction between morality and expediency. I fear that many proponents of the Price-Anderson Act have supported it for the same reasons that New England residents, decades ago, opposed construction of a lighthouse on Cape Cod. As Ralph Waldo Emerson reveals in his diary, citizens of the area feared that the tower might mean losses for the shipwreck business. Let us not follow them in confusing the business of ethics with the business of profits.

NOTES

A modified version of some of this material appears in the author's *Nuclear Power and Public Policy* (Reidel, 1979). Copyright © 1979 by K. S. Shrader-Frechette.

1. Decimus Junius Juvenal, *Satires,* VI, l. 292. Interestingly, however, Plato thought it was absurd to think that a guardian should himself need a guardian. See *The Republic,* III, 403E.

2. Other writers have made a similar point. See Garrett Hardin, "The Tragedy of the Commons," in *Technology and Society,* ed. N. de Nevers (Reading, Mass., Addison-Wesley, 1972), p. 164, and D. J. Rose, "New Laboratories for Old," in *Science and Its Public: The Changing Relationship,* ed. G. Holton and W. Blanpied (Boston, D. Reidel, 1976), pp. 151–152.

3. Alasdair MacIntyre, "Utilitarianism and Cost-Benefit Analysis: An Essay on the Relevance of Moral Philosophy to Bureaucratic Theory," in *Values in the Electric Power Industry,* ed. K. M. Sayre (Notre Dame, Ind., University of Notre Dame Press, 1977), pp. 217–218.

4. Schweitzer, *The Philosophy of Civilization,* cited by J. L. Huffman, "Individual Liberty and Environmental Regulation: Can We Protect People While Preserving the Environment?" *Environmental Law,* VII, No. 3 (Spring 1977): 447.

5. Another reason for focusing on the philosophical justification underlying the Price-Anderson Act is that this law (or any other law whose effect is the same, viz., to limit the liability of all persons who could be held responsible for a nuclear accident) establishes a necessary condition for the existence of the nuclear industry. Hence if the law is unconstitutional, unethical, illogical, or based on questionable scientific methodology, and deserves to be stricken, then the nuclear industry would not operate. On the other hand, if the law is constitutional, ethical, logical, and based on plausible scientific methodology, then one of the major arguments of nuclear opponents is clearly without merit.

6. Sheldon Novick, *The Electric War* (San Francisco, Sierra Club Books, 1976), p. 22. This fact is also corroborated by M. Willrich, *Global Politics of Nuclear Energy* (New York,

Praeger, 1971), pp. 5–6. Breakdown and documentation of the more than $100 billion cost of government development of nuclear fission technology is given by J. J. Berger, *Nuclear Power* (New York, Dell, 1977), pp. 94–147, esp. pp. 112, 144–147.

Carl Walske, who holds a Ph.D. in physics as well as the presidency of the coalition of pro-nuclear industries, the Atomic Industrial Forum, has often pointed out that military expenditures and justification were ultimately responsible for nuclear development. (Cited by Novick, *op. cit.*, pp. 32–33, in a taped interview with Walske.) In fact, no clear choice was made to develop nuclear energy as a source of power. Rather "the atomic bomb program had become . . . the country's largest industrial enterprise." In 1953, for example, the U.S. Atomic Energy Commission owned and operated three towns, employed 5 percent of the nation's construction labor force, and consumed 10 percent of the country's electric power. (Novick, *op. cit.*, p. 38.) This development of nuclear energy was considered necessary, in order to obtain needed plutonium for weapons, especially since the USSR had already tested its own nuclear devices and the United States was engaged in the beginning of the Cold War.

7. In the 1940s and 50s the United States expended billions of dollars of research and development monies on water-cooled reactors because they were not as complex to build and because their fuel was enriched uranium, already being used for making explosives. Enrichment plants existed for the bomb effort, and their continued operation could be justified if they also were used to make fuel for reactors. As a result, the U. S. nuclear technology is built around a water-cooled, enriched-uranium design. This is a critical point, because such reactors (unlike those of Canada, Britain, and the USSR) are plagued with the risk of "core meltdown," and therefore possible catastrophic accidents. (See Novick, *op. cit.*, pp. 160–161. The key benefit of reactors which do not use enriched uranium is that their fuel, natural uranium, contains only 0.7 percent uranium 235, and hence the density of power generation in their fuel is much less than that in enriched uranium. As a result, the natural uranium reactors can withstand a loss of cooling more easily than can the enriched uranium ones. Cooling helps to control the fissioning uranium or plutonium which releases energy; with such control, the fissioning would cause catastrophic damage which could go to $280 billion and destroy an area the size of Pennsylvania. See later remarks for more details regarding core melts and their effects.)

8. E. D. Muchnicki, "The Proper Role of the Public in Nuclear Power Plant Licensing Decisions," *Atomic Energy Law Journal*, XV, No. 1 (Spring 1973): 39, makes this same point. See also Novick, *op. cit.*, p 22.

9. J. R. Brydon, "Slaying the Nuclear Giants," *Pacific Law Journal*, VIII, No. 2 (July 1977): 781. Predictably, the "power companies were markedly reluctant to invest their own funds in further development, and the existing technology seemed unlikely to be profitable" (Novick, *op. cit.*, p. 40). Given the possibility of a core melt, owing to the use of enriched uranium, light-water technology, business and industrial leaders said unequivocally that they would not invest in nuclear power because "the legal consequences of a serious accident would financially destroy a company." (J. Marrone, "The Price-Anderson Act: The Insurance Industry's View," *Forum*, XII, No. 2 [Winter 1977: 607].) Nearly every company spokesperson, e.g., the vice-president of General Electric, said that "the company would withdraw from the nuclear field" unless some indemnity legislation was passed to protect the industry. (W. S. Caldwell *et al.*, "The 'Extraordinary Nuclear Occurrence' Threshold and Uncompensated Injury under the Price-Anderson Act," *Rutgers-Camden Law Journal*, VI, No. 2 [Fall 1974]: 379).

10. "AEC Staff Study of the Price-Anderson Act, Part I," *Atomic Energy Law Journal*, XVI, No. 3 (Fall 1974): 220. See also Lowenstein, *op. cit.*, p. 597.

11. Both environmental writers and spokespersons for the nuclear industry make it clear that the government is forced to affirm the Price-Anderson Act as a necessary condition for industry's employment of nuclear energy. J. W. Gofman and A. R. Tamplin, *Poisoned*

Power (Emmaus, Pa., Rodale, 1971), pp. 32, 180, 239, make this point, as do attorneys for nuclear utilities, e.g., R. Lowenstein, "The Price-Anderson Act: An Imaginative Approach to Public Liability Concerns," *Forum*, XII, No. 2 (Winter 1977): 596. See also M. Gravel, "Price-Anderson: A No-Fault Nightmare," in R. Munson (ed.), *Countdown to a Nuclear Moratorium* (Washington, Environmental Action Foundation, 1976), p. 48; A. W. Murphy and D. B. LaPierre, "Nuclear 'Moratorium' Legislation in the States and the Supremacy Clause," *Environment Law Review 1977*, ed. H. F. Sherrod (New York, Clark Boardman, 1977), 405; and M. C. Olson, *Unacceptable Risk* (New York, Bantam, 1976), p. 55. Even government proponents of nuclear energy clearly reveal that industry would not "go nuclear" without liability protection; see "AEC Staff Study of the Price-Anderson Act, Part I," p. 209, and "AEC Staff Study of the Price-Anderson Act, Part II," *Atomic Energy Law Journal*, XVI, No. 4 (Winter 1975): 298.

Industrial reluctance to enter or remain in nuclear energy production, without protection from liability claims, is easy to understand. Both nuclear reactors and nuclear weapons produce energy by fissioning plutonium or uranium, except that in a weapon the release is uncontrolled. Failure of the technological controls in a reactor would result in disastrous consequences. "A million-kilowatt nuclear power plant, after a year of operation, contains ten billion curies of radioactive material, enough (if properly distributed) to kill everyone in the United States." (Novick, *op. cit.*, pp. 152–153.) Moreover, much of this radiation remains lethal (capable of ionization causing death, cancer and genetic damage) for hundreds of thousands of years. Hence any extensive contamination with plutonium 239 is considered by scientists to be "permanent." In WASH 740, the Brookhaven Report, the (pro-nuclear) government arrived at conservative estimates for effects of a core melt and resultant releases of radioactivity. The report concluded that such an accident could lead to 45,000 immediate deaths, 100,000 injuries, innumerable cancers, contamination of an area the size of Pennsylvania and up to $280 billion in property damage. (See "Theoretical Possibilities and Consequences of Major Accidents in Large Nuclear Power Plants," USAEC Report WASH 740 [Washington, D.C., Government Printing Office, 1957] and its update, i.e., R. J. Mulvihill, D. R. Arnold, C. E. Bloomquist, B. Epstein, "Analysis of United States Power Reactor Accident Probability," PRC R-695 [Los Angeles, Planning Research Corporation, 1965]. See also J. Elder, "Nuclear Torts: The Price-Anderson Act and the Potential for Uncompensated Injury," *New England Law Review*, XI, No. 1 [Fall 1975]: 127; T. Black, "Population Criteria of Oregon and the U. S. in the Siting of Nuclear Power Plants," *Environmental Law*, VI, No. 3 [Spring 1976]: 897.) In other words, the effect would be similar to that of 100 Hiroshima bombs. This study (WASH 740) was originally requested by the Joint Commission on Atomic Energy because the Commission wanted positive safety conclusions "to reassure the private insurance companies" so that they would provide insurance for the nuclear industry. Since even the conservative statistics of the Brookhaven Report were alarming, the report was suppressed and its data were not made public until more than fifteen years later, when a suit was brought as a result of the Freedom of Information Act. (J. G. Fuller, *We Almost Lost Detroit* [New York, Ballantine, 1975], p. 60.) Because private companies would only insure the nuclear industry for an amount of up to $60 million, and given the industry's refusal to enter or remain in nuclear power generation without comprehensive liability protection, the U.S. Congress passed the Price-Anderson Act in 1957. According to Representative Price, the purpose of the Act was not only to protect the public from the dangers of nuclear accident, but also to help industry: "Clearly the enactment of the indemnity legislation was in response to direct pressure from private industry for an indirect subsidy in the form of protection against possible bankruptcy." (W. S. Caldwell, III, L. Filler, and R. S. Lee, "The 'Extraordinary Nuclear Occurrence' Threshold and Uncompensated Injury under the Price-Anderson Act," *Rutgers-Camden Law Journal*, VI, No. 2 [Fall 1974]: 379. See also Gravel, *op. cit.*, p. 45.)

12. At the time it was passed in 1957, the Price-Anderson Act was said to be needed only as a "temporary" measure; it was to remain in effect for only ten years, since the Joint Committee on Atomic Energy was sure that after ten years, "there will be enough experience gained so that the problems of reactor safety will be to a great extent solved and the insurance people will have had experience on which to base a sound program of their own." (Lowenstein, *op. cit.*, p. 597.) It was argued that the safety of nuclear plants, by 1967, would be evident and would thus "remove the fear of public liability that had previously blocked private investment." (Caldwell, et al., *op. cit.*, p. 364.)

13. At this time it was amended so that a waiver of defenses against all liable persons was required, provided that an "extraordinary nuclear occurrence" had taken place. The Atomic Energy Commission was given the power to define occasions on which such an occurrence had happened, and defenses were waived for these incidents; the "injured" person was required only to show that an accident at a nuclear power plant caused his injury, but not that the owners/operators were negligent. Amendments at this time also provided for immediate partial compensation to the victims without their having signed a release. (Marrone, *op. cit.*, pp. 608–609.)

14. On January 19, 1975, after thirty years during which the same government agency, the Atomic Energy Commission, had both promoted nuclear industry and regulated it (so as to provide for the safety of the public), the nuclear promotional and regulatory functions of the government were separated. This was accomplished through the Energy Reorganization Act, which created the Energy Research and Development Agency (to push commitment to nuclear energy) and the Nuclear Regulatory Commission (to assess the safety and security of existing and proposed nuclear installations). The Reorganization Act became necessary because of numerous lawsuits against the Atomic Energy Agency, which was repeatedly censured by the courts and the public for failing to regulate safety hazards of nuclear power plants, and for subscribing to industry's demands rather than to consideration of the public's best interests. The Energy Reorganization Act, as a response to government problems regarding the AEC's pro-industry bias, is discussed in M. H. Rowden, "Nuclear Power Regulation in the United States: A Current Domestic and International Perspective," *Atomic Energy Law Journal*, XVII, No. 2 (Summer 1975): 102 ff. See also Novick, *op. cit.*, pp. 354–355; Brydon, *op. cit.*, p. 771; V. McKim, "Social and Environmental Values in Power Plant Licensing," in Sayre, *op. cit.*, pp. 38–40; G. B. Karpinski, "Federal Preemption of State Laws Controlling Nuclear Power," *Georgetown Law Journal*, LXIV, No. 6 (July 1976): 1336; and J. G. Palfrey, "Energy and the Environment: The Special Case of Nuclear Power," *Columbia Law Review*, LXXIV, No. 8 (December 1974): 1380.

15. Currently private insurers are willing to offer $125 million in coverage to the nuclear industry, and the government provides the remaining $435 million in coverage, i.e., that up to a total of $560 million. The substitution, of industry-financed insurance for government-financed coverage, will be accomplished by a "deferred premium" plan which will operate only when an "extraordinary nuclear occurrence" takes place. In this event, each licensee will be assessed a prorated share of the damages above $125 million, provided no premium is less than $2 million or more than $5 million. The level of the premium is set, within these constraints, by the NRC. The date on which government subsidy is phased out depends on the level at which the premium is assessed and on the number of power plants operating. With all currently operating plants (60) assessed at the maximum $5 million, the total amount now available from the deferred premium plan is $300 million. This means the government would still have to supply a $125 million subsidy, since the liability limit is $560 million. (For bibliographical material regarding this amendment, see the references in the next note.)

16. For example, if 26 more plants were built by 1987, then in that year, the liability limit can be no greater than $565 million. If 27 plants are built by 1987, then in that year the liability limit cannot exceed $570 million. In any event, the ceiling of $560 million will rise

only if at least 26 more plants are built. Explanations of these two amendments to the Price-Anderson Act can be found in Lowenstein, *op. cit.*, pp. 599–603; Marrone, *op. cit.*, pp. 608–609; M. McCormack, "U. S. Congressional Attitudes and Policies Affecting Nuclear Power Development in the World," *Atomic Energy Law Journal,* XVII, No. 4 (Winter 1974): 307–309; R. Wilson, "Nuclear Liability and the Price-Anderson Act," *Forum,* XII, No. 2 (Winter 1977): 613–614; F. Schmidt and D. Bodansky, *The Energy Controversy: The Fight over Nuclear Power* (San Francisco, Albion, 1976), p. 144.

17. Caldwell, et al., *op. cit.*, p. 360.

18. Marrone, *op. cit.*, p. 606.

19. "AEC Staff Study of the Price-Anderson Act, Part I," p. 231.

20. Mulvihill, et al., *op. cit.* Elder, *op. cit.*, p. 127. See also Black, *op. cit.*, p. 897; Gofman and Tamplin, *op. cit.*, pp. 174–179, 354; Berger, *op. cit.*, pp. 44–46. Cf. U.S. Atomic Energy Commission, Summary Report, "Reactor Safety Study: An Assessment of Accident Risks in U. S. Commercial Nuclear Power Plants [Rasmussen Report, WASH 1400]," *Atomic Energy Law Journal,* XVI, No. 3 (Fall 1974): 201–202. Since nuclear plants today are approximately five times the size of those whose accidents were described in the original Brookhaven Report, it is likely that consequences of a core melt would be even more severe than those calculated in WASH 740.

21. Richard Wilson, "Nuclear Liability and the Price-Anderson Act," *Forum,* XII, No. 2 (Winter 1977): 617, who is a proponent of nuclear power and the Price-Anderson Act, provides this statistic about airline coverage.

22. The reason given by the Atomic Energy Commission for the necessity of government-subsidized insurance for the nuclear industry is that private insurance would not take the risk. See "AEC Staff Study of the Price-Anderson Act, Part I," pp. 219–220; Gofman and Tamplin, *op. cit.*, pp. 31, 313; Olson, *op. cit.*, p. 54.

23. This is a per year, per reactor probability, computed by Senator Mike Gravel (Alaska), and quoted in Berger, *op. cit.*, pp. 145–146.

24. U. S. Atomic Energy Commission, Summary Report, *op. cit.*, p. 178; see also Novick, *op. cit.*, p. 159, and U.S. Nuclear Regulatory Commission, *Reactor Safety Study—An Assessment of Accident Risks in U.S. Commercial Nuclear Power Plants,* Report No. (NUREG–75/014) WASH-1400 (Washington, D.C., Government Printing Office, 1975). This figure is also a per year, per reactor probability.

25. Former Insurance Commissioner of the State of Pennsylvania, H. S. Denenberg, calculated this annual premium; he is quoted in Olson, *op. cit.*, p. 56. See also H. S. Denenberg, *Citizen's Bill of Rights and Consumer's Guide to Nuclear Power* (Reading, Pennsylvania Insurance Department, 1973), pp. 1–5. For information on the $5 million maximum deferred premium established by the Price-Anderson Act, see Lowenstein, *op. cit.*, pp. 599–603; Marrone, *op. cit.*, pp. 608–609; Wilson, *op. cit.*, pp. 613–614.

26. See note 11.

27. Several authors have made points related to this argument of mine; see Gofman and Tamplin, *op. cit.*, p. 357; and R. Nader, "Nuclear Power: More than a Technological Issue," *Mechanical Engineering,* XCVIII, No. 2 (February 1976): 33.

28. See Wilson, *op. cit.*, pp. 617–618, and Schmidt and Bodansky, *op. cit.*, p. 144.

29. M. McCormack, "U.S. Congressional Attitudes and Policies Affecting Nuclear Power Development in the World," *Atomic Energy Law Journal,* XVII, No. 4 (Winter 1974): 311.

30. R. Drinan, "Nuclear Power and the Role of Congress," *Environmental Affairs,* IV, No. 4 (Fall 1975): 607. See also Gravel, cited by Gofman and Tamplin, *op. cit.*, p. 364.

31. Berger, *op. cit.*, pp. 112, 144–147.

32. Gofman and Tamplin *op. cit.*, p. 362.

33. U.S. Atomic Energy Commission, Summary Report, *op. cit.*, p. 178. This is the justification given by the government for the limitation on liability. The reasoning here is that a serious accident is unlikely, hence massive insurance protection is not needed for the

public; therefore, it is reasonable for the Price-Anderson Act to establish such a limitation on liability. See AEC Staff Study of the Price-Anderson Act, Part II, *op. cit.*, p. 297, where the Atomic Energy Commission makes the argument that the likelihood of a serious accident is remote.

34. See U.S. Nuclear Regulatory Commission, *op. cit.*, for the Rasmussen Report, known as WASH 1400. See also U.S. Atomic Energy Commission, Summary Report, *op. cit.*

35. Report of the APS cited in R. Doctor, et al., "The California Nuclear Safeguards Initiative," *Sierra Club Bulletin*, LXI, No. 5 (May 1976): 45. The APS critique may be found in H. W. Lewis, *et al.*, "Report to the American Physical Society by the Study Group on Light-Water Reactor Safety," *Reviews of Modern Physics*, XLVII, No. 1 (Summer 1975), S1-S124; and Study Group on Light Water Reactor Safety, "Nuclear Reactor Safety—the APS Submits Its Report," *Physics Today*, XXVIII, No. 7 (July 1975): 38–43. See also Schmidt and Bodansky, *op. cit.*, pp. 86–87.

36. Cited in R. Augustine, "NRC Finds Safety in Numbers," in Munson, *op. cit.*, p. 31. See also Berger, *op. cit.*, pp. 54–63.

37. Nader, *op. cit.*, pp. 32–33; see also Novick, *op. cit.*, pp. 156–158, 315–316.

38. C. Hohenemser, R. Kasperson, and R. Kates, "The Distrust of Nuclear Power," *Science*, CXCVI, No. 4285 (April 1, 1977): 25–34.

39. Using the Rasmussen probability (per year, per reactor) for a core melt, one can compute the probability that a core melt will occur in one of the 60 plants now operating during their thirty-year lifetime. Employing the formula for the probability of mathematically independent events, P (a core melt in at least one of 60 reactors over a thirty-year lifetime) = 1 − P (no core melt in any of the 60 reactors over a thirty-year lifetime), one obtains P (core melt) $= 1 - (1 - (1/17{,}000))^{1800} = 1 - .8995 = .100$. Thus even if reactors now under construction and now planned are not built, there is still a 10 percent probability, or a one in ten chance, of a core melt in the thirty-year lifetime of one of the reactors now operating. Using the same formula and the Rasmussen probability, one can also compute the probability of a core melt in the thirty-year lifetime of 100 plants (assuming 60 now operating and 40 under construction). This P = .1618 or 16 percent. Thus for 100 reactors, the chance of a core melt is approximately 1 in 5. Similarly the probability of a core melt in the thirty-year lifetime of 350 plants (assuming 60 now operating, 40 under construction, and 250 being planned) is .4608 or 46 percent. Thus for 350 reactors there is almost a 50/50 chance of a core melt sometime in the thirty-year lifetime of one of them.

40. Nuclear proponents such as Schmidt and Bodansky, *op. cit.*, p. 143, make it clear that the Rasmussen Report has been and ought to be used as a justification for the Price-Anderson Act.

41. See Black, *op. cit.*, p. 897; Elder, *op. cit.*, p. 127; Berger, *op. cit.*, pp. 44–46; Gofman and Tamplin, *op. cit.*, pp. 174–179, 354–355; and G. C. Coggins, "The Environmentalist's View of AEC's 'Judicial' Function," *Atomic Energy Law Journal*, XV, No. 3 (Fall 1973): 189. Elder, *op. cit.*, p. 131, and Palfrey, *op. cit.*, pp. 1398–1400 also reveal the "cover-up" of the Brookhaven Report and of problems with the emergency core-cooling system.

42. See Barry Commoner, *The Closing Circle* (New York, Bantam, 1974), p. 197.

43. W. O. Doub, "Meeting the Challenge to Nuclear Energy Head-On," *Atomic Energy Law*, XV, No. 4 (Winter 1974): 258–259.

44. Olson, *op. cit.*, pp. 72, 76, 82. See also C. L. McGuire, "Emerging State Programs To Protect the Environment: Little NEPA's and Beyond," *Environmental Affairs*, V, No. 3 (Summer 1976): 572–575; Palfrey, *op. cit.*, p. 1378; Coggins, *op. cit.*, pp. 182–184; and J. Bieber, "Calvert Cliffs Coordinating Committee v. AEC: The AEC Learns the True Meaning of the National Environmental Policy Act of 1969," *Environmental Law*, III, No. 2 (Summer 1973): 316–333.

45. Palfrey, *op. cit.*, p. 1392.

46. Gofman and Tamplin, *op. cit.*, pp. 225–226.

47. See Coggins, *op. cit.*, pp. 184, 187, and Novick, *op. cit.*, pp. 109, 256 for details of

harassment and firing of government scientists who refused to allow pro-industry censorship. In Gofman and Tamplin, *op. cit.*, esp. pp. 24–27, 135, 255–272, the authors (former AEC scientists) tell how they were commissioned to calculate radiation effects, but when they arrived at conclusions contrary to the thesis that currently allowed radiation levels specified by the AEC (NRC) were safe, they were fired and the results of their studies were suppressed.

48. McKim, *op. cit.*, pp. 46–54, tells of interviewing Argonne staff members who said that top-level Nuclear Regulatory Commissioners had tampered with environmental impact statements.

49. Novick, *op. cit.*, p. 155; J. Primack and F. Von Hippel, "Nuclear Reactor Safety," *The Bulletin of the Atomic Scientists*, XXX, No. 8 (October 1974): 7–9. See also M. Bauser, "United States Nuclear Export Policy: Developing the Peaceful Atom as a Commodity in International Trade," *Harvard International Law Journal*, XVIII, No. 2 (Spring 1977): 51.

50. Primack and Von Hippel, *op. cit.*, pp. 7, 9; Berger, *op. cit.*, pp. 46–52.

51. Both proponents and opponents of nuclear power admit that this reason for failing to make the scale-model tests is the correct one. See Schmidt and Bodansky, *op. cit.*, pp. 139–142, and Novick, *op. cit.*, p. 192.

52. This explicit assumption is made in the Rasmussen Report WASH 1400, and is cited by Doctor, Harding, and Holdren, *op. cit.*, p. 45.

53. R. Polluck, "Nuclear Emergency Plans Fail in Most States," *Critical Mass Journal*, III, No. 1 (February 1978): 1, 8.

54. D. D. Comey, "The Brown's Ferry Incident," in Munson, *op. cit.*, pp. 5–7. See also Berger, *op. cit.*, pp. 40–43. The civil defense coordinator heard about the fire two days after it occurred; the local sheriff said he was "asked to keep quiet about the incident to avoid any panic." Fifteen minutes elapsed between the start of the fire and when the alarm was called in; there was confusion as to whom to call.

55. Fuller, *op. cit.*, p. 2.

56. See notes 33 and 34.

57. M. S. Young, "A Survey of the Governmental Regulation of Nuclear Power Generation," *Marquette Law Review*, LIX, No. 4 (1976): 843–844. With only one exception, the courts have viewed the liability limitation as a technical, rather than as an ethical, issue. In 1977, the U. S. District Court in Charlotte, N.C., declared the Price-Anderson Act unconstitutional on the grounds that it would allow "the destruction of the property or the lives of those affected by nuclear catastrophe without reasonable certainty that the victims would be reasonably compensated." Although this court ruled that the act is in violation of the equal protection and due process provisions of the Fifth Amendment, because it places the costs for employing nuclear power "on an arbitrarily chosen segment of society, those injured by nuclear catastrophe," the ruling was reversed on June 26, 1978, by the U.S. Supreme Court. The highest court ruled that, "when appraised in light of the extremely remote possibility of an accident in which liability would exceed the statutory limit . . . , a $560 million ceiling is within permissible limits and not violative of due process." (Duke Power Company v. Carolina Environmental Study Group, Inc., et al. and U. S. Nuclear Regulatory Commission et al., v. Carolina Environmental Study Group, Nos. 77–262 and 77–375, *The United States Law Week*, XLVI, No. 50 [June 27, 1978]: 4845; and Berger, *op. cit.*, pp. 145–146.)

58. Muchnicki, *op. cit.*, p. 42.

59. Doub, *op. cit.*, p. 261.

60. Although not a logical error, the naturalistic fallacy is committed whenever one attempts to deduce ethical conclusions from purely empirical premises or to define ethical characteristics in nonethical terms. (G. E. Moore, *Principia Ethica* [New York, Cambridge University Press, 1951], p. 39.) It is built on the false assumption that "ethics is an empirical or positive science." (PE, p. 39.) Of course it is not, since no amount of solely factual or scientific data about *what is the case* is alone sufficient to determine ethically *what ought to*

be the case. In making this assumption, those who commit the naturalistic fallacy fall into a number of errors, the most common of which is the attempt to replace ethics with one of the natural sciences. (PE, p. 40.)

61. Muchnicki, *op. cit.,* p. 43. See also J. Lieberman, "Generic Hearings: Preparation for the Future," *Atomic Energy Law Journal,* XVI, No. 2 (Summer 1974): 142. Here Lieberman argues that the procedural rights of participants in hearings, the nature of the issues to be resolved, and the fairness of the proceeding are the major problems not only of nuclear power plant hearings, but also of "decision making in a technical society."

62. Muchnicki, *op. cit.,* p. 44.

63. Summary Report of the U. S. Atomic Energy Commission, *op. cit.,* pp. 182–183.

64. *Ibid.,* p. 179. Hans Bethe, a government consultant and proponent of nuclear energy makes the same technological/probabilistic argument. After admitting that a nuclear accident might cause 5,000 cancer deaths, he argues: "One should remember that in the U.S., there are more than 300,000 deaths every year from cancers due to other causes." (Hans Bethe, "The Necessity of Fission Power," *Scientific American,* CCXXXIV, No. 1 [January 1976]: 26.)

65. *How Probable Is a Nuclear Plant Accident?* (Washington, D.C., Office of Public Affairs, 1976), EDM-074R (9-76), p. 2.

66. See footnote 60; see also A. R. White, *G. E. Moore: A Critical Exposition* (Oxford, Basil Blackwell, 1958), 122–127; F. Snare, "Three Sceptical Theses in Ethics," *American Philosophical Quarterly,* XIV, No. 2 (April 1977): 129–130; L. Kohlberg, "From Is to Ought: How to Commit the Naturalistic Fallacy and Get Away with It in the Study of Moral Development," in *Cognitive Development and Epistemology,* ed. T. Michel (New York, Academic Press, 1971), p. 154; J. H. Olthuis, *Facts, Values and Ethics* (Assen, Van Gorcum, 1969), pp. 28–34; K. E. Goodpaster (ed.), *Perspectives on Morality: Essays by William K. Frankena* (Notre Dame, Ind., Notre Dame University Press, 1976), pp. 16–17, 210.

67. A. Einstein, "The Laws of Science and the Laws of Ethics," in *Readings in the Philosophy of Science,* ed. H. Feigl and M. Brodbeck (New York, Appleton-Century-Crofts, 1953), p. 779.

68. K. Sayre and K. Goodpaster, "An Ethical Analysis of Power Company Decision-Making," in Sayre, *op. cit.,* pp. 238–288, argue that the utility industry operates according to the concept of utility, rather than according to the concept of right. If their analysis is correct, then it would not be surprising if the Price-Anderson Act, which is essential to utility operation in the nuclear area, is also utilitarian in nature.

69. Rep. Cole made this same point during Congressional hearings on the Price-Anderson Act. His remarks are cited by Elder, *op. cit.,* p. 134. The U. S. District Court for the Western District of North Carolina agreed with this opinion in a 1977 ruling. The court said that the liability limits of the Price-Anderson Act violated the due process and the equal protection clauses of the Fifth Amendment. This ruling, however, was overturned by the U.S. Supreme Court. (See note 57.)

The limitation on liability of the Price-Anderson Act is utilitarian in ways other than its neglect of the rights of those living near a nuclear reactor. Because the Act is governed by a twenty-year statute of limitations, and since radiation-induced cancers often take thirty years before symptoms are apparent (see Olson, *op. cit.,* p. 55; see also Elder, *op. cit.,* pp. 126–132; and J. C. Bodie, "The Irradiated Plaintiff: Tort Recovery Outside Price-Anderson," *Environmental Law,* VI, No. 3 [Spring 1976]: 859–896), the Act also discriminates against injured parties whose symptoms do not appear as early as those of others. This means that a disproportionate share of the cost of the risk of Price-Anderson is, again, borne by victims.

70. For arguments to this effect, see AEC Staff Study of the Price-Anderson Act, Part II, *op. cit.,* p. 297; Schmidt and Bodansky, *op. cit.,* p. 145; Lowenstein, *op. cit.,* p. 600.

71. Cited by G. Hardin, "To Trouble a Star: The Cost of Intervention in Nature," in R. T. Roelofs, J. N. Crowley, and D. L. Hardesty (eds.), *Environment and Society* (Englewood Cliffs, N.J., Prentice-Hall, 1974), p. 125. In response to this charge of irresponsibility, proponents of nuclear industry have claimed that forcing the utility to go bankrupt, in the event of a disaster, would be of "dubious social value." (Schmidt and Bodansky, *op. cit.*, p. 146.) While it is true that such a bankruptcy would not enable the utility in question to continue to fulfill its social and legal mandate to provide power to its service area, other ethical benefits might accrue. For one thing, victims of the disaster would be able to recoup some of their losses. Moreover, the negligent utility might be prevented from operating another plant in such a way as to cause a further disaster. In other words, the bankruptcy might serve to protect other potential victims from a nuclear catastrophe. As it is, Price-Anderson now enables industry to make operating decisions only on the basis of economic and technological considerations, rather than also on the basis of liability constraints.

72. The Joint Commission on Atomic Energy admitted in 1965 that the government's indemnity was provided at a cost much lower than would be assessed commercially, were such insurance available; see AEC Staff Study of the Price-Anderson Act, Part I, p. 239. Although it is true that the 1975 amendments to the Price-Anderson Act provide for a phasing out of government subsidy, once approximately thirty more plants are built, this subsidy exists and will continue to exist for some time. Such a continuation is all the more questionable because the 1954 Atomic Energy Act has, for the last twenty-five years, aroused controversy over its being a "give-away" statute.

73. Berger, *op. cit.*, pp. 94–97, 106–112, 144–147; see also Gofman and Tamplin, *op. cit.*, pp. 177, 199; Primack and Von Hippel, *op. cit.*, p. 7; Muchnicki, *op. cit.*, p. 45. Berger, p. 112, explains that "more than a hundred billion dollars [in federal funds] has already been invested in nuclear power" subsidies since 1945.

74. These statistics are taken from M. Mesarovic and E. Pestel, *Mankind at the Turning Point: The Second Report to the Club of Rome* (New York, Signet, 1974), pp. 135–139.

75. John Rawls, proponent of the contractarian deontological theory most widely discussed currently, would probably argue that current discrepancies in patterns of consumption are not ethical. According to him (*A Theory of Justice* [Cambridge, Mass., Harvard University Press, 1971]), all persons have a duty to act according to principles of justice securing the fairest distribution of goods among all people. He argues that this duty exists by introducing the notion of the "original position." The "original position" is that place held by a person who is completely ignorant of his natural ability, his place in society, and all other details that might give him an advantage over his fellow humans. If one imagines what principles, for the distribution of the goods of society, he might hold, were he in the original position, he will arrive at truly just principles. This means, says Rawls, that persons in the original position would reject utilitarianism, because no rational being would agree to accept a societal theory that might sanction a loss for himself in order to bring about a greater net good for everyone. Hence if one adopted Rawls's "thought experiment" of the "original position" it is unlikely that one would sanction the disparity in consumption between U.S. citizens and other peoples in the world. The reason for rejecting such a utilitarian state of affairs is that, in the original position, one does not know whether he would be a U.S. citizen or not. Hence his "ignorance" provides a rational basis for adopting a principle of equity.

76. Palfrey, *op. cit.*, p. 1391.

77. Willrich, *op. cit.*, p. 181, uses this argument regarding nuclear fusion; Novick, *op. cit.*, p. 318, makes the same point in terms of solar energy.

78. Standards of the International Commission on Radiation Protection, the basis for U. S. regulations on nuclear energy, affirm that there is no safe dose of radiation, and no threshold below which no carcinogenic or genetic effects occur. ICRP standards are given in J. P. Holdren, "Hazards of the Nuclear Fuel Cycle," *The Bulletin of the Atomic Scientists*, XXX, No. 8 (October 1974): 16, and confirmed in Gofman and Tamplin, *op. cit.*, pp. 39,

293–307; Gofman and Tamplin, "Nuclear Power, Technology, and Environmental Law," *Environmental Law*, II, No. 2 (Winter 1971): 59–63; B. Commoner, *The Closing Circle* (New York, Bantam, 1974), p. 194; and O. Hansen, "Development and Application of Radiation Protection Standards," *Idaho International Law Review*, XII, No. 1 (Fall 1975): 1–32.

79. Caldwell, et al., *op. cit.*, p. 383, also makes this point; see pp. 380–383, and Elder, *op. cit.*, p. 122.

80. Higher radiation readings in the atmosphere outside the reactor building are documented in Fuller, *op. cit.*, p. 17.

81. Fuller, *op. cit.*, pp. 109–116.

82. Fuller, *op. cit.*, p. 84; see pp. 75–92 for more details on the Windscale accident.

83. See Schmidt and Bodansky, *op. cit.*, p. 88, and Fuller, *op. cit.*, pp. 17 ff.

84. Summary Report of the U. S. Atomic Energy Commission, *op. cit.*, p. 178. Government regulators make this surprising remark because the context of their discussion is the safety of *commercial* nuclear reactors. Since most reactors have been military, experimental, or government-subsidized, they are not classed purely as "commercial," and accidents in them are not called "accidents of commercial reactors."

85. Cited by Caldwell, et al., *op. cit.*, p. 366.

86. The difficulty of proving that radiation, released from a nuclear reactor, has caused a particular cancer is illustrated very well by recent attempts of U. S. veterans to prove that their cancers were caused by their participation in atomic bomb testing in the 1950s. On August 31, 1957, for example, U. S. troops were marched "to within 300 yards of ground zero" soon after the Smoky atomic blast. Although the U. S. Department of Energy has found records of 900 overexposed individuals involved in tests dating from 1951 to 1962, benefits have been granted only to ten men of the many "who claimed their disabilities were the result of radiation exposure received during the U.S. nuclear weapons tests." (M. Korchmar, "Radiation Hearings Uncover Dust," *Critical Mass Journal*, III, No. 12 [March 1978]: 5; see also Berger, *op. cit.*, pp. 65–66, 69, 71, and R. Kraus, "Environmental Carcinogenesis: Regulation on the Frontiers of Science," *Environmental Law*, VII, No. 1 [Fall 1976]: 83–135.) Because of the problems associated with all levels of radiation and the possible violations of rights involved, the U. S. House Commerce subcommittee on Health and Environment held three days of hearings in January 1978, to hear testimony by servicemen. (Korchmar, *op. cit.*, p. 5.) According to Orville Kelly, USN Commander of Eniwetok Atoll during weapons tests in the 50s, between 200,000 and 400,000 soldiers were exposed to high levels of radiation. Since radiation was not monitored, he says, the government claims it has records of only 900 overexposures. Kelly, a cancer victim and founder of "Make Today Count," has been denied benefits because the Veterans Administration says he cannot prove his cancer was caused by radiation received during the weapons tests. Kelly was within five miles of ground zero during twenty-three nuclear tests and was exposed to fallout. In the last seven years of pleading for benefits for his wife and four children, he has exhausted all legal channels of appeals. (Facts regarding the Orville Kelly case, one of many in which benefits were denied, came from a personal conversation with him in my home on August 17, 1978.)

87. Sheldon Novick raises some interesting civil liberties questions in *The Electric War: The Fight over Nuclear Power* (San Francisco, Sierra Club Books, 1976), p. 308. He points out that in Texas and California, local police forces have kept an eye on people at the request of power company officials, and that, although the authority was refused, the Virginia Electric Power Company recently asked the state legislature for police power which would allow it to arrest people and search their homes. Carl Walske, president of the Atomic Industrial Forum (and hence the top spokesperson for nuclear industry), thinks we need a new Federal police agency solely to deal with problems related to plutonium and enriched uranium. The Nuclear Regulatory Commission also contends that balancing the need for

public protection against the necessity for intrusion of government into personal liberties is "one of the most difficult and delicate questions confronting the Commission" [V. Gilinsky, "The Need for Nuclear Safeguards," *Atomic Energy Law Journal,* XVII, No. 2 [Summer 1975]: 149–150. See also R. W. Ayres, "Policing Plutonium: The Civil Liberties Fallout," *Harvard Civil Rights-Civil Liberties Law Review,* X, No. 2 [Spring 1975]: 369–443.)

Because of the susceptibility of nuclear fission plants to core meltdown, they are unique targets for sabotage and for terrorist attack. Since a core melt would take thousands of lives and destroy hundreds of square miles of the area surrounding the plant, terrorists could be almost certain of having their demands met, once they were able to take over a plant. Moreover, although there have not been any publicized accounts of such take-overs of power plants in this country, such terrorist attacks on nuclear installations have occurred in other countries. There have been celebrated cases of sabotage of nuclear plants in this country, some by utility employees, as well as several incidents during which explosives were actually placed at reactors. See J. J. Berger, *op. cit.,* pp. 187–218, esp. pp. 205–206; M. S. Young, "A Survey of the Governmental Regulation of Nuclear Power Generation," *Marquette Law Review,* LIX, No. 4 (1976): 849–850; and J. D. Schleimer, "The Day They Blew Up San Onofre," *The Bulletin of the Atomic Scientists,* XXX, No. 8 (October 1974): 24–27. Even nuclear power proponents have argued that "the main effect of the exploitation of nuclear energy for peaceful purposes will be an intensification of the problems already inherent in the structure of the global political system." (M. Willrich, *Global Politics of Nuclear Energy* [New York, Praeger, 1971], p. 186; see also pp. 37–41, 43–66, 80–83, 179–186.) Because the "peaceful" uses of nuclear facilities are such that they could be used to manufacture nuclear weapons, the International Atomic Energy Agency opened the Treaty on Nonproliferation of Nuclear Weapons for ratification in 1968. Despite this, however, the plutonium used in the Indian nuclear explosion of May 18, 1974, was produced in a reactor provided India by Canada. Hence U.S. reactors, whether used in this country or marketed abroad, have the potential for being used to fabricate nuclear weapons. See Gilinsky, *op. cit.,* pp. 144–145; R. Arfa, et al., "Nuclear Proliferation," *Brooklyn Journal of International Law,* XXX, No. 2 (Spring 1977): 258–292; M. A. Bauser, "The Atomic Energy Commission's ECCS Rule-Making," XVI, No. 1 (Spring 1974): 42–90; D. Triggs, "Prospects for Nuclear Proliferation and Its Control," *Denver Journal of International Law and Politics,* VI, No. 1 (Spring 1976): 159–190; J. E. Vaughan, "Nuclear Diversion," *Atomic Energy Law Journal,* XVII, No. 3 (Fall 1975), 180–192.

88. The transportation phase of the nuclear fuel cycle is particularly vulnerable to accident or sabotage because of the great number of nuclear materials being sent. By 1980, 300 rail shipments or 15,000 truck shipments of spent fuel will be required annually, says W. T. England, "Recent Regulatory Developments Concerning the Transportation of Nuclear Fuel and Other Radioactive Materials," *Environmental Law,* VII, No. 2 (Winter 1977): 213. Moreover, railroads all over the country are refusing to ship radioactive materials because of their hazardous nature and because they would be unable to handle the insurance claims brought against them as a result of an accident (*ibid.,* pp. 214–218). An additional problem is that, according to some authors, J. W. Gofman and A. R. Tamplin, *op. cit.,* p. 201, organized crime has made great inroads into the transportation industry; every trucking firm in one state is Mafia-owned and controlled, and according to the U.S. Attorney General, the cargo handling apparatus at Kennedy Airport is also under Mafia control.

Some authors, such as W. O. Doub and L. M. Muntzing ("A Perspective on World-Wide Nuclear Developments," *Atomic Energy Law Journal,* XVII, No. 4 [Winter 1976]: 240) maintain that solar energy is uneconomical and that its technology is not yet well developed enough for use. Other authors, such as Novick (*op. cit.,* pp. 87–88, 138–144) and Berger (*op. cit.,* pp. 244–277) maintain that in ten years solar energy could be as cheap as energy from coal, and that although solar technology is now available, it is not used because there are no commercial structures available for its utilization.

89. Answering this question depends not only on how one views the ethical answer to this question, but also on what the available energy options are. Some authors, e.g., W. O. Doub ("Meeting the Challenge to Nuclear Energy Head-On," *Atomic Energy Law Journal,* XV, No. 4 [Winter 1974]: 239) maintain that supplies of fossil fuels are so limited that there is no choice but to employ increasing amounts of nuclear energy immediately. Other authors, e.g., Novick (*op. cit.*, p. 315), Gofman and Tamplin (*op. cit.*, p. 359), and the authors of the Club of Rome Report (M. Mesarovic and E. Pestel, *op. cit.*, pp. 137 ff.) claim that there are hundreds of years of fossil fuels available, and that the government has not revealed this completely, so as to promote nuclear power, until the time that widespread solar power is available.

90. Many authors argue that nuclear energy is the cheapest source of power available, because of low fuel costs. See H. A. Bethe, "The Necessity of Fission Power," *Scientific American,* CCXXXIV, No. 1 (January 1976): 21–31; Walske, cited in Novick, *op. cit.*, p. 202; Willrich, *op. cit.*, p. 73; F. Schmidt and D. Bodansky, *op. cit.*, pp. 45–49.

Some writers point out that all government and industrial compilations of the comparative cost of nuclear energy ignore the hundreds of billions of dollars of government subsidy of the nuclear industry through its coverage of reprocessing, waste storage, research and development, insurance, and enrichment. See Berger, *op. cit.*, pp. 112, 144–145, and Novick, *op. cit.*, p. 279.

A number of experts have argued that nuclear power is economically desirable only if one makes erroneous assumptions about economies of scale, load factors, rising construction costs, and especially, if one ignores the fivefold increase in the cost of uranium for fuel, as well as the absence of reprocessing facilities. See K. Jameson, "Economies of Scale in the Electric Power Industry," in Sayre, *op. cit.*, pp. 126–130; V. McKim, *op. cit.*, pp. 70–71; Novick, *op. cit.*, pp. 300–313; Willrich, *op. cit.*, pp. 35–36; Doub and Muntzing, *op. cit.*, p. 244; P. L. Joskow, "Commercial Impossibility, the Uranium Market, and the Westinghouse Case," *Journal of Legal Studies,* VI, No. 1 (January 1977): 119–137; J. R. Brydon, "Slaying the Nuclear Giants," *Pacific Law Journal,* VIII, No. 2 (July 1977): 741–750. Moreover, claim these writers, the diseconomies of nuclear power are responsible for the cancellation of most of the reactors ordered in the last four years in the United States. See Berger, *op. cit.*, p. 112; Joskow, *op. cit.*, p. 165; Novick, *op. cit.*, pp. 138, 199, 299; and M. C. Olson, *op. cit.*, pp. 236–237.

91. Those who answer "yes" to this question include government regulators of nuclear energy, such as Anders and Ramey. See Bethe, *op. cit.*, pp. 21–31; J. T. Ramey, "Quality Assurance: Translating Doctrine into Action," *Atomic Energy Law Journal,* XV, No. 1 (Spring 1973): 6–7; W. A. Anders, "The Need for Credible Nuclear Regulation," *Atomic Energy Law Journal,* XVII, No. 2 (Summer, 1975): 124. For the opposite point of view on this issue, see Olson, *op. cit.*, p. 149; Novick, *op. cit.*, pp. 21, 228, 279–280; Willrich, *op. cit.*, pp. 70–78; Joskow, *op. cit.*, p. 121; Schmidt and Bodansky, p. 34; K. E. Boulding, "Fun and Games with the Gross National Product," in *Environment and Society,* ed. R. T. Roelofs, J. N. Crowley, and D. L. Hardesty (Englewood Cliffs, Prentice-Hall, 1974), p. 133; C. C. Howard, "Uranium Sales Contracts," *Rocky Mountain Mineral Law Institute,* XXII (1976): 397.

92. A number of persons, especially those concerned about the plight of minority groups within the economy, have answered "yes" to this question. See Willrich, *op. cit.*, pp. 13–36, 184; Schmidt and Bodansky, *op. cit.*, pp. 1–4; Ramey, p. 3; E. D. Johnson, "The Aerospace Industry," in de Nevers, *op. cit.*, p. 147; D. C. Anderson, "Policy Riddle: Ecology vs. the Economy," in Roelofs, et al., *op. cit.*, p. 110. See also Novick, *op. cit.*, pp. 241–266.

For arguments as to the detrimental effects of increased energy consumption and the desirability and possibility of conservation, see Berger, *op. cit.*, pp. 221–243; Gofman and Tamplin, *op. cit.*, pp. 214–225; Novick, *op. cit.*, pp. 233–238, 299–305; Johnson, *op. cit.*, p.

111; Doub and Muntzing, *op. cit.*, pp. 235–248; K. M. Sayre, "A Cybernetic Analysis of Certain Energy Consumption Patterns," in Sayre, *op. cit.*, pp. 1–29; National Goals Research Staff, "Environmental Policy," in Roelofs, et al., *op. cit.*, pp. 137–144; N. Notis-McConarty, "Federal Accountability," *Environmental Affairs,* V, No. 1 (Winter 1976): 130, 138; H. Sprout, "The Politics of Environmental Reform," in Roelofs, et al., *op. cit.*, pp. 172–174; P. Schrag, "Who Owns the Environment?" in Roelofs, et al., *op. cit.*, pp. 153–161. Perhaps the most famous of contemporary "no growth" arguments is found in Meadows, D. H., et al., *The Limits to Growth: A Report for the Club of Rome's Project on the Predicament of Mankind* (New York, New American Library, 1974). Other authors present excellent philosophical reasons why increasing levels of consumption are not desirable. See V. McKim, *op. cit.*, pp. 30–99; and E. Maher, "The Dynamics of Growth in the Electric Power Industry," in Sayre, *op. cit.*, pp. 149–216.

93. For philosophical and legal discussions of this question, see McKim, *op. cit.*, pp. 30–99, and C. Murdock, "Legal and Economic Aspects of the Electric Utility's 'Mandate to Serve,'" in Sayre, *op. cit.*, pp. 100–115.

94. See Gofman and Tamplin, *op. cit.*, p. 337; T. Black, "Population Criteria of Oregon and the U. S. in the Siting of Nuclear Power Plants," *Environmental Law,* VI, No. 3 (Spring 1976): 897–920; R. Doctor, et al., "The California Nuclear Safeguards Initiative," *Sierra Club Bulletin,* LXI, No. 5 (May 1976): 4–6, 44–66; G. B. Karpinski, "Federal Preemption of State Laws Controlling Nuclear Power," *Georgetown Law Journal,* LXIV, No. 6 (July 1976): 1323–1341; A. W. Murphy and D. B. LaPierre, "Nuclear 'Moratorium' Legislation in the States and the Supremacy Clause," *Columbia Law Review,* LXXVI, No. 3 (April 1976): 392–456.

For an explanation of the physical character and lethal medical effects of this radiation, and an analysis of its causal role in human disease and death, see E. Nathanson, "International Management of Radioactive Wastes," *Environmental Affairs,* V, No. 2 (Spring 1976): 363–380; Berger, *op. cit.*, pp. 92–111; Gofman and Tamplin, *op. cit.*, pp. 22, 196–203, 325; L. Breitkopf, "The National Environmental Policy Act of 1969 and Nuclear Power Plant Licensing," *DePaul Law Review,* XXVI, No. 3 (Spring 1977): 667–671; Bryden, *op. cit.*, pp. 652–679; Doub and Muntzing, *op. cit.*, p. 241; Meadows, et al., *op. cit.*, p. 81; Young, *op. cit.*, 850–851; Schmidt and Bodansky, *op. cit.*, pp. 37–39.

Olson, *op. cit.*, pp. 161 ff.; Novick, *op. cit.*, pp. 178–179; Gofman and Tamplin, *op. cit.*, pp. 190–192; and Nathanson, *op. cit.*, p. 365, describe some of the consequences of leaks in Washington, New York, South Carolina, and Kentucky.

95. A number of members of the U. S. Nuclear Regulatory Commission have stated explicitly that financial considerations have dictated the policy of allowing small amounts of radioactive discharges from power plants; see Fox, et al., "The Case for Tax-Exempt Bond Financing of Pollution-Control Facilities at Nuclear Power Plants," *Atomic Energy Law Journal,* XVII, No. 1 (Spring 1975): 43. See also E. A. Mason, "Recent Actions by the Nuclear Regulatory Commission," *Atomic Energy Law Journal,* XVII, No. 2 (Summer 1975): 164–165. See also Gofman and Tamplin, *op. cit.*, pp. 77–78, 149–157, 288–291; and Novick, *op. cit.*, pp. 95, 104, 182. For accounts of cancers and genetic damage which have already occurred as a result of "permissible" radiation levels, see Gofman and Tamplin, *op. cit.*, pp. 302–304; B. Rosenthal, "Nuclear Hazards Abound," *Critical Mass Journal,* III, No. 1 (February 1978): 6–7; T. F. Mancuso, et al., "Radiation Exposures of Hanford Workers Dying from Cancer and Other Causes," Pittsburgh, University of Pittsburgh, March 1977, Contract No. E (11-1)-3428, unpublished research report. See also note 78.

ADAMS, INHABER, AND RISK-BENEFIT ANALYSIS:

ETHICAL AND METHODOLOGICAL PROBLEMS WITH TWO RECENT ASSESSMENTS OF NUCLEAR SAFETY

K. S. Shrader-Frechette

I. INTRODUCTION

Ever since the accident at Three Mile Island, the already-raging debate over nuclear fission has escalated to astounding proportions. Proponents of atomic energy have criticized media coverage of the event and called for public education in science and technology to forestall a "Luddite revolt" against nuclear power.[1] Opponents, on the other hand, have criticized government and industry for its "abusive use of science to bias public policy decisions" in favor of fission technology.[2] On both sides of

Research in Philosophy & Technology, Volume 3, 1980, pages 343–365.
Copyright © 1980 by JAI Press Inc.
All rights of reproduction in any form reserved.
ISBN: 0-89232-102-4

the debate, there have been charges of emotionalism and scientific illiteracy. Although some scientists have argued that widespread public dismay with the technology is a consequence of the scientific ignorance of the population, others have argued that, on the question of atomic energy, "the divisions are deeper and more bitter among the scientifically literate than in the general public."[3] A score of Nobel Prize winners, in almost equal numbers, have positioned themselves in opposite camps on the question of nuclear power.

Generally speaking, the views of members of these respective camps are characterized not only by extreme disagreement over the scientific and technological facts of the matter, but also by great conflict concerning the relevant ethical questions. Regarding the factual matter of the biological consequences of exposure to radiation, for example, respected proponents of nuclear technology have argued that only 0.0002 cases of cancer per year would be induced by exposure to one rad of radiation.[4] Equally respected scientists, opponents of atomic fission, maintain that something on the order of 32,000 cases of cancer per year would be induced by this same level of exposure.[5]

Apart from these scientific disagreements, the controversy also is characterized by numerous conflicts over ethical questions. Often nuclear proponents argue that requiring "zero risk" from reactors is not an ethically acceptable position,[6] while their opponents claim that "zero risk" is ethically necessary.[7]

How does one decide who is right? Given the complexity of the question ("ought we to employ light-water reactors to generate electricity?") and given the impressive array of scholars who have aligned themselves on both sides of the issue, there is only one way to untangle the scientific and ethical considerations intertwined in the nuclear-power controversy. This is by slow and painstaking analyses of all the technical and philosophical assessments of atomic energy.

In the present essay I will take up this analytical task by focusing on two recent studies of nuclear power, both highly supportive of the technology. The first, by Herbert Inhaber, is largely technical and scientific in scope.[8] The second, by Frank Adams, addresses both technical and ethical questions.[9]

There are several reasons for choosing the Adams and Inhaber studies as a basis for analysis. Both reports represent the current state-of-the-art regarding the nuclear debate and both were published recently (Adams' work in December, 1978, and Inhaber's in February, 1979). Since the demise of the 1975 government analysis, WASH-1400, known as the "Rasmussen Report,"[10] the main work cited by scientific proponents of atomic fission has been that of Herbert Inhaber. An advisor to the Atomic Energy Control Board of Canada, Inhaber has presented his findings to numerous scholarly groups throughout the world in the last year since his research was completed. The Adams report is of interest, despite its numerous unsubstantiated and question-begging claims, because it is one of the only pronuclear works to focus precisely on the ethical aspects of the debate, rather than merely on technical considerations; to the extent that specifically *ethical* questions have been analyzed, this has been done largely by nuclear opponents.[11]

This essay is a critical assessment of recent works by two nuclear proponents. It should not be taken to mean, however, either that there are not also methodological and ethical problems in studies critical of nuclear technology, or that the difficulties identified here are an exhaustive list of those found in the work of Adams and Inhaber. Rather, the essay is one of many that must be done before a thorough assessment of all studies of atomic energy can be said to have been accomplished. The sorts of questions raised here are meant to serve both as indicators of key methodological and ethical problems in existing technology assessments and as warnings to those who uncritically accept such assessments, particularly in the case of nuclear fission.

II. THE INHABER REPORT

One work, which threatens to gain widespread, uncritical acceptance, is that by Herbert Inhaber. Authored over a three-month period,[12] this study provides precise, quantitative conclusions regarding the relative health and safety risks from alternative means of generating electricity. Because of the specific, mathematical character of its claims, the work has gained much acclaim.

Inhaber compared the risk to human health for five conventional (coal, oil, natural gas, nuclear, and hydroelectric) and six nonconventional (solar thermal electric, solar photovoltaic, solar space heating, methanol, wind, and ocean thermal) means of generating energy. Considering the entire cycle for producing electricity, he concluded that "the risk to human health from nonconventional sources can be as high as, or even higher than, that of conventional sources" when one determines relative "risk per unit energy."[13] Specifically, Inhaber deduced that total man-days lost (as a result of both public and occupational injuries and deaths), per megawatt-year output over the lifetime of the system, was very small for nuclear generated electricity as opposed to alternative energy sources. On his analysis, total man-days lost, per megawatt-year output over the lifetime of the system, is approximately 2000 for coal and oil, about 1000 for wind, solar thermal, and solar photovoltaic, roughly 400 for methanol, 200 for solar space heating, 50 for hydroelectricity, and 30 for ocean thermal, but only approximately 10 man-days lost for nuclear and 6 for natural gas. This means that, on his calculations, of eleven conventional and nonconventional energy sources, nuclear is the safest, except for natural gas.[14] "Wind, solar thermal, and solar photovoltaic have much of their risk produced by the backup they require," says Inhaber; this backup was assumed by him to be coal.[15] Hence, if one argues that energy policy ought to be determined in large part by choosing means of generating electricity which cause the least amount of human health risk, then (provided Inhaber's conclusions are accurate) on this basis one ought to give serious consideration to adopting greater use of nuclear fission.

How accurate, however, are Inhaber's conclusions, especially with regard to the alleged relative safety of atomic power? There are a number of methodological assumptions, omissions, and ethical considerations which, when properly taken into account, strongly suggest that Inhaber's deductions are incorrect. Because of the public policy significance of his views, let us examine these problem areas more closely.

III. INHABER'S OMISSIONS

The most general difficulties plaguing the Inhaber study arise from his omission of parameters constituting an important part of risk-benefit analysis. On the basis of man-days lost, for example, he concludes that nuclear-generated electricity is much safer than coal. Even if this comparison were accurate (see Section IV below for reasons why it is not), however, it alone would not constitute grounds for Inhaber's argument that greater "risks" arise from solar or coal than from nuclear-generated electricity.[16] The reason, of course, is that he has omitted consideration of many criteria in terms of which coal and solar are less risky than is atomic power. For example, nuclear-generated electricity is much more expensive than that produced by coal, according to nuclear industry, U.S. government, and university experts.[17] Inhaber admits this fact, but ignores it when he draws his conclusions.[18] In so doing, he neglects consideration of the point that cost of electricity has a definite effect on the health and well-being of energy users, particularly when those users are poor or living on fixed incomes. Hence, if more expensive means of generating electricity (e.g., nuclear) are chosen over less expensive ones (e.g., coal), and if this cost differential continues to be passed on to the consumer by the utility, then the health and well-being of certain persons in society is clearly threatened, e.g., urban poor living in poor housing during periods of extreme cold. Because Inhaber ignored cost-parameters, his risk-benefit comparison of alternative energy sources is both incorrect (by virtue of incompleteness) and broadly misleading.

Another obvious omission, also admitted by Inhaber but ignored by him in his considerations, is the political and economic benefit deriving from nonconventional, decentralized systems like solar space heating. Although he recognizes that such systems have greater reliability and also less dependence upon political and economic trends and powers,[19] Inhaber does not include these benefits in his assessment of the risks and benefits of conventional and nonconventional sources of energy. Were these paramenters, e.g., reliability, taken into account, it is possible that Inhaber's conclusion would be reversed and therefore that nonconventional means of generating electricity would be shown to be more beneficial than conventional ones.

From a philosophical point of view, Inhaber's most troubling omissions have to do with the extent to which implementation of various energy technologies promotes equity and due process. Since equal justice and due process have clear effects on human health and well-being, his omissions in this area are significant, because they suggest not only methodological reasons why his conclusions might be false, but also ethical assumptions which, if accepted, might result in grave violations of justice as a result of utilizing one, rather than another, energy technology.

For example, all of Inhaber's risk-benefit calculations are determined on the basis of the average number of deaths and injuries resulting from construction and operation of various energy sources; in other words, the risk calculated is solely medical. Medical risks, however, constitute only a part of the total class of risks borne by the public as a result of electricity generation. Some means of producing power also cause great financial risks to the public which, because of their magnitude, are likely to affect the social and psychological well-being of numerous persons. Take, for example, the risk of a core melt and a resultant catastrophic release of radiation from a nuclear power reactor. If such an accident were to occur, say U.S. government officials, it could result in $18 billion in property damages alone, apart from other billions in medical expenses.[20] Either the Price-Anderson Act is to "satisfy public liability claims" as the Atomic Energy Commission says, or it is not. If it is to meet these claims, then its liability limits must be consistent with governmental damage estimates, and not merely 3 percent (i.e., $560 million divided by $18 billion) of them. If the Act is not to compensate for damages, then Price-Anderson need not be amended, but government officials must retract the claim that it will "satisfy public liability claims." In a later section of the paper I will argue why the most ethical decision is to raise the liability limits of Price-Anderson.

A second disturbing fact is that the government estimates of the nuclear risk, presupposed in the Price-Anderson Act, are inconsistent with the estimates given by private insurers. Clearly either the $435 million government insurance subsidy (provided to the nuclear industry above the $125 million available through private insurance companies) represents a risk the government ought to take with the taxpayers' money, or it does not. If such a risk is economically desirable, then a profit can very likely be made on nuclear insurance and the taxpayers' money probably will not be lost. In this case, private industry ought to be ready to invest heavily in insuring the nuclear industry. However, facts do not bear this out. After more than twenty years during which nuclear reactors have generated power, the maximum insurance obtainable for them through private insurance pools is $125 million. This is especially interesting, since private insurance groups have provided, for example, up to $500 million in coverage for Pan American Airlines. Because of pressure from the utility industry for protection against bankruptcy in the event of such an accident, the U.S. government has, by law (the Price-Anderson Act), limited liability in such a catastrophe to $560 million (or 3 percent of total property damages, apart from other costs of the incident).[21] Although this law has been upheld (for technical reasons) by the Supreme Court, other Federal courts have ruled that it violates due process and equal protection guarantees of the Fifth and Fourteenth Amendments.[22]

In no energy technology other than nuclear is it the case that a majority of the medical and financial consequences of an accident must be borne by its victims, the public, rather than by its perpetrators, the utility. Hence the disastrous consequences of limited nuclear insurance are monumental when compared to the absence of liability limitations for accidents related to other energy technologies. In ignoring the social, psychological, and financial risks associated with atomic power and the Price-Anderson Act, Inhaber has underemphasized the actual risks associated with nuclear generation of electricity and thereby overemphasized its benefits. His omission points to both ethical and methodological flaws in his analysis.

A related ethical omission in the Inhaber report has to do with his neglecting to analyze the *equity of distribution* of risks and benefits from alternative energy sources. Instead, his study investigates merely the *magnitude* of medical risks from various means of producing power. Obviously a particular energy source might cause a smaller degree of overall medical risk than another, but might be less acceptable because those risks were distributed in a profoundly inequitable manner. Inhaber does not take account of this fact; he merely computes man-days lost from deaths and injuries arising from hazards of various energy sources.[23] Clearly, however, certain energy sources are less acceptable not only if they cause a greater number of deaths and injuries than those arising from other means of generating electricity, but also if they cause a more inequitable distribution of risks than other energy technologies.

For example, radioactive pollution caused by long-lived wastes from the nuclear fuel cycle has great potential for causing injury to future generations living hundreds of thousands of years from now. The hazard is great, because the wastes are carcinogenic, mutagenic and teratogenic (causing birth defects) for many centuries;[24] because all means of storing them in the recent past have resulted in leaks and contamination of the environment;[25] and because no solution to the problem of how to store them safely is expected by the U.S. government before 1992–1995, if then.[26]

These difficulties with radioactive waste mean that nuclear electricity, unlike that produced from other energy sources, poses a great threat to future populations who do not receive any direct benefit from present generation of power by means of fission. In the case of coal, on the other hand, although the sulfur oxides pose serious problems of air pollution to contemporary persons, their health risk is relatively short-lived, partially controllable, and is visited upon roughly the same chronological set of persons deriving benefits from coal-produced electricity.

If it is equitable and just that those who receive the benefits from generation of electricity ought also to be the ones who bear its risks, then one ought to consider the equity of distribution of the risks and benefits associated with alternative means of producing energy. Moreover, a number of philosophers have argued that considerations of equal protection are necessary to any just or ethical public policy (as well as central to the U.S. Constitution).[27] To the extent that Inhaber has ignored questions of the equity of risk transference to future generations, to that same degree his risk-benefit assessment is open to charges of methodological and ethical incompleteness.[28]

It is clear that Inhaber has glaring omissions in at least three areas of his work. He has ignored the social and economic aspects of energy risks by failing to discuss costs of generating electricity by alternative means and by ignoring the psychological, political, and financial impact associated with limited liability for energy-related accidents. From the vantage point of ethics, he has omitted consideration of the negative effects of possible violations of due process and equal protection rights and hence he has ignored important social, political, and ethical risks of various forms of energy production.

IV. INHABER'S METHODOLOGICAL ASSUMPTIONS

Apart from these difficulties with Inhaber's work, however, there are a number of purely logical and scientific problems with his methodology. These were described, in a general sense, by Rein Lemberg. According to Lemberg, who was commissioned by the Atomic Energy Control Board of Canada (for whom Inhaber did his study) to review Inhaber's work, its basic flaw is that its conclusions rest upon highly subjective assumptions about how to interpret the available mathematical data. Once other, and in some cases, more plausible assumptions are employed, the data used by Inhaber lead "to the opposite conclusion," viz., that nonconventional means of generating electricity, e.g., solar, are safer than conventional means, e.g., nuclear, and that coal-produced energy causes fewer threats to health than does that produced by nuclear fission.[29]

What are some of the methodological assumptions and interpretations which cast doubt on the validity of Inhaber's conclusions? Although it is not possible to comment on all of these problems, citing a few of them should give one an idea of the methodological difficulties besetting his work. For one thing, Inhaber assumes that in computing the risks associated with materials acquisition, component fabrication and on-site construction of energy facilities, one must include all risks of all industries making or transporting anything connected with the energy facility. This assumption, however, suggests that all these related industries would be doing no work and would have no accidents, were it not for the energy facility being built. Clearly, however, this supposition is false.[30] Instead of using *total risks* of all industries making anything connected with the energy facility, Inhaber should have used the *incremental risks* taken on specifically as a result of constructing the energy system. Only in this way could he obtain an accurate picture of energy-related risk taken on by a subcontractor, as opposed to non-energy-related risks taken on by the same subcontractor in other of his work.

This assumption (that gross, rather than incremental, numbers of injuries in material acquisition and construction ought to be used to compute energy-related risks) has an obvious effect on Inhaber's conclusions. He admits that "for most of the nonconventional systems, the cause of the [risk and injury] values is high material acquisition and construction risk."[31] Once the more plausible assumption is used, therefore, risk from construction and material acquisition for nonconventional energy systems decreases and causes the total risk associated with nonconventional energy systems to be small, relative to that of conventional energy sources.

A second doubtful methodological assumption of Inhaber's is that there is no reason to treat all energy technologies on a consistent basis. When he computes the risks associated with nonconventional energy sources, for example, he includes occupational injuries associated with construction of the facility. When he computes the risks associated with conventional energy sources, however, he does not include the occupational risks of building coal, oil, or nuclear power plants in the figures for these technologies.[32] The source of his numbers for the risks involved in these three fuel cycles is a report which includes only figures for operation and maintenance risks but not for those of materials acquisition, component manufacture or plant construction.[33] One expert points out that, had

Inhaber been consistent and included construction risks, the occupational risk associated with atomic energy could be approximately two times higher than it is.[34]

This inconsistent treatment of various energy hazards is particularly reprehensible because it has an obvious effect on Inhaber's conclusion, that nonconventional energy systems are medically more risky than conventional ones. Not only does he not include hazards from material acquisition and construction in assessing the safety of conventional energy systems, as was explained above, but also he uses inordinately high risk figures for material acquisition and construction in nonconventional energy sources. He claims that "for most of the nonconventional systems, the cause of large [risk] values is high material acquisition and construction risk," then uses these high figures, even though he later admits that "precise" numbers for these aspects of nonconventional systems "are not known."[35] Because of his omissions in the nuclear calculations and the admitted unreliability of the figures used for material and construction risks in nonconventional energy sources, Inhaber's claim about the relative safety of conventional means of generating electricity is open to question.

Another methodological assumption which throws doubt on Inhaber's conclusions is that it is permissible to use nuclear risk figures as calculated from the Rasmussen Report. Inhaber's claim, that these data are taken from the Rasmussen Report,[36] raises at least two problems. First, his "upper limit" figure for reactor accident risks is three times smaller than that actually given in the Rasmussen Report; if Inhaber actually employed the numbers he claims to have used (Smith, et al. and Rasmussen), says one scientist, the risk figures for nuclear power (because of this one reason alone) would be as much as forty times greater than that shown in his figure 6 graphs.[37] Apart from Inhaber's not doing what he claims to be doing, a second difficulty is that, because of mathematical and methodological errors, the conclusions of the Rasmussen Report have been repudiated by the U.S. government, which commissioned it, and by numerous groups of independent scientists.[38]

A fourth problem with the Inhaber study, which also suggests that his conclusions regarding nuclear risk are erroneous, is the source of his data concerning the hazards of nuclear wastes. According to him, about 6 percent of the total risks from atomic energy are a result of the dangerous and long-lived radioactive wastes.[39] Since his article in *Science* gives no citation showing how the conclusion about 6 percent risk was estimated, one is left to surmise that it was merely a guess. Also suggesting that this is purely speculative is the fact that the United States does not now have a method for storing and managing radioactive wastes; there have been continued instances of leaks and contamination since the first such wastes were generated, and the United States will not have the first experimental plan for managing these substances until 1992–1995.[40] In the light of all these unknowns, including whether a safe storage technique is even possible, it is unclear how Inhaber can arrive at such small figures for the risks associated with managing nuclear wastes for hundreds of thousands of years.

Although there is not time to cite further examples here, Inhaber's treatment of the waste problem and his questionable risk figures associated with it are typical of his work. There are literally scores of instances, like this one, where many needed data are nonexistent, yet Inhaber guesses at them, uses them to arrive at his conclusions, and then touts the conclusions as if they were without doubt.[41] Because of the unavailability of the data needed to draw true and valid conclusions, much of Inhaber's reasoning consists of the argument from ignorance, which is a logical fallacy.

Not only, however, does Inhaber underestimate the risks from nuclear energy; he also overestimates the hazards of solar and coal-generated electricity. There are several reasons why this is the case. First of all, although he maintains that his calculations are based on the risks associated with "present-day [energy] technology, models, and systems"[42] Inhaber clearly does not follow this norm with respect to coal. His occupational and public risks from the coal fuel cycle are clearly based on mining practices and SO_2 emissions (from plants) which have been illegal for some time; as one scientist points out, "present dust standards imply occupational deaths from black-lung disease as much as sixty times lower than the figure used by Inhaber." The public deaths from sulfur dioxide disease, used by him in his calculations, correspond not to emissions in force but to "emissions five times higher."[43] What all this means, of course, is that Inhaber has deliberately or indeliberately inflated the hazards resulting from coal-generated electricity.

The inflation of coal-related risk figures in Inhaber's work not only causes him to conclude, wrongly, that coal-produced electricity is more hazardous than that from nuclear sources, but also (equally wrongly) that solar generated energy has more associated risks than does nuclear. This second error occurs because Inhaber assumes: (1) that wind and solar energy require backup power;[44] and (2) that these backups are coal systems.[45] When he calculates the risk from nonconventional energy sources (wind and solar), he includes in it the risks associated with the coal backups for them, even though he admits that, for the nonconventional systems, a majority of the risk comes from the backup![46]

Obviously it is methodologically questionable whether Inhaber should have included risks from conventional energy sources as part of the computed risks of nonconventional energy systems. Moreover, this inclusion is inconsistent with his expressed purpose: to compare risks of conventional and nonconventional energy sources.[47] Instead he has compared the risks of conventional systems with those of dual systems that are both conventional and nonconventional. The implausibility of his methodology is illustrated by the fact that, if one omits the inconsistent calculations for constructing the nonconventional energy sources and if one ignores the risks associated with the conventional backup, then Inhaber's conclusion is exactly reversed. It then becomes the case that nonconventional technologies (like solar) are safer than conventional ones (like nuclear).[48] If one reinterprets Inhaber's conclusions in the light of all the omissions, inconsistencies and doubtful assumptions of which he is guilty, it is clear that the nuclear risk is indeed substantial.

V. ADAMS'S ANALYSIS OF THE NUCLEAR DEBATE

Like Inhaber, Frank Adams also claims in his work that the hazards associated with atomic energy are insignificant. Also like Inhaber, Adams accomplished his study in only three months' time.[49] Importantly, however, Adams's analysis includes material omitted from consideration by Inhaber, viz., the ethical aspect of energy risk and policymaking.

How well does Adams's work fare when one subjects it to methodological and ethical scrutiny? Although the study has several assets, these are, by far, outweighed by the liabilities. In Adams's favor, it must be said that he has attempted a task, ethical analysis of the nuclear power question, which has long been neglected by both philosophers and scientists. Members of the former group have probably neglected this topic because of the technical and scientific information necessary to handle it properly. Scientists, on the other hand, being concerned with empirical studies, have largely been either unwilling or unable to assess the ethical and theoretical issues regarding nuclear power. Adams, however, has had some training in both science (an undergraduate degree in nuclear engineering) and philosophy (study for two years). Moreover, since he has been employed in nuclear-related industries for a total of six years, he has some first-hand experience with atomic energy, an asset many scholars do not possess.

Despite these positive aspects of Adams's study, its value is undercut by three different types of problems: factual errors and omissions; false or unsubstantiated methodological assumptions; and questionable, sometimes logically fallacious, modes of ethical reasoning. I will discuss each of the problems mentioned, in the order given above. My considerations have been hindered, however, by the fact that it is often difficult to know precisely what Adams is trying to say. His study is replete, for example, not only with numerous instances of improper subject/predicate/pronoun agreement, incomplete sentences (and thoughts), and confusion of possessive case with plural number, but also with partial and question-begging treatments of the same subjects at various parts of the book (when they should be treated at one place and organized systematically).

At the most basic level, a number of factual errors and omissions need to be corrected in the Adams work. Although it would serve no purpose to cite all of these difficulties here, a few instances should suffice to illustrate the magnitude of the empirical problems in his study. His factual errors, for example, include the statement that we have more uranium reserves than coal resources, a proposition almost universally held to be false by both government and industry experts.[50] He also says that the most important (dangerous) radioactive wastes are the high-level ones,[51] even though radiologists and health physicists maintain that the high-level/low-level distinction provides no key "to potential hazards," since it is based merely on the *origin* of the residues (whether they are from the first cycle solvent extraction system or not), and since the distinction provides no key either to the longevity or dose commitments of the radioactive isotopes involved.[52] Likewise, he maintains that WASH-1400 was "an independent study" of nuclear power,[53] even though it was done by an agency whose charge was "to promote" use of atomic energy.[54] His statement that no deaths have been caused by reactor accidents[55] can be proved false by citing any one of a number of accidents; this claim has explicitly been denied by U.S. officials charged with regulating fission generation of electricity.[56]

One of the most troubling factual errors in Adams's study, however, is his claim that there is an acceptable, safe, or threshold level of radiation.[57] This belief, widely held until the fifties, has been rejected by the International Commission on Radiation Protection, the Federal Radiation Council, the Environmental Protection Agency, and the Nuclear Regulatory Commission.[58] In fact, there is no scientific or professional organization which currently subscribes to this outmoded opinion. It was rejected thirty years ago after U.S. officials discovered the carcinogenic, mutagenic, and teratogenic effects of medical, dental, and military exposures to very low (allegedly safe) levels of radiation.[59]

In addition to these and other factual errors, Adams's analysis also contains a number of omissions. He suggests that fuel reprocessing, the economic linchpin of the nuclear cycle is "just around the corner,"[60] but fails to point out: that there are no reprocessing facilities now operating; that there are no plans for any in the future; and that all three partially completed reprocessing plants were abandoned long ago because of radiation hazards and the fact that safeguarding them against such hazards would bankrupt their parent companies.[61] Likewise Adams asserts that current waste-disposal practices are "safe" for the short term,[62] but neglects to point out that every disposal facility in operation has been subject, over the short term, to continual leaking and contamination. The Maxey Flats site, for example, has had plutonium travel offsite, and the Hanford facility leaked 500,000 gallons of high-level waste to the land, river, and sea.[63] None of these problems has been solved and the U.S. government estimates that accidental contamination will continue to occur, at a rate of two to three releases per year, per storage site.[64] Likewise, Adams naïvely asserts that all features of reactor safety and operation have been "tested" and "checked."[65] He fails to point out, however, at the time at which he makes this statement (he admits it later), that there have been no full-scale tests of the emergency core-cooling system. This system is the single most important safety feature in reactor design; it is supposed to prevent core melt and the consequent release of radiation which, according to U.S. government documents, could kill immediately more than 45,000 people and destroy an area the size of Pennsylvania.[66]

VI. ADAMS'S QUESTIONABLE METHODOLOGY

Adams's factual errors and omissions would not be so troublesome if his methodology were more careful. As it is, however, these errors, combined with his logical and epistemological problems, together result in a serious misrepresentation of the risks of atomic energy. Let us examine several of these methodological difficulties.

The source of many of Adams's errors and omissions, and itself a methodological problem, is the fact that he does not have a good background in the relevant literature. Many of his citations are to popularized nuclear industry magazines; key government and university documents, on both sides of the atomic energy debate, are missing from his citations and notes. On the pro-nuclear side, numerous important AEC, ERDA, and NRC documents, such as WASH-1224, and ERDA-1536, 1537, and 1538, for example, were apparently not considered. On the con-nuclear side, equally numerous and important books and articles were ignored; some of the most important of these are all of Gofman's and Tamplin's works, Nader's work, various EPA documents (such as those on radioactive waste), and Saunders Miller's key book, *The Economics of Nuclear and Coal Power* (New York: Praeger, 1976).

Perhaps the most serious of the methodological flaws in his work is that Adams repeatedly begs important questions rather than argues regarding them. When discussing the Rasmussen Report, WASH-1400 (whose summary conclusions were rejected by the Nuclear Regulatory Commission, and whose methodology and conclusions were questioned by the Environmental Protection Agency, the American Physical Society, and numerous other groups),[67] for example, Adams defends its highly criticized mathematical methodology with the unsubstantiated remark: "Most experts seem to agree that all of the important accident sequences or categories of sequences have been analyzed."[68] This is precisely what the NRC, EPA, APS, et al., denied. Other unsubstantiated assertions are that "the major problem types [for nuclear reactors] will show up in the first few years of an operation";[69] that a nuclear war has a higher probability of occurrence than a major nuclear accident;[70] that fatalities resulting from use of nuclear power are "less than or equal to fatalities from a coal plant";[71] that one ought not to be "concerned about a few more bombs in a few more hands" as a result of nuclear proliferation;[72] that transportation accidents involving nuclear materials are unlikely;[73] that fuel recycling and reprocessing have been stopped by political interests, rather than because of medical, financial, and technological problems;[74] that a major melt-down "probably" will not occur;[75] that major use of solar power will not be possible till 2010;[76] that fossil fuels are more polluting than nuclear ones;[77] that casks for transporting radioactive wastes "are safe" (whatever that means);[78] and that radioactive emissions from a nuclear power plant are "insignificant."[79]

In lieu of substantiating remarks such as those just cited, Adams characteristically resorts to emotive outbursts. He accuses opponents of "just plain ignorance,"[80] says they have "unreasonable fears" about nuclear energy, and that they persist in "crys [sic] of 'wolf.' "[81]

Behind the unsubstantiated *ad hominems* that Adams directs at those who question atomic energy is his apparent inability to engage in logical or conceptual *analysis*. When he does give a citation to back up his remarks, usually he develops the point so poorly that it is not clear whether it supports his case or not. For example, Adams tells us that low-level radiation is safe, because "there was a report of a study of the Hanford [waste facility] workers" which showed that their exposures produced no "extra cancer."[82] Although Adams gives a page citation to a popular magazine published by the nuclear industry, he does not explain: (1) who conducted the study; (2) the title of the study; (3) the length (time period) of the study; (4) the controls on the study; and (5) the "normal" rate of cancer incidence for persons of the age, history, and geographical location of the Hanford workers. Item (3) is especially important, since cancers from low-level radiation are often latent for thirty to forty years.[83] Hence if the study was done for any period less than forty years (which is likely, since Hanford has only been in operation for a fraction of this length of time), the total number of "additional" cancers would not be detected.

Although there are numerous other examples of Adams's inability to analyze and to argue effectively,[84] let us leave this point and illustrate another repeated methodological error in his book. This is his tendency to use partial inductions and to assume that, because something has or has not happened in the past, it will or will not happen in the future. For example, he claims that because the United States has protected its nuclear weapons from misuse, therefore nuclear reactors and their fuel cycle can be protected from terrorists and saboteurs.[85] While this may be the case, past experience in a related area is obviously not sufficient to ensure nuclear security in the future. Likewise, for example, Adams asserts that because the hundreds of bombs and threats directed at nuclear reactors in the past have resulted in "no significant damage," therefore threats of terrorism and sabotage pose no danger to atomic plants in the future.[86] Obviously, again, the premise in his argument does not constitute anything near sufficient grounds for drawing his conclusions.

Adams's problem with inductive generalizations suggests that he has a habit of relying on vague, general, intuitive, or naïve judgments about nuclear safety rather than on detailed assessments of it. For example, despite the fact that there have been no full-scale tests of the emergency core-cooling system, Adams maintains that this constitutes neither an empirical nor a logical obstacle to asserting that reactors are safe. He says, simply, "This is a point where common sense and engineering judgment are sufficient."[87] If this common-sensical belief (that the emergency core-cooling system will work) is correct, however, then nothing is to be lost by testing the emergency core-cooling system. Moreover, it is incredible that Adams is willing to rely on "common sense and engineering judgment," rather than on good empirical data, to insure that 45,000 people are protected from immediate death caused by a loss-of-coolant catastrophe.

VII. ADAMS'S DISCUSSION OF ETHICS

Many of Adams's naïve, unsubstantiated assumptions about the safety and workability of nuclear reactors seem based on the erroneous supposition that because regulations prohibit certain practices, therefore these dangerous occurrences are not likely to take place. Adams seems to discount the possibility of inadvertent, financially motivated, or malicious attempts to subvert regulations governing atomic plants. He says, for example, that decisions regarding energy policy are made "primarily on safety considerations"; that the containment structure "forms a virtually leak-tight barrier to prevent the escape of fission products"; that the nuclear industry reports all its safety problems to the NRC, and that, although quality control in a nuclear plant is done typically by persons with a high-school education, "radiation protection" is achieved because these personnel have one and one-half years of "extensive training" on the job.[88] Nowhere in his entire work does Adams admit that safety considerations might take a back seat to economic ones, or simply to human carelessness. In failing to discuss his assumption that nuclear plants always operate as specified, regulated, or planned, he leaves one with the impression, again, that he has not really analyzed all the possible consequences of implementing nuclear technology.

This same naïve, unsubstantiated faith in humans' perfect actions is also connected with Adams's naïve, unsubstantiated faith in technology. Rather than advocating precise assessments of each technology to determine its merits, and hence whether it ought to be implemented, Adams simply says that "a Christian response" to technology is "to use the technology while reducing the risks of the technology."[89] Why, *a priori,* ought "a Christian response" to be one of using technology? Are there no technologies which ought never to be used? Why ought all technologies, regardless of merit, to be employed?

As these questions reveal, at the heart of Adams's undocumented, error-laden, incomplete, and emotive work, there is as much *ideology* as *philosophy*. If his study is viewed in this way, then his lack of analysis and question-begging approach become understandable. His pro-technology ideological bias explains why he can admit that it is impossible to assess the risks of nuclear power,[90] but at the same time, claim that Christians ought to use technology and that we must make a hard choice in favor of a "theology of risk."[91]

Why ought one to risk numerous human lives as a result of a nuclear accident, especially if (as he says) nuclear risks cannot be assessed? It seems equally rational and ethical *not* to take a risk when its probability and consequences are not known, as Adams claims. Rather, his entire "theology of risk" justification for employing nuclear power, when one cannot *rationally* assess it, is based on his personal, subjective interpretations of numerous Scriptural passages and what he calls "the will of God." On the contrary, it seems that, where thousands of human lives are at stake, one ought to be wary of risking them for reasons which one admits cannot be rationally assessed, but only obscurely "discerned" in the "prophetic living word of God."[92] What is troublesome with his approach is Adams's confidence that he knows the will of God. How many evils have been perpetrated on humankind in the name of "the will of God"?

In all fairness to Adams, it must be admitted that his somewhat subjective remarks, on how religious considerations and the "theology of risk" justify employment of nuclear power, comprise 80 percent of his chapter on "Ethical and Moral Considerations."[93] He does spend three pages of the chapter (pp. 104, 107, 108) discussing properly *ethical* (as opposed to religious or theological) considerations regarding nuclear power. (Note, however, that in this book, entitled *Technical and Ethical Aspects of the*

Nuclear Power Debate, only three pages are devoted to ethical discussion; the bulk of the work covers what Adams calls "technical aspects of the debate.")

In his brief discussion of the ethical aspects of the nuclear-power controversy, Adams treats two topics: the energy needs of Third World nations and the rights of future generations to use fossil fuels which have been conserved for them.[94] Both of these issues are extremely important, both deserve serious and careful analysis, and on both points Adams reveals that his heart is in the right place. He appears to be a caring, compassionate person who wishes to defend the rights of all persons, whether present or future ones and whether from developed or underdeveloped countries, to an equal share of the goods of the earth. How well does Adams fare in his defense of this just and important cause?

Unfortunately, he spends two pages quoting Scripture, the Church Fathers, papal documents, and the Second Vatican Council of the Roman Catholic Church. While brilliant and persuasive insights are contained in many of these religious teachings, because they constitute numerous "arguments from authority," it is doubtful if they would help any but the already committed Roman Catholic to come to terms with the two ethical points in whose support they are enlisted. The problem is not only that Adams uses an appeal to authority when he promised us an "ethical consideration." Rather, the difficulty is also that these citations leave Adams little space for *arguing* why members of present generations have obligations to future ones and why developed countries have obligations to help underdeveloped ones. Both these claims can be rationally defended and several philosophers have done a good job of doing so.[95] Adams, however, neither cites these important analyses nor presents alternative ones of his own. Instead, he resorts to question-begging, apodictic assertions that nuclear power ought to be used in Third World countries and that it will enable us to conserve oil and gas for the future.

When Adams gives us his unanalyzed conclusions without argument, he claims to be using "a simplified form of John Rawls's method."[96] Rather than employing Rawls's mode of analysis, however, Adams simply reprints several long selections from Rawls's *Theory of Justice*; the four quotations comprise part of Rawls's explanation of the "just saving principle," according to which each generation is enjoined to pass on to subsequent ones "a fair equivalent in real capital" of resources, culture and civilization. After citing these Rawlsian remarks, Adams concludes (but does not *argue*) that "technological advancement, as well as learning in general, is a method of savings for the future . . . it would seem that one can pursue a new technology for the benefits and count on mankind's ability to develop solutions to the problems."[97]

As is obvious, Adams has *assumed* that advancement of nuclear technology is a method of saving for the future. He is guilty here of *petitio principii,* begging the question; he is probably also incorrect in his conclusion since substantiation of his point here requires him to explain: (1) how we save for the future when we use uranium in reactors; (2) whether nuclear pollution, e.g., radioactive waste (which is carcinogenic, mutagenic and teratogenic for hundreds of thousands of years) is a "saving for the future" since it leaves them a dangerous burden from the energy we have used; and (3) why conservation and use of solar energy do not constitute desirable alternatives (to nuclear power) for "saving for the future."

Even if Adams effectively argued regarding points (1)–(3), however, he would still need to employ all of Rawls's principles and techniques, and not just single one out of context and then use it for a hasty conclusion. Rawls is one of the most profound and rational ethicians the world has been blessed to know; Adams does him and scholars a disservice by employing his remarks in so brief and slipshod a fashion. One of the points Adams might discuss, in a more thorough employment of Rawls's method, is the extent to which technologies such as nuclear power inflict *inequitable risks* upon present and future populations.[98] Rawls's philosophy is grounded on egalitarian principles; so long as Adams does not examine questions of fuel supply, low-level radiation, the Price-Anderson Act, core-melt catastrophe, and radioactive waste, insofar as they promote or inhibit equal justice and equal protection, then it is questionable whether he ought to claim to be following Rawls's method.

VIII. CONCLUSION

What the errors, omissions, faulty methodological assumptions, and question-begging of Adams's work suggest is that, despite his concern for the poor of the world and for future generations, he has provided a rational foundation neither for protecting their interests nor for defending the employment of nuclear fission technology. Moreover, both the studies of Inhaber and Adams contain so many factual errors and omissions that it is impossible to say, apart from ethical considerations, that they have technically done an adequate job of discussing atomic power.

The allegedly factual summaries of both authors leave so much to be desired that one wonders if the quality of the nuclear debate was not hindered more than helped by their three-month studies. Even worse, since both authors have some scientific and technical training, their backgrounds might provide grounds for the uninformed to assume they are correct and hence to underestimate the hazards of atomic energy. In this regard, one is reminded of the "official" reports published (even in journals such as *Science*) after the Three Mile Island accident. Scientist-authors of some of these bulletins stated that members of the public near the accident site were exposed to the same amount of radiation as one might receive in a chest X-ray. As more careful scientists were quick to point out,[99] however, the analogy between Three Mile Island and X-ray exposures is erroneous because the former was (and is) "whole body," while the latter is not. Moreover, even were the previous error corrected, one would have to obtain numerous X-rays (one for each day's exposure at Three Mile Island) in order for the two radiation doses to be equivalent. Hence the scientific reports should have said that *daily* radiation exposure (from the accident), to a member of the public, was comparable to one whole-body X-ray.

Just as the public is not served by such half truths as those perpetrated about Three Mile Island, so also it is not served by discussions such as those of Inhaber and Adams. They bring the nuclear debate down to the level of sophistry, a charge which can be laid at the doors both of some proponents and some opponents of atomic energy. Society deserves better than that from its alleged scholars.

NOTES

1. R. L. Meehan, "Nuclear Safety: Is Scientific Literacy the Answer?" *Science,* 204, No. 4393 (11 May 1979): 571; hereafter cited as: Literacy.

2. B. L. Welch, "Review of Rasmussen Report," *Science,* 203, No. 4386 (23 March 1979): 1198; hereafter cited as: Review.

3. Meehan, Literacy: 571.

4. U.S. Atomic Energy Commission, *Comparative Risk-Cost-Benefit Study of Alternative Sources of Electrical Energy,* WASH-1224 (Washington, D.C., Government Printing Office, December 1974), pp. 3–95; hereafter cited as: WASH-1224. According to the *Code of Federal Regulations* (Washington, D.C., U.S. Government Printing Office, 1978), 10, Part 20, p. 184 (hereafter cited as: 10 CFR 20), one "rad" of radiation is "a measure of the dose of any ionizing radiation to body tissues in terms of the energy absorbed per unit mass of the tissue." For more information on the ethical problems raised by the controversy over standards for low-level radiation, see K. S. Shrader-Frechette, *Nuclear Power and Public Policy* (Boston, D. Reidel, 1979), Chapter Two.

5. Arthur R. Tamplin and John W. Gofman, *'Population Control' Through Nuclear Pollution* (Chicago, Nelson-Hall, 1970), pp. 4, 85; hereafter cited as: *Nuclear Pollution.* Damage estimates are made by plotting the curve obtained when various health effects ("response") are interpreted as a function of radiation dose. Since dose-response coefficients for low-level radiation are obtained by hypothesis from data on high levels of exposure (i.e., Hiroshima and Nagasaki), there is no general agreement regarding health effects of low-level radiation. Government scientists assume that the dose-response curve follows a linear pattern according to which there is a small, constant risk per rad, independent of dose rate. Many prominent academicians such as Gofman and Tamplin, on the other hand, believe that there is a rapid rise in the dose-response function (such that there are great damages even from low radiation exposures) followed by a leveling of the curve at higher exposures.

6. U.S. Environmental Protection Agency, "Criteria for Radioactive Wastes," *Federal Register,* 43, No. 221 (November 15, 1978): 53266; hereafter cited as: Radioactive.

7. Tamplin and Gofman, *Nuclear Pollution,* p. 51.

8. Herbert Inhaber, "Risk with Energy from Conventional and Nonconventional Sources," *Science,* 203, No. 4382 (23 February 1979): 718–723; hereafter cited as: Risk. See also note 41.

9. Frank Adams, S.J., *Technical and Ethical Aspects of the Nuclear Power Debate* (St. Louis, Mo., Institute for the Theological Encounter of Science and Technology, 1978); hereafter cited as *Aspects.*

10. See U.S. Nuclear Regulatory Commission, *Reactor Safety Study—An Assessment of Accident Risks in U.S. Commercial Nuclear Power Plants,* Report No. (NUREG-75/014) WASH-1400 (Washington, D.C., Government Printing Office, 1975); hereafter cited as: WASH-1400. WASH-1400 has been criticized severely on mathematical and methodological grounds by the American Physical Society, the Environmental Protection Agency, the Union of Concerned Scientists, and numerous other groups. (See NRC, WASH-1400, Appendices I and XI, for the APS and UCS critiques, and R. Doctor, et al., "The California Nuclear Safeguards Initiative," *Sierra Club Bulletin,* LXI, No. 5 [May 1976]: 45; hereafter cited as: Safeguards. See also H. W. Lewis, et al., "Report to the American Physical Society by the Study Group on Light-Water Reactor Safety," *Reviews of Modern Physics,* XLVII, No. 1 [Summer 1975]: S1-S124, and R. Nader, "Nuclear Power: More Than a Technological Issue," *Mechanical Engineering,* XCVIII, No. 2 [February 1976]: 33; hereafter cited as: Issue.) Finally, in the wake of so many scientists' denying the validity of the 1975 Rasmussen Report, the Nuclear Regulatory Commission withdrew its support from it on January 19, 1979. There is currently no complete study of nuclear power safety which bears the approval of the U.S. government, despite the fact that there are approximately 70 commercial reactors now operating and 70 more under construction in the United States. For an ethical and methodological analysis of the Rasmussen Report, WASH-1400, see K. S. Shrader-Frechette, *Nuclear Power and Public Policy* (Boston, D. Reidel, 1979), Chapters Four and Six: hereafter cited as: *Nuclear Power.* These chapters also provide an analysis of other important government and private assessments of nuclear safety.

11. See, for example, Nader, Issue, p. 33; Sheldon Novick, *The Electric War* (San Francisco, Sierra, 1976), Tamplin and Gofman, *Nuclear Pollution,* and K. S. Shrader-Frechette, *Nuclear Power.*

12. Rein Lemberg, "Energy: Calculating the Risks," *Science,* 204, No. 4392 (4 May 1979): 454; hereafter cited as: Energy. Lemberg, who reviewed the Inhaber work for the Canadian Atomic Energy Board, was told by Inhaber that his study was prepared over a total of three months' time.

13. Inhaber, Risk, p. 718.

14. *Ibid.,* p. 722.

15. *Ibid.,* p. 721.

16. *Ibid.,* pp. 718, 722.

17. See U.S. AEC, WASH-1224; Amory B. Lovins, "Energy: Bechtel Cost Data," *Science,* 204, No. 4389 (13 April 1979): 124–127; hereafter cited as: Bechtel; and K. S. Shrader-Frechette, *Nuclear Power,* Chapters Three and Five. The Bechtel (nuclear industry) and Lovins data, more up-to-date than that of WASH-1224, reveal that capital-cost calculations are \$3179 to \$5000 per kWe delivered for nuclear, and \$2476 per kWe delivered for coal (Bechtel, p. 126).

18. Inhaber, Risk, p. 719.

19. *Ibid.,* p. 719.

20. "Theoretical Possibilities and Consequences of Major Accidents in Large Nuclear Power Plants," U.S. AEC Report WASH-740 (Washington, D.C., Government Printing Office, 1957); see also R. J. Mulvihill, D. R. Arnold, C. E. Bloomquist, B. Epstein, "Analysis of United States Power Reactor Accident Probability," PRC R-695 (Los Angeles, Planning Research Corporation, 1965); hereafter cited as: WASH-740-UPDATE. The for-

mer study will hereafter be cited as: WASH-740. See also J. J. Berger, *Nuclear Power* (New York, Dell, 1977), p. 45; hereafter cited as: *Nuclear Power.*

21. Even government proponents of nuclear energy clearly reveal that industry would not "go nuclear" without liability protection; see "AEC Staff Study of the Price-Anderson Act, Part I," p. 209, and "AEC Staff Study of the Price-Anderson Act, Part II," *Atomic Energy Law Journal,* XVI, No. 4 (Winter 1975): 298. For an account of industrial reluctance to go nuclear without Price-Anderson support, see K. S. Shrader-Frechette, "Technology, Public Policy, and the Price-Anderson Act," this volume. Also J. Elder, "Nuclear Torts: The Price-Anderson Act and the Potential for Uncompensated Injury," *New England Law Review,* XI, No. 1 (Fall 1975): 127; hereafter cited as: Injury. This reason for limiting liability is substantiated by the U.S. government in "AEC Staff Study of the Price-Anderson Act, Part I," *Atomic Energy Law Journal,* XVI, No. 3 (Fall 1974): 219–220. Although there are provisions for small increases (in the future) in total liability limits, these increases still provide that nuclear utilities are not liable for a majority of the financial consequences of a catastrophic accident. For discussion of these future changes, see R. Lowenstein, "The Price-Anderson Act," *Forum,* XII, No. 2 (Winter 1977): 599–603, and J. Marrone, "The Price-Anderson Act," *Forum,* XII, No. 2 (Winter, 1977): 608–609.

22. See Duke Power Company v. Carolina Environmental Study Group, Inc., et al. and U.S. Nuclear Regulatory Commission, et al., v. Carolina Environmental Study Group, Nos. 77–262 and 77–375, *The United States Law Week,* XLVI, No. 50 (June 27, 1978): 4845; and Berger, *Nuclear Power,* pp. 145–146. For further discussion of the ethical, methodological, and logical problems inherent in the Price-Anderson limit on liability, see K. S. Shrader-Frechette, *Nuclear Power,* Chapter Four.

23. See Inhaber, Risk, p. 722.

24. Interagency Review Group on Nuclear Waste Management, *Report to the President,* TID-28817 (Springfield, Virginia, National Technical Information Service, October, 1978), p. iii; hereafter cited as: IRG, *Report.*

25. Plutonium has traveled off-site from storage facilities, and more than 500,000 gallons of high-level waste have leaked from containers and have contaminated surrounding areas, e.g., in Hanford, Washington. For more information regarding the hazardous nature of these leaks, see B. Kiernan, et al., Legislative Research Commission, *Report of the Special Advisory Committee on Nuclear Waste Disposal,* No. 142 (Frankfort, Kentucky, Legislative Research Commission, October, 1977), pp. ix–17; hereafter cited as: LRC; and U.S. Energy Research and Development Administration, *Final Environmental Statement: Waste Management Operations, Hanford Reservation, Richland, Washington,* Vol. 2, ERDA-1538 (Springfield, Virginia, National Technical Information Service, October 1975), pp. 11.1-H-1 through 11.1-H-4. See also G. Hart, "Address to the Forum," U.S. Environmental Protection Agency, *Proceedings of a Public Forum on . . . Radioactive Wastes,* ORP/CSD-78-2 (Washington, D.C., Government Printing Office, May 1978), p. 6; hereafter cited as: ORP/CSD-78-2.

26. IRG, *Report,* p. xxiii. For further discussion of ethical problems associated with radioactive wastes, see K. S. Shrader-Frechette, *Nuclear Power,* Chapter Three.

27. See, for example, John Rawls, *A Theory of Justice* (Cambridge, Mass., Harvard University Press, 1971), pp. 14–15. See also Charles Fried, *An Anatomy of Values* (Cambridge, Mass., Harvard University Press, 1978), pp. 116–117, 126–127; Alan Donagan, *The Theory of Morality* (Chicago, University of Chicago Press, 1977), pp. 221–239; and Marshall Cohen, "The Social Contract Explained and Defended," *The New York Times Book Review,* Section 7 (July 16, 1972): 1, 16–17.

28. For arguments in favor of the rights of future generations, see Environmental Protection Agency, "Criteria for Radioactive Wastes," *Federal Register,* 43, No. 221, Part 9 (November 15, 1978): 53267; hereafter cited as: Criteria. See also R. M. Green, "Intergenerational Distributive Justice and Environmental Responsibility," *BioScience,* 27,

No. 4 (April 1977): 261; Joel Feinberg, "The Rights of Animals and Unborn Generations," in T. A. Mappes and J. S. Zembaty (eds.), *Social Ethics* (New York, McGraw-Hill, 1977), pp. 358–359, and Daniel Callahan, "What Obligations Do We Have to Future Generations?" *The American Ecclesiastical Review*, 164, No. 4 (April 1971): 265–268.

29. Rein Lemberg, Energy, p. 454.

30. This point is also made by Lemberg, Energy, p. 454; see Inhaber, Risk, p. 721.

31. *Ibid.*, p. 721.

32. *Ibid.*, p. 722.

33. This report is by K. R. Smith, J. Weyant, J. P. Holdren, *Evaluation of Conventional Power Systems* (Berkeley, University of California, July 1975), Report ERG 75-5, Energy and Resources Group; hereafter cited as: ERG 75-5.

34. John P. Holdren, Kirk R. Smith, Gregory Morris, "Energy: Calculating the Risks (II)," *Science*, 204, No. 4393: 564; hereafter cited as: Energy.

35. Inhaber, Risk, pp. 721, 723.

36. *Ibid.*, p. 719.

37. Holdren, et al., Energy, p. 564; see also Inhaber, Risk, p. 722.

38. See note 10.

39. Inhaber, Risk, p. 721.

40. See notes 24, 25, 26.

41. For Inhaber's entire report, see H. Inhaber, *Risk of Energy Production* (Ottawa, Ontario, Atomic Energy Control Board, 1978), Report AECB 1119, 3rd ed.; hereafter cited as: AECB 1119. Numerous other "guesses" and omissions in Inhaber's study are cited in Holdren, et al., Energy, pp. 564 ff. and Lemberg, Energy, p. 454.

42. Inhaber, Risk, p. 719.

43. Holdren, et al., Energy, p. 564. See Inhaber, Risk, p. 722.

44. *Ibid.*, p. 720.

45. *Ibid.*, p. 721.

46. *Ibid.*, p. 721.

47. *Ibid.*, p. 718.

48. This point is also made by Lemberg, Energy, p. 454.

49. Adams, *Aspects*, Foreword (p. 1).

50. *Ibid.*, p. 108. See U.S. AEC, WASH-1224 for confirmation that the statement is erroneous.

51. *Ibid.*, p. 10.

52. B. Kiernan, et al., LRC, p. 3, and U.S. Environmental Protection Agency, *Considerations of Environmental Protection Criteria for Radioactive Waste* (Washington, D.C., US EPA, February, 1978) pp. 1, 9.

53. Adams, *Aspects*, p. 49.

54. See note 10.

55. Adams, *Aspects*, p. 74.

56. H. T. Peterson (Nuclear Regulatory Commission), "Nuclear Industry: Safety Record," *Science*, 205, No. 4406 (10 August 1979): 544.

57. Adams, *Aspects*, pp. 8, 75.

58. See notes 4 and 5. See also US AEC, WASH-1224; US AEC, *Nuclear Power and the Environment* (Washington, D.C., Government Printing Office, 1969), p. 41; and US EPA, *Considerations*, p. 30.

59. See notes 24 and 58. See also K. S. Shrader-Frechette, *Nuclear Power*, Chapter Two; US AEC, WASH-1224, p. 43; O. Hansen, "Development and Application of Radiation Protection Standards," *Idaho International Review*, 12, No. 1 (Fall 1975): 1–32; and M. S. Young, "A Survey of the Governmental Regulation of Nuclear Power Generation," *Marquette Law Review*, 59, No. 4 (1976): 842–843.

60. Adams, *Aspects*, p. 10.

61. US AEC, WASH-1224, pp. 3–64 and 3–65. Berger, *Nuclear Power,* pp. 78–84, 94–97.
62. Adams, *Aspects,* p. 10.
63. See notes 24, 25, and 26.
64. US ERDA, ERDA-1538, vol. 1, p. 111.2-2.
65. Adams, *Aspects,* p. 21.
66. US AEC, WASH-740-UPDATE. See also Berger, *Nuclear Power,* p. 45.
67. See note 10.
68. Adams, *Aspects,* p. 51.
69. *Ibid.,* p. 56.
70. *Ibid.,* p. 60.
71. *Ibid.,* p. 74.
72. *Ibid.,* p. 93.
73. *Ibid.,* p. 96.
74. *Ibid.,* p. 100.
75. *Ibid.,* p. 101.
76. *Ibid.,* p. 83.
77. *Ibid.,* p. 80.
78. *Ibid.,* p. 72.
79. *Ibid.,* p. 73.
80. *Ibid.,* p. 2.
81. *Ibid.,* pp. 119, 89.
82. *Ibid.,* p. 70.
83. J. C. Bodie, "The Irradiated Plaintiff: Tort Recovery Outside Price-Anderson," *Environmental Law,* VI, No. 3 (Spring 1976): 859–896; and Elder, Injury, pp. 126–132.
84. See Adams, *Aspects,* pp. 87, 89, 102, 104, 107, for example.
85. *Ibid.,* p. 96.
86. *Ibid.,* p. 58.
87. *Ibid.,* p. 55.
88. *Ibid.,* pp. 100, 19, 17, 22.
89. *Ibid.,* p. 112.
90. *Ibid.,* p. 76.
91. *Ibid.,* p. 122.
92. *Ibid.,* p. 113.
93. This chapter is pp. 104–119, whereas the treatment of religious considerations and the "Theology of Risk" section is pp. 105, 106, 109–119 of Adams, *Aspects.*
94. *Ibid.,* pp. 104–109.
95. See note 28.
96. Adams, *Aspects,* p. 104. For a longer analysis of numerous methodological, ethical, and social problems with employing nuclear fission technology to generate electricity, see K. S. Shrader-Frechette, *Nuclear Power.*
97. Adams, *Aspects,* p. 109.
98. See, for example, K. S. Shrader-Frechette, *Nuclear Power,* Chapter Two, for such an analysis.
99. H. W. Ibser, "Three Mile Island," *Science,* 204, No. 4395 (25 May 1979): 795.

THE RELATIONSHIP OF MORAL AND STRATEGIC ARGUMENTS IN THE DEFENSE DEBATE

J. Bryan Hehir

INTRODUCTION

The purpose of this paper is to analyze how strategic and moral issues are joined in the debate about U.S. nuclear policy. The specific focus of analysis will be the SALT II debate; this choice omits several important elements of U.S. defense policy, but it also facilitates the task of defining the relationship of strategic and moral factors in the policy debate. The methodological premise of the paper is that moral argument does not add something new to the policy debate, but is a process of making explicit and available for analysis the ethical logic which is already present in policy positions.[1]

The argument of the paper moves from a sketch of relevant strategic and moral data to examining a spectrum of representative positions in the

Research in Philosophy & Technology, Volume 3, 1980, pages 367–383.
Copyright © 1980 by JAI Press Inc.
All rights of reproduction in any form reserved.
ISBN: 0-89232-102-4

SALT II debate. Each position on SALT II will be stated, then the dominant ethical issues raised by the position will be discussed. To say that every policy position contains an ethical logic is not to expect that the moral position will be consciously addressed. The task of the paper will be to explicate the strategic and moral positions, then in a final section to assess the role and weight of moral argument in the strategic debate.

I. ESTABLISHING THE GROUNDS OF DEBATE

To evaluate the SALT II debate in political and moral terms it is necessary to establish the central elements of the problem. From the perspective of ethical theory there are a variety of positions one could assume to assess the SALT II debate. In this paper the position chosen (because it is my own, because I know it best and because I believe it to be the most useful instrument of *policy* analysis) is the classical Just-War Tradition. The moral theory of just war is a complex and multifaceted mode of ethical analysis. It poses a series of tests for any use of force which examines the *cause* for which force is used, the *motives* of actors, the *authority* legitimating their action, the *necessity* of the resort to force and the *alternative* forms of action available. Although all of these (*jus ad bellum*) considerations are significant for a just-war analysis of policy, the principles which are central to an analysis of nuclear policy are those of proportionality and noncombatant immunity.[2]

To grasp the nature of moral restraint imposed on war by these *jus in bello* principles it is necessary to examine their content and relationship in assessing policy. The principle of noncombatant immunity has as its objective the protection of civilians from directly intended attack. The moral logic of the principle is rooted in the essence of just-war theory, and in the era of technological warfare it has moved progressively to the center of attention. The principle specifies when killing in warfare is unjustified. Ralph B. Potter illustrates the link between just-war theory as a whole and the specific rule of noncombatant immunity:

> We live by the presumption that men must do no harm to their neighbors. That presumption can be overridden when it is necessary to restrain wrong-doers from inflicting harm. Only this necessity can grant to any man an excuse to kill. It follows that those who need not be restrained must not be killed.[3]

Noncombatants are by definition those who do not need to be restrained or killed; hence they are to be protected even in the midst of warfare. The moral principle protects them by distinguishing between justifiable defense and murder. To fall under the prohibition of the principle, the attack on civilian life must be directly intended, not the unintended by-product of a legitimately targeted attack on a combatant or military target.

The principle of proportionality can be applied to the enterprise of a war as a whole or to means used within a war. It is the means question which surfaces in the debate about nuclear policy. Proportionality of means is addressed to the tactics of war, counting the costs of specific measures even in a justified war. Paul Ramsey, who has written more extensively on *jus in bello* than any other ethicist, describes the principle of proportionality as a framework for moral judgments. It is a framework which requires a judgment of political prudence to give it content, a judgment which assesses the relative *moral* costs and benefits of a given tactic or policy.

In assessing the morality of strategic policy, the two principles are used in tandem, bringing two restraints to bear upon policy. Proportionality assesses the scope of intended damage; it judges consequences. Noncombatant immunity assesses the kind of damage intended; its object is to determine the legitimacy of the killing done in the name of justice. Ramsey states the relationship between the two principles:

> An act may be gravely wrong for one thing, because its primary thrust is to kill the innocent: that is murder. The same act, or another act, may be gravely wrong for another thing, because of serious disproportion between its good and evil effects.[4]

The two principles bring to bear upon policy two styles of moral reasoning. The strength of the just-war tradition is to utilize both forms of ethical restraint; the constant temptation is to collapse the principles, thereby losing some of the restraining force which reason can place on power. The proportionality principle, using a cost/benefit style of analysis, is by definition consequentialist. The principle blends easily with the style of policy discourse, but, in Ramsey's words, "is open to the greatest uncertainty." Proportionality is not a firm line in the face of escalation of means.

In contrast, the principle of noncombatant immunity assesses the nature of each act of killing. Its deontological character gives it the capacity to isolate an action or a policy and call it into question in spite of the good consequences it may achieve. Because this moral principle can point at a given action or policy and declare it to be murder, rather than justifiable killing, it can render judgment much sooner than proportionality. Together the principles question not only how many may be killed in war but who is being killed, not only what the cost of targets is, but which targets are being deliberately attacked. These two measures of the morality of policy will form the instrument of ethical analysis for evaluating positions in the SALT II debate.

In political-strategic terms the SALT II debate was generated by the expectation that a SALT II "package" would be concluded and sent to the Senate for consideration in 1979. The SALT I package, signed and ratified in 1972, contained a Treaty severely limiting ballistic missile defenses in the U.S. and U.S.S.R., and an Interim Agreement limiting offensive nuclear weapons for five years. In 1974 SALT I was supplemented by the Vladivostok Accords which established (very high) ceilings for strategic nuclear delivery vehicles (SNDVs) and for multiple-independently-targeted warheads (MIRVs).

The SALT II negotiations, conducted in the final months of the Ford Administration and continued by President Carter, are designed to produce the first actual reduction in nuclear delivery systems and to prepare the way for SALT III negotiations which, it is hoped and promised, will produce "deep cuts" in the nuclear arms race. The SALT II package is expected to contain a *Treaty* lasting until 1985, a *Protocol* lasting for three years, and a *Statement of Principles* to guide SALT III negotiations.[5]

The heart of the negotiations is the Treaty which is designed to place both quantitative and qualitative restraints on the U.S.-U.S.S.R. competition. Among its principal elements are the following measures:

—a ceiling on aggregate number of SNDVs: 2250;
—a subceiling on MIRVed missiles and bombers equipped with air-launched cruise missiles (ALCM): 1320;
—within the 1320, a further subceiling on land and sea based MIRVed missiles: 1200;
—finally within the 1200 a subceiling on land-based intercontinental ballistic missiles (ICBMs): 800.

Accompanying the Treaty is a Protocol designed "to codify some of the restrictions on weapons modernization sought by the Administration . . . and to meet Soviet concerns over the American sea- and ground-launched cruise missiles."[6] Finally the Statement of Principles is designed to set the pattern for future negotiations involving questions such as further qualitative controls and the possible inclusion of weapons systems which in SALT I and II fell outside the perimeter of negotiations.

Although the negotiations have not been concluded as yet, and there is no possibility the Senate would consider a SALT II package before the new Congress convenes, a serious, sustained debate about the wisdom, utility and morality of the proposed agreements is already underway in the United States. In assessing the debate the authoritative *National Journal* defined its two purposes as trying to influence the Administration's assessment of what kind of treaty could pass the Senate, and trying to influence the Senate vote itself. As Congress approaches the end of its present session, most analysts believe that SALT II will face strong opposition in the Senate. The comparison is regularly drawn with the Panama Canal Treaties which were ratified by a margin of one vote. SALT II, in addition to facing the intrinsic complexity of gaining a two-thirds vote in the Senate, has two liabilities not present in the Panama debate. First, the issue is much more complicated, admitting honest intellectual differences about facts and interpretation of data in a manner distinctly different from the Panama debate. Second, the perceived stakes and risks are much greater in SALT II.

Since no actual text is yet before the public and the Senate, much of the speculation regarding the fate of SALT II is still just speculation. However, the present debate, drawn in broad strokes and directed at issues which undoubtedly will shadow the SALT II vote, will set the framework conceptually and politically within which the SALT II agreements will be evaluated. By examining the positions of key actors in the debate, it is possible to sort out the factors which will weigh heavily in the public discussion and policy decision-making.

II. ANALYZING THE ACTORS AND THE ARGUMENTS

To assess the political and moral content of the arguments in the SALT II debate, it is useful to establish a spectrum of views set in relationship to the Administration position, then to answer three questions for each actor in the debate: first, what is their position on SALT II; second, what is the political-strategic view supporting the position; third, what are the relevant moral issues raised, implied or contained in the strategic position.

A. The Administration Position

The Carter Administration is committed to a position of successfully concluding the SALT II negotiations with the Soviet Union and of advocating acceptance of the SALT II package by the Senate. The principal spokesmen for the Administration, in addition to the President, are the Secretary of State and Mr. Paul Warnke, Director of the U.S. Arms Control and Disarmament Agency and Chief Negotiator in the SALT II process. Since Mr. Warnke has thus far presented the Administration case publicly in the greatest detail, his views will be used here for analysis.

The Political-Strategic Position: The political-strategic argument which is the foundation of the Administration's case for SALT II moves in three steps. First, a description of the present U.S.-U.S.S.R. relationship as "a stable strategic balance." The balance rests upon the strategic theory and policy most clearly articulated by former Secretary of Defense McNamara as "mutual assured destruction." The Warnke support of this position is summarized in a recent speech: "It is the very fact of sure retaliation that today provides the only sure deterrent to nuclear aggression."[7] Second, from this premise follows the conclusion that U.S. strategic interests are best served by preserving the existing military balance and insulating it against chance, miscalculation or misperception on the part of either superpower. Third, SALT II is seen as a crucially important step in preserving a stable strategic balance.

The attributes of the SALT II package are not only that it caps the nuclear arms race at approximate equality, but that it solidifies the strategic balance through a mix of measures which include Soviet reductions in delivery vehicles and qualitative controls on especially provocative weapons systems (land-based ICBMs). The Warnke argument is a mix of *Realpolitik* (the agreement is in our interest) and enlightened global leadership (the international interest is also served). Finally, the SALT II agreements, since they serve our interest in a crucial area of foreign policy, should be evaluated on their own merits; they ought not to be "linked" to other issues (e.g., human rights in the Soviet Union) or policy objectives (Soviet actions in Africa) or other aspects of defense policy (countering Soviet conventional strength in Central Europe). The summation of the Warnke position is contained in the following passage:

> So we have today a coincidence of interests that logically leads the Soviets to view with favor a SALT Agreement, even though it protects the security of the United States. And whatever our distaste for certain Soviet behavior, our own self-interest requires that we seize this opportunity to advance our own security and to make more remote the prospect of nuclear catastrophe for us and for the world as a whole.[8]

The Ethical Issues: The ethical questions which are embedded in the Administration position center on the imperative to control the arms race. As in any policy position there are multiple values being pursued, but there is a primacy given to controlling the dynamic of the arms race. The first imperative of control is insuring that nuclear weapons not be used. Hence, in the Administration position there is a clear option to stabilize the deterrence system rather than to shape a force structure designed to pursue limited nuclear options or war-fighting capability. The Administration view implicitly adopts the position that fighting a nuclear war is not a rational political end. Hence, our strategic interest is defined in terms of nonuse, and SALT II is designed to reinforce strategic stability and the confidence each side has in the mutual assured destruction doctrine.

Viewed in terms of the just-war categories this priority on deterrence receives mixed reviews. On one hand, it could be argued that on the basis of both proportionality and noncombatant immunity, the stress on deterrence is the best option available in the existing configuration of nuclear balance. Prevention of use preserves both principles; any nuclear war is considered irrational and immoral. On the other hand, the acceptance of deterrence based on mutual assured destructive capacity means the acceptance of a declared intent to strike civilian centers if deterrence fails. The credibility of deterrence is locked into the declaration of this intent and its belief by one's adversary. This strategy incorporates the classical dilemma of whether one can threaten to do evil in order to attain a good. The Warnke position does not explicitly address the dilemma, but the shape of the Administration policy means a position has been taken: the evil of use recognized and resisted; the intent to do evil ignored.

The second dimension of policy which highlights the centrality of control in the Administration position is the way in which vertical and horizontal proliferation are joined. The primacy of control means insulating the strategic balance not only from the escalatory moves of the superpowers, but also from the expansion of the nuclear club. The Administration position is an endorsement of the logic of Article VI of The Nonproliferation Treaty (NPT). Under pressure from the nonaligned nations, many of which were essential for an effective ban on proliferation, the superpowers committed themselves to move toward control and reduction of the vertical nuclear arms race in return for pledges of absention on the part of potential nuclear nations. In Mr. Warnke's view the conclusion of SALT II and the movement toward SALT III is crucial because: "I think it's unrealistic to assume that, if we and the Soviet Union are unable to agree on nuclear arms control and continue instead to develop more and more nuclear weapons, other countries are going to forego the acquisition of nuclear weapons themselves."[9]

This position is a different version of linkage. It argues that prevention of proliferation, a major objective for both superpowers, is tied to a principle of fairness. The nuclear powers cannot be unrestrained while continuing to preach restraint. The Warnke position is a prescriptive norm drawn accurately from the policy of the NPT. While accurate, it may now be in danger of being inadequate. The decade of experience with NPT has been a frustrating one for the nonnuclear states. At the U.N. Special Session on Disarmament (SSD) in 1979 the frustration was expressed not only in terms of a lack of progress on the Article VI pledge, but in a claim that the NPT "regime" is fundamentally unfair. The charge of unfairness reaches beyond nuclear weapons to fears of the nonaligned (most of whom are also developing countries) that the NPT regime will be used to restrict access to nuclear power and technology for peaceful purposes. The fairness charge, however, also relates to the marginal moves made by the superpowers since 1968 to implement the pledge of disarmament.

In the moral calculus of global arms control Warnke has identified a key principle. The problem is not simply how to implement the fairness principle, but how do the nuclear states take meaningful steps to reestablish the belief that the NPT system is itself a fair bargain for nonnuclear states.

The Administration position, therefore, is centered on the value of control and uses a mix of self-interest and world-order arguments. The implicit ethical logic is easily grasped, acknowledging the incalculable risks of using nuclear weapons while avoiding the more stubborn question of the moral quality of deterrence. The moral position is not immune to criticism but it provides coherent support for the political-strategic view with which it is joined.

B. The Challengers

The strategic case against SALT is epitomized in the writings of Paul Nitze.[10] A former Secretary of the Navy, Assistant Secretary of Defense and member of the SALT I negotiating team, Nitze now serves as Chairman of Policy Studies for the Committee on the Present Danger. The Committee as a whole, and Mr. Nitze in a unique way, are the most articulate opponents of the emerging SALT II package. Nitze's extensive writings offer a clear example of what the strategic opposition to SALT II looks like in its most sophisticated form.

Political-Strategic View: In Nitze's view the purposes of the SALT process should be—in descending order of priority: (1) to seek limits on arms which maintain crisis stability (i.e., no incentive for a first strike) and reduce the threat of nuclear war; (2) to seek both the reality and the appearance of essential equivalency in the U.S.-U.S.S.R. strategic balance; (3) to provide a basis for reducing the arms competition. In contrast to these objectives Nitze sees the evolution of the SALT process resulting in the erosion of the deterrent relationship and the growth of Soviet strategic superiority. His most recent assessment is that "Recent U.S. program decisions plus the terms of the probable SALT agreement make it virtually impossible to maintain crisis stability, rough equivalence or to reverse presently unfavorable trends during the period of the agreements."[11]

These unfavorable trends which point to Soviet ascendancy in the arms race are at the heart of Nitze's skepticism/opposition toward SALT II. The core of his strategic analysis is a sophisticated assessment of U.S.-U.S.S.R. strategic capabilities. Nitze uses a combination of tests to estimate relative U.S.-U.S.S.R. strength in three situations: before a nuclear exchange; after a Soviet counterforce strike; and after a counterforce exchange. The pessimistic conclusion of the assessment is that:

> Trends shown in these charts by all methods and in all indices move in a direction favorable to the Soviet Union from the mid-1960s through the mid-1980s. Today, after a strategic nuclear counterforce exchange under normal conditions the Soviet Union would hold superiority in all indices except numbers of warheads, and even that sole remaining U.S. advantage would be gone within two or three years.[12]

The trends, in Nitze's view, are not only unfavorable but destabilizing, i.e., they may provide an incentive for the Soviets in a crisis situation to initiate a counterforce exchange which will leave them in a superior strategic position. In this sense Nitze sees Soviet decisions tending toward a "war-winning nuclear capability." Such a capability is foreshadowed by two factors, neither of which, in Nitze's estimate, is addressed adequately by SALT II. The first is a significant Soviet superiority in throw-weight for their SNDVs; the second, a continuing program of civil defense in the U.S.S.R. which reduces the effectiveness of the U.S. retaliatory threat.

The acquisition of a "war-winning" capability does not mean the Soviets would initiate a nuclear war, but it does mean that they would enter a crisis with options unavailable to the United States and in noncrisis periods they could use their perceived strategic superiority for political purposes. The politically significant consequences of such superiority are most relevant in the areas of Central Europe and the Middle East; in these theaters of competition perceived strategic superiority can be translated, Nitze argues, into political influence.[13]

In light of this assessment of the political-strategic balance, Nitze's prescription is that critical preconditions must be established for successful arms control negotiations to proceed: "If, and only if, the United States now takes action to redress the impending strategic imbalance, can the Soviet Union be persuaded to abandon its quest for superiority and to resume the path of meaningful limitations and reductions through negotiation."[14] Such action by the U.S. would principally be designed to increase accuracy of U.S. missiles and to decrease the vulnerability of the land-based ICBMs. The latter would include going ahead with Trident II and the MX (mobile) missile, and perhaps reconsidering the B-1 decision.

The Ethical Issues: The strikingly different political-strategic view of the Nitze position from that of the Administration yields a different moral agenda. The value placed on controlling the arms race as a strategic and moral imperative is replaced in the Nitze position by the quest for security as a primary value. The Warnke position is that strategic stability and equality is to be achieved *through* control of the technological dynamic of the arms race. Nitze argues that strategic stability must first be reestablished, then one can pursue control of the balance.

Nitze agrees that the deterrence relationship is the basic fact of strategic existence because "the failure to deter would be of enormous cost to the United States and to the world."[15] He is concerned, however, that U.S. policy relies too exclusively on a deterrence policy with little or no regard for strategic choices which must be made if deterrence fails. In Nitze's view, "Americans think in terms of deterring nuclear war almost exclusively. The Soviet leaders think much more of what might happen in such a war." Nitze argues for a posture which he describes as "denial of a war-winning capability to the other side."[16] There is some affinity here with former Defense Secretary Schlesinger's desire to shape U.S. policy and force structure so that the deterrent would be supplemented with a capability for "limited nuclear options." Nitze does not advocate a "war-fighting" capability and in that sense his position is distinct from the Schlesinger view. His concern is to link deterrence with other strategic choices making deterrence more credible and providing the United States with a response to a counterforce strike which is less than a spasm reaction against Soviet urban/industrial targets.

In advocating this position ("deterrence-plus") Nitze joins the ethical debate with the Administration. The dilemma of the Warnke position is that it *threatens* massive violation of the principles of noncombatant immunity and proportionality in order to prevent any *use* of nuclear weapons. The dilemma of the Nitze position is that in seeking to make deterrence more credible by making it more usable, it may in fact render resort to limited nuclear engagement more likely.

The Nitze-Schlesinger view is that the mutual assured destruction position devoid of supporting limited options is not only strategically unsound but morally corrupting since it relies exclusively on holding civilians hostage. This threat is a direct violation of noncombatant immunity and is precisely the dimension of ambiguity cited in the Administration position earlier in this paper. The Nitze position, however, is not free of ambiguity. Even though it is a straightforward application of noncombatant immunity to the nuclear dilemma, arguing for targeting military objectives first and holding civilian centers as a last resort, it deals lightly with two questions. First, whether nuclear weapons are usable at all within any recognized moral framework. Second, whether limited options language and strategy tend to diminish the psychological, strategic and policy barriers—firebreaks—which have slowly, arduously and delicately been constructed in the age of deterrence.

The Nitze position does not address the first question; the implicit logic of the position is that it is possible to fight certain forms of nuclear engagement within the bounds of the twin principles of noncombatant immunity and proportionality. This position has supporters among both strategists and ethicists, but it also meets significant opposition from practitioners of the just-war doctrine who find the nature and scope of nuclear weapons and the design of nuclear policy such that the just-war answer in the nuclear age is a form of nuclear pacifism. Since Nitze's answer to the first question is that nuclear weapons are morally usable, the second question is less moral than strategic for him. His response, prefigured above, is that multiple options render deterrence more credible, hence the chance of crossing the nuclear firebreak is lessened by threatening limited as well as massive retaliation.

C. The Advocates

The advocates of SALT II are nongovernmental authors or groups who support the proposed agreement but who also support steps beyond the Administration position and/or use different arguments to support arms control policies. The counterpoint to the Nitze position and the goad to the Administration view is represented most extensively by Dr. Herbert Scoville, Vice President of the Arms Control Association.[17] For Scoville, SALT II is necessary but in fact very limited and very late. He supports the SALT II package, but his real interest is in moving rapidly beyond SALT II.

Political-Strategic View: Scoville's perception of the political-strategic situation is strikingly different from Nitze's and provides an illustration of how the debate about SALT II will be structured. Scoville accepts the Warnke view that a stable strategic balance exists. On this premise he then outlines three objectives for SALT II: (1) to turn the arms race around and to reduce the number of SNDVs; (2) to bring under control programs for replacing existing weapons systems with more advanced versions; (3) to restrict creation of new systems "which could destabilize the current strategic balance."[18] These objectives clash starkly with Nitze's not only in their order of priority, but in the strategic perception supporting them. Scoville is an unequivocal supporter of the position that security today is assured only through control of nuclear weapons. Rather than placing the SALT negotiations within the framework of U.S.-Soviet relations and seeing it as part of the strategy of détente, Scoville identifies the very limited and specific objective of SALT as "improvement of security by controlling nuclear weapons."[19]

This view of SALT, like the Administration position, is based on the premise that the strategic balance is presently stable. Nitze denies this premise, arguing that recent Soviet strategic planning has destabilized the balance so that it presently cannot serve as the foundation for a new arms agreement. Scoville does not deny that the Soviets have made decisions about force structure since the initiation of SALT which have been counterproductive from the perspective of arms control. But he argues that the moves have not placed the United States at any fundamental disadvantage and he also points to a series of decisions taken by the United States in the same period to bolster its strategic forces.

For Scoville the most important task of SALT II is to prevent decisions which are destabilizing; for Nitze the gravest threat of the moment is that the United States is on the debit side of an unstable balance. This divergence about what constitutes the threat of the moment leads Scoville to an even stronger disagreement with Nitze about the desirability of enhancing U.S. counterforce capabilities. The dominant conclusion in Nitze's analysis of U.S.-Soviet capabilities is that under prevailing conditions the Soviets would emerge from a counterforce exchange in a superior position. This calls for increasing U.S. counterforce options and decreasing the vulnerability of land-based ICBMs. Scoville denies that the emerging vulnerability of the ICBMs really presents the Soviets with an attractive option for risking nuclear exchange. In addition he views moves by either side to increase counterforce capabilities as fundamentally mistaken.

In the final analysis Scoville gives strong support to what he regards as a very limited agreement. Although late and limited it is not, he argues, a one-sided treaty or inimical to U.S. strategic interests. It may not be worthy of great praise, but it is worthy of support.

Ethical Issues: Scoville's political-strategic view is very similar to the Warnke position. In contrast to the Administration, however, Scoville is quite explicit about the ethical choices he makes to support his options for primacy of control in the arms race and for reliance upon mutual assured destruction strategy as a necessary part of a process for controlling and reducing the arms race.

The Nitze (and Schlesinger) position makes explicit appeal to the principle of noncombatant immunity in support of a counterforce component of policy. Scoville is just as explicit in rejecting the rationale for counterforce strategy by relying on the ethical logic of the proportionality principle. This ethical norm, identified earlier in this paper as a utilitarian principle, focuses on the consequences of political and strategic decisions. It is precisely the perceived consequences of trying to shift the strategic balance in the direction of counterforce capabilities which brings Scoville to oppose the move.

In Scoville's view limited nuclear exchange which then presents the possibility of rational political bargaining, à la Nitze's scenario, is a chimera. For Scoville, "the overriding objective is to prevent nuclear wars not fight them."[20] Nitze would agree; but he would probably not agree with the following Scoville proposition:

> Therefore, the primary criterion for any new strategic policy must be the assurance that it will in no way increase the likelihood that nuclear war will be unleashed. A

> limited nuclear conflict presents a major risk of uncontrollable escalation to widespread nuclear devastation so that almost no gain is worth risking an increase in the probability that it will occur.[21]

This "primary criterion" of policy means that in the process of controlling the arms race the emphasis must be on preventing any use of nuclear weapons. The logic of this position is that moves to equalize the Soviet counterforce capability with a similar U.S. potential carry more risks than they convey benefits and should not be pursued. The basic moral perspective supporting Scoville's analysis is that any conceivable nuclear exchange violates the principle of proportionality, overwhelming any political gain by its destructive results. For Scoville this is the controlling moral principle, and appeals to counterforce arguments, either in the name of increasing strategic stability or protecting noncombatant immunity, are unpersuasive since they contain the inherent risk that "the improved ability to respond at lower levels of violence increases the risk that nuclear war will erupt."[22] This risk, in Scoville's view, cannot be tolerated, and his political and moral case is shaped to render the intolerable unthinkable in U.S. policy.

D. Two Nonstrategic Arguments About SALT II

To fill out the picture of key actors in the SALT II debate two other positions, neither of which is argued in strategic terms, can be briefly summarized.

1. A Political Argument About SALT II

This position, represented in the views of Senator Daniel P. Moynihan (D-N.Y.), has a different theme from Nitze's detailed strategic critique. The argument is broader than SALT or even broader than U.S.-Soviet relations. Indeed, it may not ever be used against SALT II since Moynihan is not announced in opposition to an agreement. The substance of the argument, however, *could* be used by Moynihan or others in the SALT debate and it does combine a mix of political and moral themes.

Senator Moynihan sees the fundamental political and moral problem in world affairs as the conflict between liberal democracy and totalitarianism. The U.S.-U.S.S.R. relationship is central to this cosmic drama but it does not exhaust its dimensions. In Moynihan's view, "If one only identifies totalitarianism with the sort of papal claims of Moscow, one misreads history and wrongly analyzes the present situation."[23] While Moynihan locates much of the problem in an inability of elites in the West to recognize and refute the totalitarian challenge, the question of Soviet intentions and capabilities is a central piece of his policy perspective.

If the fundamental value conflict in world politics is freedom vs. totalitarianism then the Soviets at a political-strategic level embody that challenge for the United States in a unique way. Moynihan, from a different perspective and for different reasons, agrees with Nitze that the perception of Soviet power as increasingly dominant is growing.

> The dominance is clearly perceived in Western Europe, and it is beginning to be perceived here. . . . I said in a speech last weekend that it seemed to me that much of the difficulty we have understanding our Middle East situation is that for the first time, American policy is beginning to accommodate to the assumption that the Soviets are now, or soon will be, the superior military power.[24]

Both the larger value question and the specific assessment of U.S.-Soviet relationships are pertinent to a judgment on SALT II. It is not clear that in Moynihan's case these broader political and moral themes will lead to a negative conclusion about SALT II. But the nature of the argument made here could easily shape the background of the specific strategic debate and the contour of the argument promotes skepticism about seeking a relationship with the Soviets on the basis of strategic control rather than strategic competition.

2. A Geo-Political Argument About Security

A distinctly different perspective and framework of moral and political argument is found in the position of Richard Barnet of the Institute for Policy Studies in Washington.[25] Barnet would support SALT II, it appears, but would do so with minimal enthusiasm because of the limited nature of the agreement. SALT II looks limited to him not simply because of the specifics of the package but because of the way he defines the security problem today.

Barnet seeks to redefine the concept of security in light of a new context of international relations: "The end of colonialism, the new awareness that vital resources of the globe are limited, and the realization that the political and economic problems of the United States (and every other country) cannot be solved except in the context of global solutions are transforming the dream of world community into a political imperative."[26] In this context a definition of security which is tied exclusively or predominately to weapons systems is in Barnet's view fundamentally mistaken. The new security problems touch the stability of the domestic economy, the access of necessary raw materials and the quality of urban life as much as they relate to external threats.

The SALT II agreement is limited, argues Barnet, because it is based on an inadequate conception of security which inherently preserves the superpower competition in military hardware. The critique of SALT II is less an argument against it than it is a proposal for much more broadly conceived initiatives to supplement the SALT process. Barnet proposes a series of "independent" U.S. initiatives, moves which the United States could take without detriment to its security but which offer the promise of a breakthrough, conceptually and politically, in the arms race. These include proposals by the United States to cease all weapons testing; to cease testing of SNDVs, and to suspend all weapons acquisition for three years. The proposals would be independently made but would look toward a reciprocal response from the Soviets.

This redefinition of security requires both a political re-evaluation of what constitutes the fundamental themes of world politics, and a moral or "spiritual" shift which allows us "to put our trust in something other than weapons stockpiles which are always too small and are always obsolete."[27] Barnet tries to fuse political analysis of transnational forces and problems which analysts today recognize as posing new problems for any state with a call for a new value analysis which includes strategic security but displaces it from the center of the political picture.

III. THE ROLE OF MORAL ARGUMENT IN THE STRATEGIC DEBATE

In addition to identifying the implicit or explicit moral positions in the SALT II debate, it is possible to assess the role of moral argument in a policy decision like SALT II. The understanding held in the Just-War Tradition of the contribution moral analysis makes to policy is expressed by a modern expositor: "Primarily, its value resides in its capacity to set the right terms for rational debate on public policies bearing on the problem of war and peace in this age, characterized by international conflict and advanced technology."[28]

Supporting this view of morality and policy are the assumptions that every policy position incorporates some moral calculus, but the moral logic of a policy is interwoven with the empirical data of the problem. In short, there is no "nonmoral" approach to policy and there cannot be a "purely moral" answer to policy questions.[29] In assessing how the ethical and empirical factors of policy are related in the strategic debate three perspectives are helpful: the way moral argument complicates policy debate; the way it clarifies policy debate; and how it can affect the outcome of debate.

In one dimension of the SALT II debate the introduction of explicitly moral categories serves to complicate an otherwise consistent position. The case is the Warnke-Scoville argument for clear and explicit reliance on deterrence, based on massive retaliation, for the purpose of placing any consideration of *using* nuclear weapons beyond the pale of rationality. In strategic terms the position is dangerous and fragile, but it has been effective in ruling out any use of nuclear weapons for three decades. A moral analysis of deterrence theory can acknowledge the value in prohibiting use, but it cannot simply condone the declared intent, always present in deterrence theory, to kill a major portion of the adversary's civilian population as a conscious choice of policy. Deterrence policy may be both effective and in fact the best choice among alternatives, but it can never be held with moral integrity.

Moral analysis, on the other hand, can help to sort out and clarify some strategic choices. It can distinguish between the moral quality of a *threat* as opposed to the increased danger of *use* and give priority to opposing use. Similarly in the Warnke-Nitze clash over primacy of control or of security as the first value to be sought in arms limitation, the ethical theory provides a framework for adjudicating the content of the terms in the present state of the arms race.

The weight which moral argument has in policy debate is a function of how explicitly the ethical presuppositions of various policy choices are analyzed and how they are then used to judge other factors in the policy decision. In some of the positions analyzed in this paper, ethical choices have been made but are not articulated in explaining the policy; in other positions the ethical content of the policy is explicitly used to convince others of its validity. For moral argument to have any decisive function in a debate like SALT II, all the moral choices in each of the positions must be put on display and tested explicitly. The purpose of this paper has been to indicate how such testing might be done.

NOTES

1. Cf. Ralph B. Potter, Jr., *War and Moral Discourse* (Atlanta, Ga.: John Knox, 1969).
2. For a treatment of just-war doctrine cf.: James T. Johnson, *Ideology, Reason and the Limitation of War: Religious and Secular Concepts* (Princeton, N.J.: Princeton University Press, 1975); Ralph B. Potter, Jr., *The Moral Logic of War* (Philadelphia: Department of Church and Society, Board of Christian Education, United Presbyterian Church, n.d.); LeRoy Walters, "Five Classic Just War Theories" (Unpublished Ph.D. dissertation, Yale University, 1971); Paul Ramsey, *The Just War; Force and Political Responsibility* (New York: Scribner, 1969).
3. Ralph B. Potter, Jr., *The Moral Logic of War,* p. 20.
4. Paul Ramsey, *The Just War,* p. 347.
5. For the most thorough published review of the elements of SALT II, cf. R. Burt, "The Scope and Limits of SALT," *Foreign Affairs* 56 (1978): pp. 751–770; also R. Burt, "Asking SALT to Do Too Much," *Washington Review* 1 (1978): 19–33.
6. R. Burt, "The Scope and Limits of SALT": 758.
7. Paul Warnke, "Strengthening United States Security Through SALT," Arms Control and Disarmament Agency (Washington, D.C., 1978); also "Arms Control: A Global Imperative," Arms Control and Disarmament Agency (Washington, D.C., 1978).
8. Paul Warnke, "Strengthening U.S. Security," p. 10.
9. *Ibid.*, p. 10.
10. Cf. Paul H. Nitze, "The Strategic Balance Between Hope and Skepticism," *Foreign*

Policy, no. 17 (Winter, 1974–1975): 136–156; "Assuring Strategic Stability in an Era of Détente," *Foreign Affairs* 544 (January 1976): 207–232; "Deterring Our Deterrent," *Foreign Policy,* no. 25 (Winter, 1976–1977): 195–210; Press Release (March 28, 1978), Committee on the Present Danger.

11. Press Release, Committee on the Present Danger, p. 10.

12. Paul H. Nitze, "Deterring Our Deterrent," p. 201.

13. Paul H. Nitze, "The Global Military Balance," in Grayson Kirk and Nils Wessell, eds., *The Soviet Threat: Myths and Realities* (New York: Praeger, 1978), pp. 4–14.

14. Paul H. Nitze, "Assuring Strategic Stability": 208.

15. *Ibid.,* 222.

16. *Ibid.,* 212.

17. Cf. Herbert Scoville, "The Salt Negotiations," *Scientific American* 237, no. 2 (August 1977); "SALT II Means More Security," *Christian Science Monitor* (April 24, 1978); and "The True Utility of SALT," *Nation* (May 27, 1978).

18. Herbert Scoville, "The True Utility of SALT."

19. Herbert Scoville, "SALT II Means More Security."

20. Herbert Scoville, "Flexible Madness," *Foreign Policy,* no. 14 (Spring 1974): 176.

21. *Ibid.,* 168.

22. *Ibid.,* 174–175.

23. Daniel P. Moynihan, "The American Political Elite," *Washington Review* 1 (1978): 35.

24. "Is There a Crisis of Spirit in the West? A Conversation with Dr. Henry A. Kissinger and Senator Daniel P. Moynihan," *Public Opinion,* vol. 1, no. 2 (May-June 1978): 7.

25. Richard Barnet, "A Time to Stop" (Washington, D.C.: Institute for Policy Studies, Xerox copy).

26. *Ibid.,* p. 2.

27. *Ibid.,* p. 6.

28. John Courtney Murray, *We Hold These Truths; Catholic Reflections on the American Proposition* (New York: Sheed and Ward, 1960), p. 269.

29. Cf. Ralph B. Potter, *War and Moral Discourse,* p. 26.

APPROPRIATE TECHNOLOGY: FACTS AND VALUES

J. van Brakel

I. INTRODUCTION

In the latest edition of the *Encyclopaedia Britannica* the entry "philosophy and (of) technology" does not appear. The subject is partly covered in a brief article on "conceptions of technology," dealing with such concepts as "faith in progress" or the anti-technology trend. Two paragraphs discuss the promise of technology for underdeveloped regions. It is noted that:

> Two important principles have emerged: first, that the technology of the developed nations is not automatically ideal, or even always suitable, for the underdeveloped nations, and, second, that the underdeveloped nations need not necessarily pass through the same succession of stages that the vanguard of developed nations has undergone.[1]

Research in Philosophy & Technology, Volume 3, 1980, pages 385–402.
Copyright © 1980 by JAI Press Inc.
All rights of reproduction in any form reserved.
ISBN: 0-89232-102-4

Contemplating these two principles, it would seem that "technology for/in/and development" should be a fruitful source of problems for the philosophy of technology, or at least a useful and necessary area to apply any theories or analytical tools that have emerged in the philosophy of technology. However, the attention given to "technology and development" in "philosophy and technology" is negligible.

That this is the case is easily checked by consulting such sources as the *Bibliography of the Philosophy of Technology*[2] or the various newsletters that cover this subject. There is a lot of interest in "humanities perspectives on technology" or the "impact of technology on society," but, as far as the perspective of society is concerned, it seems that the latter is confined to OECD-members, if not the United States of North America.[3] In the rare cases that some attention is given to technology and development this is only as an afterthought and does not come across as being based on a serious study of the subject matter. For example, in a book by Calder we find a chapter on the "pigmented poor," which concludes:

> Given the prospect of a century of creative work to a self-evidently desirable end, the enrichment of the whole world, the present habit whereby research and technology attend to the needs of the poor only as an afterthought is hard to tolerate, let alone defend.[4]

There is abundant evidence in the literature to show that, even if one favors an optimistic view of the promise of technology, Calder's assessment is slightly simplistic. I have to admit that this quotation is ten years old, and that one can still find people who agree with it. However, the point is that no significant contribution has been made by "philosophy and technology" in analyzing the problems of "technology and development."[5]

Returning now to the first quotation in this section, we may regard the two principles referred to as concerned with the statics and the dynamics of "technology and development." The present article has as its subject a review of the recent use of such adjectives as "appropriate," "intermediate," and "alternative" in connection with the statics and dynamics of technology and development. As will appear, the conceptual analysis of such terms is simple and straightforward; they are not like "democratic" or "beautiful." Consider:

> Simple, inexpensive machinery that can be operated and maintained by persons of little education is far more appropriate to these agrarian nations than computers and automatic devices (*Encyclopaedia Britannica*).

or:

> In any case, imitation of the present rich countries may be quite inappropriate for countries with different climates, cultures and interests (Calder).

The meaning of "appropriate" in these quotations will give few problems. However, if one looks into the discussions about the (pragmatic) truth of these and similar statements a host of philosophical, political and other problems turn up.

Although in the present article none of these philosophical problems will be discussed in detail, an attempt is made to provide some conceptual clarity which may give a better grasp of underlying problems. Apart from other prejudices, the major assumption is that the distinction between facts and values, as well as that between means and ends, is useful. This does not exclude the possibility that (some) values are in some way open to rational justification; neither does it identify the means/ends dichotomy with some form of technocracy.[6]

II. PRELIMINARIES: TECHNOLOGY AND PRODUCTION SYSTEMS

A lot of semantic, and sometimes ideological, confusion with respect to the terms "technology," "technological," etc., is due to the fact that in the English language the word "technology" is often used in the sense of "production system," or even "machines." This is the case, in particular, among economists and other social scientists. It is preferable to use the term "technology" only when "knowledge of production systems" is meant. Although the related term "appropriate technology" occasionally means "appropriate knowledge," it usually means "appropriate production systems."

A production system is already a quite complex entity. It may be described in terms of:

(a) the product it should produce (fulfilling some need);
(b) the raw materials, the mechanical, physical, and/or chemical operations, and the machines and instruments used to process the raw materials into the final product;
(c) the individuals and social structures inside and the institutions outside the production system that are necessary to have it produce;
(d) the interaction of the production system with the physical and social environment (including macro- and long-term effects).

Technology includes *all* knowledge about production systems. Technology is the know-*how* to use various production techniques in order to produce,[7] plus the know-*that* and *how* a particular production system affects both the individuals taking part in it and the environment.[8] The technology that is developed (or emerges) at a particular time and place is subject, as a matter of course, to sociocultural influences.

The following terminology may elucidate some important techno-economical characteristics of production systems. A production system is technically possible if it can produce the product needed. It is technically efficient if there are no other production systems that produce more output using the same amount of production factors.[9] A production system is economically efficient if—given the prices of the production factors—there is no other production system that can produce the same output at lower cost. It is economically feasible if it is economically efficient at some reasonable shadow prices.[10] What is reasonable is subject to discussion. Because production systems are complicated entities, interacting with their environment in complex ways, it is a technocratic fallacy that the economic feasibility of alternative production systems—that all fulfill the same need—can be evaluated by technical means only.

III. OBSERVATIONS: TECHNOLOGY FOR APPROPRIATE DEVELOPMENT

1. On Adjectives

The number of adjectives attached to "technology" (read "production systems") to indicate that something other than "normal" technology is meant is still growing. Among others, I have come across the following:

(a) labor-intensive, low-capital, capital-saving, low-cost;
(b) small-scale, small-capacity, intermediate;
(c) rural, survival, self-help, barefoot, Third World, peoples, indigenous;
(d) alternative or alternate, liberating, progressive, radical;
(e) clean, eco-, soft;
(f) appropriate, correct, optimal, adapted, mixed, pluralistic.

Although the classification in six groups, as given here, is rather arbitrary, it may give some first indication of the differences in emphasis. Of course, in the literature these adjectives are not always used with a clearly indicated meaning. In this review only a brief indication of the intended connotation can be given.[11]

The terms given under (a) refer to the evaluation of the production factors capital and labor. In theory, precise definitions can be given. In practice, there are major problems: the lack of homogeneity of production factors (hence measurement problems) and their evaluation relative to such other production factors as raw materials, managerial skills, and product quality.[12]

The terms under (b) refer to the scale of operation; "intermediate" sometimes has a wider connotation than "medium-scale" (see below). Taking into account that what is large for Gambia may be small to Brazil, and that a large bakery may be small compared to a small car factory, these are also basically descriptive terms. Again in assessing the economic feasibility of a production system, the choice of scale of production is interrelated with many other production factors.[13]

The adjectives given under (c) basically express a humanistic or holistic concern with the people or societies the technology should serve. The first six adjectives mainly advocate the interest of those considered most in need of development: "for the hundreds of millions of farmers left out of the development process" (Hoda, 1974); "technology must benefit the largest number of people possible [which means] distribution of surplus production at the lowest possible level" (Garg, 1975). Perhaps the term "indigenous" should have a class for itself:

> An indigenous technology means a technology that really takes root and once rooted begins to grow its own branches, gathering nutrition from the fertility of the soil in which it grows. In short, we are looking for that kind of influence which generates confidence and spontaneity in a people's own ability to succeed. We are looking for know-how shared by the largest number of people in the community (Onyemelukwe, 1974, p. 28).

The notions given under (d) express a political opinion and are defined in opposition to the prevailing use of technology by the prevailing powers in society. For example: "The word 'radical' literally means 'going to the root,' and accordingly 'radical technology' implies a fundamental re-examination of the role of technology in modern society. It also implies a commitment to the ideals of the political Left" (Harper, 1976). The terms are used both in connection with technology in industrialized as well as in underdeveloped countries.[14]

The terms given under (e) are mainly used with reference to industrialized countries. They are related to such notions as "limits to growth" and "Nature knows best" and may be traced back to *Mother Earth* and the *Blueprint for Survival*: "As the new technology developed, it has become increasingly clear that the old Faustian dream of endless 'progress' towards man's total domination of the earth has not been fulfilled" (Mackillop, 1975).[15]

The adjectives mentioned under (f) are the most general and the least descriptive. The terms "optimal," "mixed," and "pluralistic" emphasize that, a priori, any production system may be the best:

> The basic thesis in this is that each requirement has to be considered individually. The "optimal technology" in each situation is then *not* determined by any general economic or sociological theory; it is *not* limited to any degree of sophistication, scale of operation, labour intensity or cost; it is *not* seeking in any sense a second-rate solution; but it is *the best* technology that can be advanced or devised to meet the individual requirement, taking full account of all the relevant local political and other factors, particularly the need to maximise social and economic benefit to the community and to make best use of local entrepreneurship, management personnel, skilled labour, unskilled labour and such funds as may be available.[16]

The terms "appropriate" and "intermediate" are most difficult to classify, because their meaning may include several of the ideas expressed in the more descriptive adjectives. The original meaning of intermediate presupposes a particular development model:

> If effective help is to be brought to those who need it most, a technology is required which would range in some intermediate position between the £1 technology—the traditional indigenous technology—and the £1000 technology—the technology of modern industry. . . . At such a level of capitalisation, very large numbers of workplaces could be created within a fairly short time. . . . The intermediate technology would also fit much more smoothly into the relatively unsophisticated environment in which it is to be utilised (Schumacher, 1973, p. 150f).

The term most in use nowadays is "appropriate." The following quotations cover the greater part of its vague connotation:

> Briefly, the most appropriate technology is one that directly meets people's needs, is consistent with local values and environmental concerns, and can be controlled, managed, and maintained by local materials, resources and expertise (VITA, 1976).

> Appropriate technology may be defined as the set of techniques which makes optimum use of available resources in a given environment. For each process or project, it is the technology which maximizes social welfare if factors and products are shadow prices.[17]

> Appropriate technology can be looked upon as an approach to supplying technology to a society that requires it. The choice of a technology that is appropriate should be made on the basis of the types of production, availability of manpower, size of market, and other social, economic, and cultural considerations (DSE, 1972).

> Appropriate industrial technology is not necessarily aligned to labour-intensive or small-scale production. Depending on the conditions it could well be capital-intensive, advanced in nature, and for large-scale production. With changing conditions the appropriate technology would also change. These changing conditions could be the result of the development process itself. The particular form appropriate technology takes can thus change in the course of time. In this sense appropriate technology is a dynamic concept (UNIDO, 1977).

2. History of "Intermediate" and "Appropriate"

The term "intermediate" was coined by Schumacher while consultant to the Government of India in 1963. Intermediate technology should provide a way of dealing with mass rural emigration, mass unemployment and mass undernourishment: industries should be set up of a type suited to rural areas, located in and supplying the needs of rural communities. In 1964 the first conference on intermediate technology ever held took place in Hyderabad (India). Significantly the conference proceedings were entitled "Appropriate Technology for Indian Industry":

> Right at the time of this invitation we had signalled we did not like the term "intermediate." We hope that the term "appropriate" technologies will now replace

> it. "Intermediate" suggests a stage on the way to something better and is therefore confusing. The real problem is to develop in India technologies that are appropriate to India, i.e., technologies that are the very best in the context of India's circumstances, situations and resources. Technologies appropriate to India—with its low wage rates, millions of unemployed people and the continuing shortage of capital, foreign exchange, certain raw materials and specialised skills, which accompany rapid development—are necessarily different from technologies appropriate in countries where capital is plentiful, labour highly paid and large markets are open to an ever changing array of consumer and capital goods.[18]

From the very beginning the term "intermediate" acquired the connotation of inferior in developing countries. Nevertheless for quite some time "intermediate" was used more frequently than "appropriate" because Schumacher founded in 1966 an organization called "Intermediate Technology Development Group" (ITDG) which quickly gained momentum. In December 1967 its first major publication was launched, *Tools for Progress*. In January 1968 the first conference on intermediate technology in the United Kingdom was held in Oxford, whose purpose was to consider "the further development in the United Kingdom of appropriate technologies for, and their communication to developing countries." In the same year a similar conference was held in Rome.[19]

The breakthrough of the appropriate technology movement can be situated in 1972. On instigation of ITDG a *Technology Consultancy Centre* was established at the University of Science and Technology in Kumasi (Ghana), and an international conference on appropriate technology sponsored by the *Deutsche Stiftung für Entwicklungshilfe* was held there (DSE, 1972). In the same year the OECD organized a conference in Paris on the "choice and adaptation of technology in developing countries," where much attention was given to appropriate technology (OECD, 1974). The UNESCO adopted a resolution on appropriate technology.

Already in 1971 the *World Plan of Action* had been published (UN, 1971), in which an important place was reserved for appropriate technology. Although the plan has never, as such, been accepted by the United Nations General Assembly, in subsequent years all UN organizations involved in one way or another with production systems had to adopt "appropriate technology" as a major part of their policies. The *International Labour Organisation* adopted the notion in 1974. Their philosophy was further specified in the "basic needs" concept, launched in 1976.[20] UNIDO adopted the term "appropriate" as part of their policy in 1975 in the "Lima Declaration and Plan of Action on Industrial Development and Cooperation."[21]

In 1974 the OECD held another major conference on the subject.[22] ITDG started its influential journal *Appropriate Technology* also in 1974. In 1975 the U.S. House Committee on International Relations authorized the U.S. Agency for International Development to support "an expanded and coordinated private effort to promote the development and dissemination of technologies appropriate for developing countries" (US-AID, 1976).

This should suffice to indicate that by now every respectable institution involved in some way or other with technology and development will express that it is after appropriate technology, whatever that may mean. Since about 1977 articles are appearing from "progressive" sources to draw attention to the fact that appropriate technology is no miracle panacea for development and does not lead to any improvement in the conditions of those it is alleged to help:

> Although a lot of noise has been made in its favour, appropriate technology remains conspicuous only for its failure to have an impact on rural development in India.[23]

3. Criteria for Appropriateness

Numerous detailed definitions of "appropriate" have been given in the literature. The longest definition is probably the one given by Hvelplund (1974), who uses 15 criteria and asks for at least 200 words to summarize it. The problem with the long lists of "good things" is that it does not follow in any way whatsoever how incompatible criteria are to be combined. The problem with short lists of criteria is that it is never clear whether the list is meant or thought to be exhaustive. For example, if it is not mentioned that an appropriate production system should be economically feasible, does it follow that this is an irrelevant criterion?

Most of the criteria that have been mentioned in the literature can be classified as follows:

I. Production factors:
   1. labor-intensive;
   2. low capital cost: (a) per workplace, (b) per unit of output, (c) per machine;
   3. low energy costs;
   4. simplicity: (a) in manufacture, (b) in operation, (c) in maintenance and repair, (d) in organization;
   5. locally oriented with respect to (a) scale of market, (b) available skills and entrepreneurs, (c) use of raw materials, (d) use of energy resources;

II. Process of change aspects:
   6. geographically dispersed, rural development;
   7. ecologically sound;
   8. commercially viable (short/long term, increasing GNP);
   9. fulfilling basic needs (products for low-income groups);
   10. self-help, control over production factors;
   11. "organic growth" of production systems.[24]

The criteria or features under II imply and set boundary conditions for the criteria given under I. Most pairs of criteria could be mutually inconsistent; some frequently are.[25] Operationalization of the criteria under II often makes it impossible to meet one or more of the criteria given under I.

The justification of the criteria is based, on the one hand, on macroeconomic considerations, such as rural development, employment, import substitution or national energy policy; on the other hand, it is based on an idea as to how the given society should be in the future: self-sustaining, with local control over all aspects of the production systems.

IV. ANALYSIS: FACTS AND VALUES

1. Relative Value Judgments

Adjectives such as "appropriate," as commonly understood, have no descriptive content. Usually what is said is a tautology, whereas the writer tacitly refers to some value judgment. For example, as Mendoza (1977) says in, appropriately, the *Journal of Speculative Anthropology,* "Alternative technology is applicable to the development of the third world because it provides permanent, non-polluting development for local communities of people." Such statements are not statements of fact supported by evidence, as the form suggests (A is applicable to B, because . . .), but a conjunction of a formal definition (i.e., the terms in the definition may or may not refer to existing things) and a value statement (i.e., that local communities should have . . .).

A production system is appropriate, relative to a given set of "ultimate" value judgments (providing the choice of product and an evaluation of the effect any production system has on its environment), when the best production system is chosen and put into operation according to expectations—given a correct analysis of the boundary conditions and their possible change. That, given a number of boundary conditions and specified goals, one should choose the right (= correct = appropriate = good = adapted = optimal) production system (to be developed) is not a very original idea. The question is:

What are the boundary conditions, and can they be changed?
Which goals have to be chosen, and by whom?
What then is the best production system, and
How can it be put into operation?

Although it is never possible to completely separate questions of fact and value, it seems appropriate not to call values or goals appropriate if one adheres to them. The meaning of "appropriate" is best restricted to apply to relative value judgments,[26] i.e., judgments as to what means to use to arrive at a certain end. If used that way, I think there is no redundancy in talking about appropriate production systems, because it is by no means trivial that one succeeds in selecting the appropriate system in a given context. The loaded meaning, however, is then gone; it is not necessary to stipulate that one is in favor of appropriate production systems, which is trivial. But we can sensibly discuss whether in a particular situation—say, wanting to transport large quantities of agricultural products from the hinterland to the city—hovercraft vehicles are more appropriate than building roads and bridges.

2. Processes of Change

From the assessment given in the previous subsection, it does not follow that it is a simple procedure to arrive at the appropriate production systems in a particular society. Consider the following stages in the development of a production system:

(a) a need of (a part of) a society is identified;
(b) goals and differences between the goals of different groups are identified;
(c) relevant areas of technology and available hardware are identified;
(d) alternative ideas for products (methods, goods, or services) to fulfill needs/goals are developed;
(e) specifications of alternative production systems are developed including their technical characteristics, economic feasibilities, institutional requirements, and socio-cultural implications;
(f) alternatives (e) are evaluated in terms of (b) and a choice is made;
(g) a prototype or raw material is developed or acquired and tested in an actual social situation;
(h) on the basis of (g) the production system is again evaluated in terms of (b);
(i) inputs of the production system become or are made available on a commercial scale and the production system(s) is adopted throughout society wherever appropriate.

Of course, not every aspect is covered in the above classification of stages, but it will be detailed enough to give some idea of the complexities involved in arriving at the right choice.[27] This is not the place to analyze this any further. However, one aspect that is fundamental for the understanding of the relation between technology and development should be emphasized.

In connection with production systems, two major processes of change should be considered simultaneously. First, the development in time of the production system itself (compare the stages above). Second, the fact that the introduction or adaptation of a production system induces a process of change in its (physical and social) environment. Ideally these two processes should be integrated, such that the environment (in which the production system eventually will take part) induces the development or the transfer and adaptation of the production system. Of course there are problems: (i) the foregoing is rather vague (in particular as regards which people to include in the environment); (ii) it blurs the fact that the social part of the environment is not homogeneous (different groups have different interests); (iii) external intervention is very often both impossible to prevent and may seem appropriate; and (iv) success is under no circumstances guaranteed because not all consequences of actions chosen are foreseeable. Nevertheless, I think the concept of integrating the two processes of change is useful, because the further away one is from the ideal, the more complicated the process of change in the environment will be; and the more difficult it becomes to judge what would seem to be appropriate.

It is this type of analysis which leads to the concept of indigenous technology:

> The idea of an indigenous technology has of course vast social, political and cultural implications. It is an expression of confidence in the belief that no people whatever their culture are incapable of initiating and carrying on a development effort. It means that instead of the present attempt to manipulate and change the socio-cultural characteristics of a people to fit the claimed norms of Western industrial societies, the people themselves now sow the seeds of industrialisation in their community with their own hands.[28]

3. Denotation of "Appropriate"

Much has been said up till now about the connotation of "appropriate"; i.e., which conceptual ideas are part of the concept of "appropriateness" when applied to production systems. It would be interesting to make a careful study of which production systems have actually been called appropriate and to analyze them in terms of the questions outlined in the previous subsection. Without digging too deep, there are a number of ways in which the evaluation as to whether a particular production system is to be considered appropriate presents us with severe difficulties.

(a) Checklists of criteria do not work. In section III.3. it was already noted that the criteria that have been proposed in the literature are often mutually inconsistent. One might attempt to gather a number of definitely appropriate production systems and then try to analyze what they have in common. One would find, of course, that no one characteristic is universally applicable: similarities crop up and disappear. For example, consider an anaerobic digestor (processing agricultural waste materials into methane gas and "synthetic" manure) for a farmer in rural Scotland and for a farmer in rural India; or a distillation unit for an isolated oil well in Alaska and for a local gin producer in Kenya. All four are small-scale and could be appropriate, but for each pair all other characteristics differ vastly. On the other hand, consider making a long canal plus irrigation facilities; there are numerous ways of doing that, appropriately, but it will remain a large-scale project. Similar examples can be constructed to falsify the general applicability of all other descriptive criteria.

(b) The examples under (a) illustrate the influence of the nature of the needs to be satisfied and of the environment in which this has to take place. Assuming that these are fixed, there is the problem that a production system may be considered appropriate on different grounds, depending on the interests and preoccupations of those making the evaluation. For example, somebody concerned with poverty may be in favor of certain appropriate systems for the development of rural areas in developing countries. He or she may find the United States government in complete agreement with this line of action, but the interest of the latter may be in the fact that rural development programs decrease the investment in urban export industries that may compete with the American industry on the world market, whereas rural development may well increase the export market for the United States.

(c) The evaluating entities mentioned under (b) are both external. Whether one thinks one is led by humanistic values or exploitative goals, this kind of evaluation of "appropriateness" is made from a standpoint of cultural dominance. This aspect has been analyzed in great detail in the literature and need not be covered here.[29]

(d) Finally, no matter who is making the evaluation, there is the problem of what to choose as the boundary conditions of the system to be evaluated. On theoretical grounds one has to say that one should always evaluate the whole world, if not the universe—but, as Laplace and others noted, this is rather impractical. If one looks into the activities of people and institutions involved in the appropriate technology movement, it is clear that no serious attempt is made to evaluate the macro-economic effects and the efficiency of the investment in promoting "appropriate" production systems. The question is not only how much real money and "free" time of volunteers goes into creating one appropriate production system, which can and does survive in the environment it was designed for. It is also how many units have to be established to have a significant macro-effect in terms of output and employment, and in terms of bringing about a change in the society concerned.

Many examples can be given of clearly inappropriate production systems. I do not know of one good example of an appropriate production system.

4. The All-Embracing Technocracy

If, for the present purpose, we consider technocracy as the situation in which all goals and values have become subject to some objective development of science, technology and society,[30] we may say that, on average, "appropriate technology" is just a new metaphysics of technocracy, basically expressing an absolute faith in the power of science and technology. At the conference in Hyderabad, mentioned in section III.2, Malhotra (1964) warned that "there is a danger that intermediate technology may be treated as an abstraction like 'truth' and modern Jesting Pilate may ask 'What is intermediate technology?' and may not wait for an answer." The same applies to any other adjectives, capitals, or the invention of terms like "systems" or "unified" approach to express one's concern with important problems. It all boils down to the New Technology, or whatever, being an abstraction capable of solving all problems. Perhaps the following two examples may illustrate this point.

In his contribution to the first conference on appropriate technology for the United Kingdom, Professor M. W. Thring, Sc.D., C.Eng., F.I.Mech.E., F.I.E.E., M.I.Chem.E., F.Inst.P., F.Inst.F., F.R.Ae.S., wrote:

> So far UK technology has been based on short-term local interests. . . . The paper will outline the Creative Society described in my books . . . in which all these misuses are overcome by having a society based primarily on human creativeness in which Engineering is devoted to serving the ordinary world citizen. . . . Nevertheless, it is only the Engineer who can see what is possible and hence he has a very great responsibility to educate the Public and the Politicians to the essential long-term world-wide Ethos.

Now I do not want to raise questions such as: "Did not a lot of human creativeness go into the industrial revolution?" or: "Is not engineering serving the ordinary world citizen by providing transistor radios?"—because Thring probably explains in his books that he means something different. What is of prime interest is the form of the statement. In 1822 Comte wrote:

> A social system in its decline, a new system arrived at maturity and approaching its completion—such is the fundamental character which the general progress of civilisation has assigned to the present epoch. . . . The existing crisis is manifestly common to the several nations of Western Europe. . . . The true Organic Doctrine can alone produce the harmony so imperatively demanded by the condition of European civilisation . . . the necessity for confiding to Scientific Men the preliminary theoretical labours recognized as indispensable for reorganizing society is solidly based upon . . .: (1) scientific men are by the character of their intellectual capacity and cultivation alone competent to execute these works; . . . (3) they exclusively possess the moral authority. . . .[31]

Apparently nothing very much has changed.

As a second illustration I quote from a report of the Secretary General of the United Nations to the Commission for Social Development (UN, 1973), in which it is argued that the concept of "unified approach" should be widely diffused:

> The unified approach aims at providing a conceptual framework for understanding the development process. Development is not just the achievement of a satisfactory rate of economic growth measured by the conventional yardstick of gross national product. Nor is it merely a matter of evidence of growth in various economic sectors plus progress in particular social fields. Rather it encompasses and is influenced by a whole complex of forces that pervade a society—forces that are political, social and technological as much as economic. The unified approach is an attempt to recognize these forces and to understand the manner in which they interact on each other in the process of growth and change. In spite of the foreseeable obstacles and difficulties, it is imperative to advance towards a realization of a unified approach since it appears to represent the only possible means of overcoming the problems emerging from the failures and deficiencies that have prevailed up to the present time.

It is clear who is the mother of the unified approach:

> An integrated systems approach cannot only tell us that values are objective factors in behavior and development, but can show which values lead to the actualization of human potentials, and which bring about conditions that frustrate and thwart the individual. . . . Systems theorists investigate dynamic, goal-oriented systems, with definite if flexible programs for coping with their environment and assuring their growth and development.[32]

Is this not what we want and need for developing societies in the right direction? At least we who write and read about philosophy and technology, and technology and development, should be glad that there is systems theory, because "it offers scholars, educators, engineers, and artists new and harmonious ways of looking at the world."[33]

NOTES

1. *Encyclopaedia Britannica,* 15th edition, entry "Technology, conceptions of."
2. *Bibliography of the Philosophy of Technology,* compiled by C. Mitcham and R. Mackey, Chicago: University of Chicago Press (1973).
3. The same restricted meaning of society is employed in activities related to "Science, Technology, and Society," as is apparent from the contents of newsletters and courses on this subject.
4. N. Calder, *Technopolis: Social Control of the Uses of Science,* New York: Simon and Schuster (1971).
5. Due to the sheer increase of academic interest in (philosophy of) science, technology, and society, there has been some increase in the interest in technology and development from these circles over the last ten years, but still in no proportion to other fields of interests. Contributions do not come above the level of a superficial review of current literature (for example Sardar and Rosser-Owen, 1977). In Ellul's recently published *Le Système techni-*

cien some attention is given to the developing nations. I would appreciate being informed about other references to developing countries in philosophical publications.

6. That is to say, without taking sides in, for example, the Popper/Albert versus Adorno/Habermas debate. Statements expressing values or ends may or may not have a truth value, but in any case their epistemological status is different from that of descriptive statements.

7. Economists tend to use "technology" (read "production systems") and techniques interchangeably; for example, "choice of techniques" and "choice of technology" are usually considered synonyms. This is confusing. In one production system, say, soap manufacture, a large number of techniques are used (drying, labor management, advertising). One technique is used in many different production systems: drying of soap, bricks, peanuts.

8. Environment refers to *all* surrounding conditions and influences that may affect the production system: physical, ecological, social, and cultural conditions.

9. Production factors are the (categories of) things to develop, make, and operate a production function, for example: capital, labor, raw materials, managerial skills, know-how.

10. In making a social-cost/benefit analysis we are concerned with the economy as a whole. Instead of calculating in terms of costs and revenues for a private firm in a cost-benefit analysis the social benefit of a project is compared with the opportunity costs or social costs of the project. In making such an analysis, prices of various goods may be assumed that are different from the existing prices in the economy concerned; these are called shadow prices. "Social-cost/benefit analysis is often seen as a panacea. The technical problems in applying it are, however, enormous. It has its limitations and must be applied with great care" (Griffin and Enos, 1970).

11. See for more detailed accounts, *inter alia,* Jéquier (1976) and van Brakel (1978).

12. A detailed analysis of the underlying complexities can be found in Sen (1975). In this connection it may be stressed that the epistemological analysis of measurement in economics, or for that matter in any other science, is a highly neglected subject.

13. See on current prejudices with respect to economies of scale ("large scale is always cheaper") Onyemelukwe (1974, pp. 41–49), Wuyts (1974), van Brakel (1978, pp. 29–40), Khan (1973).

14. This is not self-evident. Usually it is stated that one of the more important features of radical or alternative technology is that it is nonalienating. It is by no means clear that this means the same for the industrialized countries (where the technology has been developed in at least some kind of interaction with Western workers) and underdeveloped countries, which receive technology and production systems that are (more basically?) alien to their culture and history.

15. A recent philosophical publication on this subject is J. Passmore, *Man's Responsibility for Nature,* London (1974).

16. Spensley (1974); he is director of the British Tropical Products Institute, one of the largest establishments that is solely engaged in R&D for developing countries.

17. Morawetz (1974, p. 517). See, for a critique of this rather technical definition, Stewart (1977, p. 95).

18. Quoted from the address of R.N. Jai, Principal Director of the SIET institute in Hyderabad, which organized the conference.

19. See Caritas (1968). The conference was addressed by the pope.

20. The basic needs strategy was adopted in the "Tripartite World Conference on Employment, Income Distribution and Social Progress" (ILO, 1976, 1977).

21. Although resolution No. 2 of the Lima conference produced an action program in 1977 (UNIDO, 1977) on the creation, transfer, and use of appropriate industrial technology, implementation of this program has hardly started.

22. In fact, two: "Seminar on Development Projects Designed to Reach the Lowest Income Groups"; and "Study Group on Low-Cost Technology and Rural Industrialization."

23. Joseph Hanlon in the *New Scientist* of May 19, May 26, June 2, and June 9, 1977. For neo-Marxists appropriate technology has always just been another variant of the capitalistic ideology, because "It ascribes the same neutrality to technology in the development process . . . it is not claimed that past policies have been proved wrong, but that economists have been insufficiently sophisticated in their evaluation of technological needs, failing to take into account a wide range of social and cultural, as well as economic factors" (Dickson, 1974, pp. 151f).

24. The criteria are derived from UN (1971), Stewart (1972), Khan (1973), Dos Santos (1974), Hvelplund (1974), Behari (1974), Collombon (1975), Singh (in Behari, 1975), Reddy (in Behari, 1975), Askin (in US-AID, 1977), and others.

25. The criteria 2,a-c, as well as 4,a-d, will often be inconsistent, hence "low capital cost" and "simplicity" are without further elucidation void of meaning. See for a detailed analysis Stewart (1972, 1977).

26. In the sense of Wittgenstein; see Wittgenstein's lecture on ethics, *Phil. Rev.* 74 (1965): 3–26, and J. Walker, "Wittgenstein's earlier ethics," *Am. Phil. Rev.* 5 (1968): 219–232.

27. The desperate attempts of those engaged in such fashionable subjects as technology assessment and technological forecasting to say something sensible may support this. What about this one? It is from a book entitled *The Methodology of Technology Assessment:* "Step 14. Subtract ratings in the 'capability' profile column from corresponding ratings in the 'relevance' profile column. This then is the 'GAP' profile rating which is by and large identical with the profile of institutional needs in science technology" (Cetron and Connor, 1972).

28. Onyemelukwe (1974, p. 29). It would be a most interesting research project to analyze the combination of the ideas underlying the concept of indigenous technology and Habermas' theoretical writings on symbolic interaction and related subjects. Cf. also: "Unless a people's basic knowledge system is voluntarily changed, no fundamental self-sustaining development can occur" (Friday, 1978).

29. See, out of many, Cooper (1974), Streeten (1974), and Mazrui (1975).

30. For example, in the way this is analyzed in J. Habermas, *Technik und Wissenschaft als "Ideologie,"* Frankfurt a/m: Suhrkamp (1968).

31. Fletcher, *The Crisis of Industrial Civilisation: The Early Essays of Auguste Comte,* London: Heinemann (1974, pp. 11–134).

32. E. Laszlo, Introduction in *The Relevance of General Systems Theory,* E. Laszlo, ed., New York: Braziller (1972). Also in the "Science, Technology and Society" subculture, (general) systems theory is taken over, because "Current socio-technological problems cannot be treated intelligently without consideration of the total system within which the problem arises" (quoted from the announcement of a course by J. Truxal in Curriculum Newsletter of the Lehigh University HPT Program, June 1978).

33. In the development policy establishment, "appropriate technology" has by now been given its appropriate place: "What the A.T. movement and its affiliates should be seen to be is a step in the development of thinking that leads one to the adoption of a wider 'systems approach' to the choice of technology in development projects" (McInerney, 1977).

REFERENCES

[Note: Only references on technology for appropriate development are included.]

Behari, B., (ed.). *Appropriate Technology for Balanced Regional Development* (conference proceedings), vol. I & II. New Delhi: Government of India, Appropriate Technology Cell (1975).

Brakel, J. van. *Chemical Technology for Appropriate Development*. Delft: Delft University Press (1978).

Caritas. *Populorum Progressio por le Tiers-Monde*. Rome: Caritas International (1968).

Cetron, M. J. and Connor, L. W. "A Method for Planning and Assessing Technology against Relevant National Goals in Developing Countries." In: *The Methodology of Technology Assessment*, M. J. Cetron and B. Bartocha (eds.), New York: Gordon & Breach (1972), pp. 1–32.

Collombon, J.-M. "Le Processus d'appropriation technologique." *Options Méditerranéennes* 27 (1975): 104–115.

Cooper, C. "Science, Technology and Production in Underdeveloped Countries." In: *Science, Technology and Development*, C. Cooper (ed.), London: Cass (1974), pp. 1–8.

Crane, D. "Technological Innovation in Developing Countries: A Review of the Literature," *Research Policy* 6 (1977): 374–395.

Dickson, D. *Alternative Technology and the Politics of Technical Change*. Glasgow: Fontana/Collins (1974).

Dos Santos, A. *Technologie peu couteuse: Revue de quelques problèmes avec référence spécial aux pays moins développés d'Afrîque*. [CD/TI (74) 16] Paris: OECD Development Centre (1974).

DSE. *Development and Dissemination of Appropriate Technology in Rural Areas*. Workshop report of the international workshop: 17–28 July 1972, Kumasi (Ghana). Berlin: Deutsche Stiftung für Entwicklungshilfe (1972).

Friday, J. "Science and Development: The Case of Papua New Guinea." Mimeo, based on a lecture given at the Royal Institution of Great Britain on January 18, 1978.

Garg, M. K. "Some Lessons from the Indian Experience." In: OECD (1975), see below.

Griffin, K. B. and Enos, J. L. *Planning Development*. Reading, Mass.: Addison-Wesley (1970).

Harper, P. "Radical Technology." *Undercurrents* 15 (1976):11.

Hoda, M. M. Various papers in OECD (1974b), below.

Hvelplund, F. *The Appropriate Technology Process* [the technology of liberation]. In OECD (1974b).

ILO. *Employment, Growth and Basic Needs: A One-World Problem*. Geneva: International Labour Organisation (1976); also: *Meeting Basic Needs* (1977).

Jéquier, N., *La Technologie appropriée: Problèmes et promesses*. Paris: OECD (1976).

Khan, A. U. "Appropriate Technology: Do We Transfer, Adapt or Develop?" Paper presented at the Ford Foundation Seminar on Technology and Employment, New Delhi (India), 1973.

MacKillop, X. *Why Soft Technology?* London: Methuen (1974).

Malhotra, D. K. "Technologies for Small Industries in Rural Areas." In: SIET (1964), pp. 19–22.

Marsden, K. "Progressive Technologies for Developing Countries." *International Labour Review* 99 (1969): 621–646.

Mazrui, A. A. *The African University as a Multinational Corporation: Comparative Problems of Penetration and Dependency*. Discussion Paper No. 95. University of Sussex: Institute of Development Studies (1975).

McGarry, M. G. "Appropriate Technology in Civil Engineering." Paper presented at the 1977 Annual Conference of the ASCE, San Francisco.

McInerney, J. P. " 'Appropriate' Technology: No Miracle Panacea for Economic Development." *Report* (News of the World Bank Group), May/June 1977, pp. 4–6.

Mendoza, E. L. "The Need for a New Development Economics." *ANTHRO-TECH: A Journal of Speculative Anthropology* 1 (1977): 12–15.

Morawetz, D. "Employment Implications of Industrialisation in Developing Countries: A Survey." *The Economic Journal* 84 (1974): 491–542.

OECD (1974a). *Choice and Adaptation of Technology in Developing Countries: An Overview of Major Policy Issues*. Paris: OECD Development Centre.

OECD (1974b). *Working Papers for Study Group on Low-Cost Technology and Rural Industrialisation, Paris 17–20 Sept. 1974*. Paris: OECD Development Centre.

OECD (1975). *Low-Cost Technology—An Inquiry into Outstanding Policy Issues*. [CD/TI (75) 1] Paris: OECD Development Centre.

Onyemelukwe, C. C. *Economic Underdevelopment: An Inside View*. London: Longman (1974).

Sardar, Z. and Rosser-Owen, D. G. "Science Policy and Developing Countries." In: *Science, Technology and Society: A Cross-Disciplinary Perspective*, I. Spiegel-Rösing and D. de Solla Price (eds.), London: Sage (1977).

Schumacher, E. F. *Small Is Beautiful*. London: Blond & Briggs (1973).

Sen, A. *Employment, Technology and Development*. Oxford: Clarendon Press (1975).

SIET. *Appropriate Technologies for Indian Industries*. Papers of a meeting. Hyderabad: SIET (1964).

Spensley, P. C. "Improving Technology Post-Harvest." *Chem. and Ind.*, (Jan. 1974): 17–21.

Stewart, F. *Intermediate Technology: A Definitional Discussion*. Background paper no. 11 for Study group on choice and adaptation of technology in developing countries. Paris: OECD Development Centre (1972).

Stewart, F. *Technology and Underdevelopment*. London: Macmillan (1977).

Streeten, P. "The Limits of Development Research." *World Development* 2 (1974): 11–34.

Thring, M. W. "UK Technology for the XXIst Century." Preprint for conference, "Appropriate Technology for the UK." New Castle-upon-Tyne: University (1976).

UN. *World Plan of Action*. [No. E.71.II.A.18] New York: United Nations (1971).

UN. *Report of a Unified Approach to Development Analysis and Planning, UN Commission for Social Development*. [No. E/CN.5/490] New York: United Nations (1973).

UNIDO. *Co-operative Programme of Action on Appropriate Industrial Technology*. Industrial Development Board, 11th session. [No. ID/B 188] Vienna: UNIDO.

US-AID. *Proposal for a Program in Appropriate Technology*. Washington, D.C.: Government Printing Office (1976); reprinted with additions, 1977.

Wuyts, M. E. *The Choice of Techniques: Some Critical Questions*. Paper 74.3, Economic Research Bureau. University of Dar-es-Salaam (1974).

Research in Philosophy and Technology

A Research Annual

Series Editor: **Paul T. Durbin**
Philosophy Department and Center for Science and Culture
University of Delaware.

Bibliography Editor: **Carl Mitcham**
St. Catharine College, Kentucky.

Official Publication of the Philosophy & Technology Group

Review: ". . . Reflecting a sudden burst of interest in this subject over the past decade, the essays attempt to answer the question 'What would constitute and adequate philosophy of technology?' . . . the collection offers and introduction to the subject that will be useful to specialists and nonspecialists alike." —*Science*

Volume 1. Published 1978 Cloth
ISBN 0-89232-022-2 390 pages

CONTENTS: Introduction. Introduction to Volume 1. Approaches to Philosophy of Technology Old and New.

Part I: METHOD, DESCRIPTIVE FRAMEWORKS, AND A PRACTICAL PROGRAM FOR PHILOSOPHY OF TECHNOLOGY. Culture and Technology, *Joseph Margolis, Temple University.* **The Problem of Scale in Human Life: A Framework for Analysis,** *Robert E. McGinn, Stanford University.* **Technology, Mass Movements, and Rapid Social Change: A Program for the Future of Philosophy of Technology,** *Joseph Agassi, Boston University and Tel Aviv University.*

Part II: THE UNIVERSITY OF DELAWARE CONFERENCE, 1975. Toward a Social Philosophy of Technology, *Paul T. Durbin, University of Delaware.* **The Evaluation of Technology,** *Albert Borgmann, University of Montana.* **Values in Science,** *Willis H. Truitt, University of South Florida.* **Technology as Ideology,** *Kai Nielsen, The University of Calgary.* **Humanization of Technology: Slogan or Ethical Imperative?,** *Edmund Byrne, Indiana University-Purdue University Indianapolis.* **What is Technology?,** *Robert E. McGinn, Stanford University.* **Shifting from Physical to Social Technology,** *Joseph Agassi, Boston University and Tel Aviv University.* **Types of Technology,** *Carl Mitcham, St. Catharine College.*

Part III: REVIEW AND BIBLIOGRAPHY. Review and Bibliographical Survey, *Carl Mitcham with the assistance of Jim Grote, St. Catharine College.* **Current Bibliography in the Philosophy of Technology: 1973-1974.**

Volume 2. **Published 1979** **Cloth**
ISBN 0-89232-101-6 **403 pages**

CONTENTS: Introduction to Volume 2: The International Cast of the Philosophy and Technology Movement.

Part I: EUROPEAN CONTRIBUTIONS. "Euthyphronics" and the Problem of Adapting Technical Progress to Man, *Jozef Banka, Katowice, Poland.* **Toward an Interdisciplinary and Pragmatic Philosophy of Technology: Technology as a Focus for Interdisciplinary Reflection and Systems Research,** *Hans Lenk and Gunther Ropohl, University of Karlsruhe.* **Fear and Insight in French Philosophy of Technology,** *Daniel Cerezuelle, University of Grenoble.*

Part II: CONFERENCE PAPERS. 1976-1977. Freedom and Determinism in a Technological Setting, *Albert Borgmann, University of Montana.* **The Normative Side of Technology: Philosophy and the Public Interest,** *Edmund Byrne, Indiana University — Indianapolis.* **Art in a Technological Society,** *Philip R. Fandozzi, University of Montana.* **Marx, Machinery, and Alienation,** *Bernard Gendron, University of Wisconsin — Milwaukee, and Nancy Holmstrom, University of Wisconsin — Madison.* **Technological Culture and the End of Philosophy,** *Michael Zimmerman, Tulane University.* **Comment of Zimmerman's "Technological Culture and the End of Philosophy",** *Robert E. McGinn, Stanford University.* **Reply to McGinn,** *Michael Zimmerman, Tulane University.*

Part III: HISTORICO PHILOSOPHICAL STUDIES. The Concepts of Nature and Technique According to the Greeks, *Wolfgang Schadewaldt, University of Tuebingen.* **On the Nature of Nature,** *Jacob Klein, St. John's College — Annapolis.* **Documentation: Analyses of Machines in the French Intellectual Tradition (Espinas, Lafitte, Weil),** *Edited by Carl Mitcham, St. Catharine College.* **On the Antagonism Between Philosophy and Technology in Germany and Austria,** *Werner Koenne, Austrian Elektrizitat Wirtschafts-Aktiengellschaft, Vienna.* **Heidegger and Marcuse: Technology as Ideology,** *Michael Zimmerman, Tulane University.*

Part IV: REVIEW AND BIBLIOGRAPHY. Philosophy of Technology and the Verein Deutcher Ingenieure, *Alois Huning, Wulfrath, Germany.* **Philosophy of Technology in France in the Twentieth Century: Overview and Current Bibliography,** *Jean-Claude Beaune, University of Clermont.* **Symposium: Hendrik van Reissen and Dutch Neo-Calvinist Philosophy of Technology,** *Edited and Translated by Donald Morton, Free University, Amsterdam.* **Technology Assessment: A New Type of Science?,** *Frederick A. Rossini, Georgia Institute of Technology.* **Technology Assessment: Supplementary Bibliography,** *Carl Mitcham and Jim Grote, St. Catharine College.* **Kenneth Sayre on Information Theoretic Models of Mind,** *K.S. Schrader-Frechette, University of Louisville.*

Appendix: Author Index to Current Bibliography in the Philosophy of Technology: 1973-1974. Research in Philosophy and Technology, Volume One, *Carl Mitcham and Jim Grote, St. Catharine College.*

Volume 4. **January 1981** **Cloth**
ISBN 0-89232-181-4 **Ca. 400 pages**

CONTENTS: Introduction. Current Bibliography in the Philosophy of Technology: 1975-1976, *Carl Mitcham and Jim Grote, St. Catharine College.* **(Analytic Table of Contents follows Introduction.) Author Index for 1975-1976. Bibliography,** *Carl Mitcham and Jim Grote, St. Catharine College.* **Appendix: Bibliography of the Philosophy of Technology (University of Chicago Press 1973.)**

INSTITUTIONAL STANDING ORDERS *will be granted a 10% discount and be filled automatically upon publication. Please indicate initial volume of standing order*
INDIVIDUAL ORDERS *must be prepaid by personal check or credit card. Please include $1.50 per volume for postage and handling.*
Please encourage your library to subscribe to this series.

JAI PRESS INC., P.O. Box 1678, 165 West Putnam Avenue, Greenwich, Connecticut 06830.

Telephone: 203-661-7602 **Cable Address: JAIPUBL**

a93